Cities and Nature

Cities and Nature fosters high-quality multi-disciplinary research addressing the interface between cities and the natural environment. It provides a valuable source of relevant knowledge for researchers, planners and policy-makers. The series welcomes empirically based, cutting-edge and theoretical research in urban geography, urban planning, environmental planning, urban ecology, regional science and economics. It publishes peer-reviewed edited and authored volumes on topics dealing with the urban and the environment nexus, including: spatial dynamics of urban built areas, urban and peri-urban agriculture, urban greening and green infrastructure, environmental planning, urban forests, urban ecology, regional dynamics and landscape fragmentation.

Indexed in Scopus!

Kwi-Gon Kim · Catherine Atkin
Editors

Cross-Cultural Perspectives on Climate Change Adaptation

Adapting to Flood Risk

Springer

Editors
Kwi-Gon Kim
Seoul National University
Seoul, South Korea

Catherine Atkin
Stanford CodeX Climate Data Policy
Initiative
Stanford, USA

ISSN 2520-8306 ISSN 2520-8314 (electronic)
Cities and Nature
ISBN 978-3-031-50364-1 ISBN 978-3-031-50365-8 (eBook)
https://doi.org/10.1007/978-3-031-50365-8

This Springer imprint is published by the registered company Springer Nature Switzerland AG
The registered company address is: Gewerbestrasse 11, 6330 Cham, Switzerland

Foreword

Cities and communities across the Global South face a complex web of crises and challenges as climate change drives shifts in weather, causing incalculable human suffering, biodiversity loss, and undermining urban economies and systems of support. Countries in Africa, Latin America, the Caribbean, Pacific Islands, and Asia are experiencing a disproportionate increase in climate change-related disasters—including floods, heatwaves, droughts, hurricanes, and wildfires. Since 1991, their populations have suffered over 90% of the global total of people affected by extreme weather and climate-related events.

I have seen the devastating effects of extreme heat and flooding first-hand as climate-related disasters increase in frequency and severity in the places where I live and work. What is more untenable is the uncertainty with which changes occur and the suddenness with which disaster strikes. According to the United Nations Office for Disaster Risk Reductions (UNDRR), Africa is the continent that has been most affected by climate-related disasters over the past two decades. Flooding—the focus of this book—is one of the most common climate-related disasters in Africa, with an average of 41 million people affected by floods each year between 2001 and 2019. It is also a key example of a systems problem driven by limited drainage and sanitation infrastructure, poor maintenance of natural infiltration areas, and non-existent or limited waste management programs—and causing downstream waste issues, storage issues for food, the spread of disease, and loss of reliable shelter, to name a few concerns. Therefore, addressing flooding and broader climate-related disasters requires cross-sector awareness and response.

Given this new reality, the imperative for concerted global, regional, and local action to address adaptation and build resilience has never been greater. The good news is that the focus on adaptation and resilience is growing globally, including new adaptation initiatives and a long overdue creation of a "loss and damage" fund at COP-27 in Sharm El-Sheikh, Egypt. This increased focus on adaptation has been hard-won and is an important step forward. Still, we are far from the desired outcomes of having substantial resources specifically directed toward adaptation. According to the OECD, in 2019, only 25% of overall climate finance was directed to adaptation. The African Development Bank has indicated that only 3% of climate finance is directed toward Africa. These resources will not enable the necessary transformative change. We must produce more evidence of real opportunities for impact, change the normative logic of financial risk to accept broader investable projects, and concentrate political pressure at all levels of government toward resourcing such transformation.

The IPCC outlines the key drivers of effective adaptation in its recent 6th Assessment Report:

> The impacts of climate change in the global south are disproportionately affecting vulnerable populations, including indigenous peoples, small island states, and low-lying coastal areas, making adaptation an urgent priority. Effective adaptation requires a combination of societal, technological, and policy responses, including investment in resilient infrastructure, risk reduction, and adaptation planning at all levels of governance.

Equally crucial to the increased global attention is the fact that leaders and communities across the Global South are marshaling resources and human capital to advance comprehensive adaptation and resilience efforts. Much of the focus of my work is building city collaborations and mutual exchange platforms globally and in Africa that support local efforts to adapt and build resilient communities. Members of ICLEI's network of local and subnational governments, who are also part of our CitiesWithNature, CityFood, ICLEI Circulars, Solutions Gateway, and LearnWithICLEIAfrica communities, are making remarkable progress in the face of significant sustainability and climate challenges. However, I am aware that many of their efforts to undertake comprehensive climate risk management are hindered by disparities in access to resources, outmoded approaches to financing and technical assistance, and a lack of focus on local capacity building and authentic community engagement.

We need to gain deeper insights into the on-the-ground challenges and opportunities as we work on these fronts at the global, country, regional, and city levels. "Cross-cultural Perspectives on Climate Change Adaptation: Adapting to Flood Risk" is a critical offering at this pivotal moment as it addresses gaps in advancing equitable climate risk management, offers new insights, explores current challenges, puts forward new tools and approaches, and suggests a national and global agenda to move forward. Highlights of the essential contributions the book makes to the current global adaptation conversation include:

- **Research, analysis, and insights from the perspective of in-country and on-the-ground public sector professionals:** The centerpiece of the book is case studies by thirteen authors who undertake climate risk assessment and green recovery planning in select locations across Africa, South and Southeast Asia, South America, and the Middle East. This unique compilation of cross-cultural explorations brings forth the voices of in-country authors who are professionals in government planning and services.
- **Comprehensive analysis and understanding of local context and community voice that must inform adaptation and resilience efforts:** The case studies tell the story of communities experiencing flooding disasters across the Global South. The authors offer insights into the variables in the local context, including community demographics, local needs, and the economic, governance, and policy landscape that must inform disaster risk assessment and green recovery planning.

In addition, deep dives into the power of community co-design and engagement of impacted persons bring into focus this too often overlooked dimension of sustainable change.

- **Practical tools and concrete guidance implementers and planners need to take action:** Comprehensive climate risk assessment is the foundation of smart city and community planning, but many existing tools and frameworks lack specificity. Those in government and NGO sectors engaging in climate risk management will benefit from the exploration of the current status and utility of climate risk management frameworks and new methodologies and tools that give on-the-ground implementers step-by-step guidance to carry out their efforts.
- **Exploration of how digital and technology solutions can be leveraged for implementation at scale:** Many Global South communities do not have access to the data and digital tools needed to understand climate risks and to evaluate and choose between hazard risk reduction measures. Authors of chapters in the fourth part of the book—including authors from ICLEI and the Green Digital Finance Alliance—explore digital tools and technologies like cloud computing, AI, and blockchain and showcase how solutions such as digital twins and the metaverse can accelerate adaptation and resilience efforts.
- **Articulation of a national and global policy agenda:** Cities and communities need the support of a robust enabling environment to adapt to climate change hazards and build resilient communities. The book offers a blueprint for how international communities, including UN agencies, can help finance, design, and implement adaptation plans and calls for developing a new co-benefit funding initiative that combines adaptation with mitigation funding for integrated climate solutions.

The book's offerings are a powerful contribution to the broader discourse and bring forth new resources and tools that are crucially needed to support meaningful change on the ground in cities worldwide. It offers a candid assessment of where we are, identifies gaps in our current theory and practice of climate risk management, and picks out the critical governance, policy, and local capabilities that can strengthen local efforts. I also commend the editors and authors for taking a positive and asset-based approach to their work that recognizes the power of local knowledge and resists the deficit frame that often pervades the conversation. The case study authors, who, like many in their countries, are constrained by resources and data, share critical knowledge and insights. Books like this are part of an exciting and critically important emerging approach to research and knowledge creation that centers communities and honors the richness of cultures and local contexts.

As you read this book, I encourage you to challenge yourself to take the knowledge, insights, and tools it offers to become part of the global adaptation and resilience movement and find ways to support local communities across the Global South that will be drivers and architects of brighter futures for their communities.

Cape Town, South Africa

Kobie Brand
Deputy Secretary General
of ICLEI—Local Governments for
Sustainability, and Regional Director
of ICLEI Africa

Preface

Global warming has changed everything, everywhere, forever. What we already know from reading the paper and watching the television is underscored by the constant stream of reports detailing the acceleration of climate-related disasters—including floods, droughts, wildfires and hurricanes—across the globe and statistics on the disproportionate impact on Global South communities. We know that more than 20 million people are being forced from their homes each year and that the economic costs are staggering with estimates that developing countries that already spend USD 70 billion a year to adapt to climate change will see that burden quadrupled by 2050.

The good news is that we are finally seeing a shift in the conversation and an increasing focus on adaptation and resilience because leaders and communities in the Global South facing the ravages of climate change have pressed for it. Now the resources need to flow from those that caused climate change to those that are enduring it. The systems of on-the-ground sense and solution making need to be activated and scaled to protect people, animals, and the environment from climate-drive impacts.

My co-editor, Catherine Atkin, Esq., and I have come together on this project united by our commitment to addressing climate change in ways that also ensure global communities thrive. Through a confluence of factors and the opportunity to work together, this book project was born.

After three decades at the national and international levels working at the intersection of the planning, environmental, climate and technology fields with a focus on planning sustainable and climate smart cities, I began to teach professionals from Global South as part of the KOICA-Hallym master's degree Program on Climate Change Policy in Korea in 2019. In my engagement with these students from public agencies and NGOs across the Global South, I am continually struck by the power of their insights. We need to build the capacity of professionals who live and work in these countries to analyze the impacts of climate change on their communities, undertake climate risk assessment, and to chart a course for change with green recovery planning. I have also had the chance to understand more deeply the realities they face, to see first-hand the limitations of existing models and tools, and to think about how they and other professionals can be better supported.

I had the opportunity to first collaborate with Ms. Atkin, an attorney working at the intersection of climate data, innovation, and finance and Chair of the Stanford CodeX Climate Data Policy Initiative, through our work with the UNFCCC Global Innovation Hub that was launched at COP-26 in Glasgow. We were both drawn to this effort because of its commitment to advancing sustainable and low-carbon innovations and value chains with a focus on cities in the Global South and worked together on a demonstration project in Goyang City, Korea, to develop and test consumption-based GHG accounting innovations.

As we collaborated on this project, we also had the opportunity to discuss how to best support equitable low-emissions development given the reality that for many places in the Global South, their most pressing issue is how to adapt to climate change. We identified some of the critical challenges and gaps that hinder the comprehensive climate risk management—including the hazard risk assessment, green recovery planning and implementation—that is so desperately needed in countries facing the ravages of climate change.

We are very excited to have had the opportunity to undertake this book project so that we can not only explore these issues but also be part of solution making as adaptation and resilience take center stage. I am also grateful to Springer Publisher for allowing me to build on the foundation of two earlier Springer Publisher books I authored and edited including *Low-Carbon Smart Cities: Tools for Climate Resilience Planning* in 2018 and *Planning Climate Smart and Wise Cities: A Multidisciplinary Approach* in 2022.

The heart of the book is the rich array of case studies that my students undertook focusing on risk assessment and green recovery planning in areas of their countries facing flooding hazards. The thirteen case studies from Global South countries in Arica, South America, South and Southeast Asia, and the Middle East expose the reader to rare insights into what is currently happening on the ground in these countries.

I also want to give special thanks to my co-editor, Catherine Atkin, for her efforts to showcase the work of global experts who wrote additional chapters on a variety of high-value topics. The thirteen case studies, combined with these additional contributions, offer the reader new thinking and approaches on a variety of topics and issues including: adaptation capacity, local context variables and community voice that must inform adaptation and resilience efforts, practical tools and concrete guidance implementers and planners need to take action, digital and technology solutions that can be leveraged for greater impact, new integrated frameworks and strategies that advance co-benefit solutions and the elements of local, national, and global policy agenda to catalyze the comprehensive green recovery planning and implementation.

We hope you find the book provocative and useful and that it generates additional exploration. We have structured the book with a focus on adding value to on-the-ground researchers and implementers, and policymakers at the local, regional, national, and international levels. While this book answers some questions, it also uncovers areas where more learning and analysis is required. It sources new tools and frameworks that should continue to be improved and developed as they are used on the ground. Most importantly, we hope that it is of use in Global South communities who are on the frontlines of this work and whose voices and needs must be heard.

Kwi-Gon Kim
Seoul National University
Seoul, South Korea

Acknowledgments

As editors of this book, we thank all the contributors for their willingness to participate in this exciting endeavor. We are particularly grateful to the case study authors from across the globe that shared the papers they wrote for "Climate Disaster and Restoration," a seminar taught by Prof. Kwi-Gon Kim, editor of this book, in 2020 and 2021 as part of the KOICA-Hallym master's degree program in Response to Climate Change.

Bringing together the many case studies with chapters written by global experts in planning, climate risk management and policy as well as technologists required intensive efforts on the part of all involved to create a cohesive narrative and reader experience.

We want to thank Springer for the opportunity to undertake this project and their team for their critical guidance and feedback and their tireless efforts to support the finalization of this book.

A complete list of contributors and their affiliation is provided following the table of contents.

June 2024

Kwi-Gon Kim
Catherine Atkin

Contents

Contributors

Jannika Aalto Green Digital Finance Alliance (GDFA), Geneva, Switzerland

Catherine Atkin Stanford CodeX Climate Data Policy Initiative, Stanford, USA

Ezekiel Atanga Azuure Ministry of Energy, Accra, Ghana

Azhari Putri Cempaka Indonesia Agency for Meteorology, Climatology, and Geophysics, Central Jakarta, Indonesia

Navin Chea Royal University of Agriculture (RUA), Phnom Penh, Cambodia

Si Chen Open Source Strategies, Inc., Pasadena, CA, USA

David Corbett ICLEI—Local Governments for Sustainability, Bonn, Germany

Neil Paul Fara-on Department of the Interior and Local Government, Quezon City, Philippines

Katherine Foster Green Digital Finance Alliance (GDFA), Geneva, Switzerland

Tewelde Alema Fukur Environment, Forest and Climate Change Commission, Addis Ababa, Ethiopia

Chris Velmida Perfas Gianzon Department of the Interior and Local Government Region VI, Iloilo City, Philippines

Kambiz Karimi National Environmental Protection Bureau, Kabul, Afghanistan

Keo Hak Provincial Department of Rural Development Svay Rieng (PDRD-SVR), Ministry of Rural Development (MRD), Svay Rieng, Cambodia

Kwi-Gon Kim Seoul National University, Seoul, South Korea

Saadatu Gambo Madaki Department of Research, Innovation and Outreach, National Council on Climate Change, Abuja, Nigeria

Saharnaz Mirzazad California Governor's Office of Planning and Research, Sacramento, USA

Del Piero Raphael Arana Ruedas Universidad Continental, Huancayo, Junin, Peru

Mohammad Nazir Safi National Environmental Protection Agency, Kabul, Afghanistan;
Hallym University, Chuncheon-si, Gangwon-do, Republic of Korea

Pourya Salehi ICLEI—Local Governments for Sustainability, Bonn, Germany

Siddique Md Abu Bakar Ministry of Local Government, Rural Development and Cooperatives, Dhaka, Bangladesh

Pietro Visetti Green Digital Finance Alliance (GDFA), Geneva, Switzerland

Yawe Samuel Rakai District Local Government, Kampala, Uganda

Introduction

Catherine Atkin and Kwi-Gon Kim

Abstract In this chapter, the book co-authors provide an overview to this groundbreaking book and describe the ways in which readers of the book—including planners, program implementers, policymakers, and academics—will gain insights and new tools to advance flood risk management, including risk assessment and green recovery planning across the Global South. The centerpiece of the book are the thirteen case study papers focused on Global South communities in Africa, South America, South and Southeast Asia and the Middle East focused on flooding cities and regions by authors who are in-country professionals from government agencies and NGOs. Chapters "Climate Disaster Risk Management—Risk Assessment and Green Recovery Planning: Core Concepts and Identification of Current Issues"–"Overview and Exploration of Nature-Based Flooding Solutions: The Green Transition" review core concepts associated with flooding and climate-change driven disasters, climate risk assessment and green recovery principles, identifies new and emerging issues and solutions in flood risk management and presents a new Climate Risk Assessment and Green Recovery Planning framework and methodology created by Dr. Kwi-Gon Kim. In addition, Chaps. "Leveraging Leading Edge Digital Models, Developed Under Urban Computing Principles, to Support Flood Risk Assessment in Cities" and "Distributed Ledger Technology (DLT) to Drive Flood Reduction Solution Markets and Financing",explore the role of technology to accelerate impact and scale and explores digitalization and next-gen technologies and solutions including AI, IoT, Distributed Ledger Technology, Digital Twins and the Metaverse. Chapters "Case Study Findings and Synthesis of Learnings"–"Looking Forward: Building Vibrant, Climate Adaptive, and Resilient Cities and Communities" provide an analysis of the thirteen case studies, features a new integrated framework for achieving adaptation and mitigation co-benefits, explores the power of community-led initiatives to drive transformational impact and shares summary insights and new directions from the editors.

C. Atkin (✉)
Stanford CodeX Climate Data Policy Initiative, Stanford, USA

K.-G. Kim
Seoul National University, Seoul, South Korea
e-mail: kwigon@snu.ac.kr

K. Kim and C. Atkin (eds.), *Cross-Cultural Perspectives on Climate Change Adaptation*, Cities and Nature, https://doi.org/10.1007/978-3-031-50365-8_1

Keywords Flood risk assessment · Climate risk management · Disaster risk management · Community-driven resilience building · Digitalization of climate risk management · Adaptation co-benefit models · Nature-based hazard reduction measures · Comprehensive urban planning

Weather-related hazards driven by climate change are forever altering our natural ecosystems and the lives for the people that inhabit them. It has been estimated that the number of climate-related disasters has tripled in the last 30 years, and that more than 20 million people are being forced from their homes by climate change each year. The human and economic costs are staggering: the UN Environmental Programme (UNEP) estimates that developing countries are already spending 70 billion USD a year to adapt to climate change. This figure is expected to rise to between 140 and 300 billion USD in 2030 and 280–500 billion USD in 2050.

While communities across the global are negatively affected, the urban and rural poor in Global South countries are disproportionately experiencing the devastating impacts of extreme heat, drought, sea-level rise and flooding hazards. This book focuses on cities and communities, particularly low-lying urban centers, in the Global South where flooding hazards are intensifying and the risk of negative impacts on people, infrastructure and the environment is greatest. As more and more countries fall into the "extreme risk" category based on their climate index, unless adaptation planning and resilience building are strengthened, flooding events will increasingly ravage global communities.

Climate risk management is an emerging discipline in higher education, research and practice. The need for integration between the planning, environment, climate change and disaster management sectors is increasingly recognized as a critical next step in efforts to combat climate change and build vibrant and resilient communities. As we increase our needed investments in adaptation, we must give local communities the tools—and build their capacity—to address these challenges themselves to ensure equitable outcomes for nations and localities across the globe most affected by climate change.

This book breaks important new ground in the field of adaptation by featuring:

Voices from the Global South and culturally inclusive tools for addressing climate disaster and green recovery. This book centers and showcases the voices of local, in-country professionals. All too often the current conversation about climate action and adaptation is told from the perspective of those outside the countries most impacted by climate change. This is a seminal book in the field of climate adaptation and urban studies as it discusses problems and solutions and identifies gaps from the perspective of those who live and work in the Global South.

Locally drive and context-specific insights and solution-making: Climate and disaster risk management efforts have too often been "top down," applying national and global data sets and models regardless of community context and favoring outside expertise for risk assessment, planning and implementation. This book shows the limitations of legacy approaches and provides insights into new frameworks and models that value localized data, knowledge creation and community engagement that can unlock greater impact.

The power of integrated adaptation and mitigation co-benefit models: Current city-level adaptation and mitigation efforts too often take place in siloes. This book provides new integrated models that can drive the development of smart hazard reduction measures that unlock powerful decarbonization and sustainability co-benefits and calls for an approach to adaptation and resilience planning that brings together city planners and others focused on community services and infrastructure with staff focused on mitigation.

New technologies and the digitalization of climate solutions. The book identifies new approaches to the digitalization of disaster risk management processes as well as opportunities to leverage the power of new technologies like Digital Twins, the Metaverse, and Distributed Ledger Technology (DLT) to unlock new data insights and technology-driven hazard reduction solutions.

1 Case Studies from Eleven Global South Countries

The centerpiece of this book is the thirteen case studies and research presentations focused on flood risk assessment and green hazard reduction in eleven countries across the Global South. These papers were written by professionals and practitioners as part of student coursework for "Climate Disaster and Restoration," a seminar taught by Professor Kwi-Gon Kim, co-editor of this book, in 2020 and 2021 as part of KOICA-Hallym Master's Degree program in Response to Climate Change. The program at Hallym University in Korea was created to nurture field-oriented climate change policy experts from the Global South and is funded by the Korean International Cooperation Agency (KOICA).

These materials taken as a body of work offer planners, program implementers, and academics a rare glimpse of the current state of play when it comes to flood risk management, including risk assessment and green recovery planning across the Global South. Case study authors show how flood risks imperil communities and use their risk assessments to identify key structural and non-structural hazard reduction strategies that can be employed to protect inhabitants and improve the resiliency and sustainability of local ecosystems. Case study authors also draw on an understanding of local context variables to identify ways to create and strengthen the social, economic, institutional and technological determinants of adaptive capacity needed to ensure communities can effectively address climate risk hazards.

While this book focuses on telling the story of what is happening in communities across the Global South, the rich case study insights coupled with the cutting-edge models and innovative approaches featured in chapters by a variety of global experts make this book extremely relevant for readers from across the globe. The case studies and additional contributions should be read as sources of insight and most importantly as provocations for future action in climate risk management and green recovery. How are the currently available climate risk management and planning tools and frameworks used by authors and what does their application tell us about opportunities and limitations of their current application and deployment? What emerging best practices and innovations are on the horizon in adaptation and resilience building—including integrated frameworks, technologies and inclusive practices—that can accelerate and deepen efforts to address flooding and other climate-related hazards that imperil communities?

2 Book Structure

Chapters "Climate Disaster Risk Management—Risk Assessment and Green Recovery Planning: Core Concepts and Identification of Current Issues" and "International Initiatives and Current Context for Adaptation Planning and Green Recovery" provide an overview to core concepts associated with flooding and climate change driven disasters, climate risk assessment and green recovery principles as well as identifies new and emerging issues and solutions in flood risk management.

Chapters "Nine-Step Conceptual Framework for Climate Disaster Risk Assessment and Green Recovery Planning" and "Overview of Case Studies: Common Elements and Framework for Analysis" introduce Dr. Kwi-Gon Kim's Climate Risk Assessment and Green Recovery Planning framework and methodology for taking a suitability and risk-based indicator approachas well as an analysis of the power and co-benefits of nature-based hazard reduction solutions and provides an overview to the case studies featured in the book.

Chapters "Flood Risk Analysis: Causes and Landscape-Based Mitigation Strategies in Dire-Dawa City, Ethiopia"–"Proposed Study Design for Spatio-Temporal Drought Assessment Using Standardized Precipitation Evapotranspiration Index (SPEI) and Adaptation Over Mantaro Valley, Peru" showcase the thirteen case study papers by in-country professional from government agencies and NGOs from the Global South focused on flooding in cities and regions across Africa, South America, South and Southeast Asia and the Middle East.

Chapters "Leveraging Leading Edge Digital Models, Developed Under Urban Computing Principles, to Support Flood Risk Assessment in Cities" and "Distributed Ledger Technology (DLT) to Drive Flood Reduction Solution Markets and Financing" explore how technology can be leveraged to accelerate impact and scale and explores digitalization and next-gen technologies and solutions including AI, IoT, Distributed Ledger Technology, Digital Twins and the Metaverse including contributions from experts from leading global institutions ICLEI and the Green Digital Finance Alliance.

Chapters "Case Study Findings and Synthesis of Learnings"–"Looking Forward: Building Vibrant, Climate Adaptive, and Resilient Cities and Communities" include an analysis of the thirteen case studies and features a new integrated framework for achieving adaptation and mitigation co-benefit, looks at the power of community-led initiatives to drive transformational impact and shares summary insights and outlines new directions forward from the editors.

Climate Change and Flood Disaster: Core Concepts, Emerging Issues and Global Context for Change

Climate Disaster Risk Management—Risk Assessment and Green Recovery Planning: Core Concepts and Identification of Current Issues

Kwi-Gon Kim

Abstract There is the urgent need to adapt to climate change. Flooding affects 106 million people annually and causes the largest economic damage (USD 31 billion, annually). 6000 people are annually killed by flooding. This chapter addresses concept and definition of disasters, and flood disaster-related newly emerging issues from a theoretical and empirical perspective. More in specific, the chapter reviews foundational principles of resilient systems, increase in frequency and intensity of flood occurrence, differential effects based on geography and wealth, bifurcation of mitigation and adaptation planning and implementation, capacity and technology gaps, and climate justice issues discussed at COP27. The chapter concludes that an integrated framework and methodology for key issues are needed to help cities and communities develop a comprehensive, risk- and scientific, evidence-based flood adaptation, and recovery strategies and actions from a multiplicity of perspectives.

Keywords 17 SDGS · Hazard · Vulnerability · Geological disasters · Hydrologic disasters · Meteorological disasters · Indicator-based approach · Climate index · Uncertainties · Probability · Cross-cultural practices · Flash floods · Urban flooding · Rural flooding · Riverine floods · Climate justice · Green recovery principles

1 Concept and Definition of Disasters, and Green Recovery Principles

Climate disasters and disease pandemics will increase rapidly as a compound risk. EU's President calls for ambitious climate change and health policy at the State of Union 2020. Disaster management is an emerging discipline in higher education,

---

K.-G. Kim (✉)
Seoul National University, Seoul, South Korea
e-mail: kwigon@snu.ac.kr

K. Kim and C. Atkin (eds.), *Cross-Cultural Perspectives on Climate Change Adaptation*, Cities and Nature, https://doi.org/10.1007/978-3-031-50365-8_2

research and practice. The interrelation between environment, climate change and disasters has attained significant recognition.

A disaster is a sudden, calamitous event that seriously disrupts the functioning of a community or society and causes human, material and economic or environmental losses that exceed the community's or society's ability to cope using its own resources, as can be seen in the equation below. Though often caused by nature, disasters can have human origins (International Federation of Red Cross).

$$(\mathbf{VULNERABILITY} + \text{HAZARD})/\mathbf{CAPACITY} = \mathbf{DISASTER}$$

A disaster occurs when a hazard impacts on vulnerable people

The combination of hazards, vulnerability and inability to reduce the potential negative consequences of risk results in disaster.

When disaster strikes, government and citizens in at-risk, countries have protocols to respond, including shelters to protect displaced and affected populations and emergency plans for geographically vulnerable locations. Leaders and citizens know they must always be ready to respond and begin anew the process of recovery. This year, however, countries need a new plan because the economic recession and the COVID-19 pandemic are part of the equation (WORLD BANK BLOGS 2020).

In the field of disaster risk management, when a new risk appears in already risky situation, we call it a compound risk. The COVID-19 pandemic means communities need to prepare even more thoroughly for disasters, as resources are limited or have been reallocated to respond to the health crisis. In simple terms, the economic and health impacts of COVID-19 are making already disaster-prone countries more vulnerable than ever before.

Temperature and weather irregularity caused by climate change increase the possibility of flooding and droughts and impact the quality and availability of global fresh water supply. As the water supply is affected, farmers will need to irrigate crops.

Floods are considered as one of the most frequently occurring natural hazards worldwide and are occurring more frequently in recent decades. Flooding affects 106 million people annually and causes the largest economic damage (USD 31 billion, annually). 6,000 people are annually killed by flooding (PBL Netherlands Environmental Assessment Agency 2018).

An example of principles of resilient systems is shown in Box 1.

Box 1 Wildavsky's principles of resilient system

Box 1.2 *Wildavsky's principles of resilient systems*

- *The homeostasis principle.* Systems are maintained by feedbacks between component parts which signal changes and can enable learning. Resilience is enhanced when feedbacks are transmitted effectively.
- *The omnivory principle.* External shocks are mitigated by diversifying resource requirements and their means of delivery. Failures to source or distribute a resource can then be compensated for by alternatives.
- *The high flux principle.* The faster the movement of resources through a system, the more resources will be available at any given time to help cope with perturbation.
- *The flatness principle.* Overly hierarchical systems are less flexible and hence less able to cope with surprise and adjust behaviour. Top-heavy systems will be less resilient.
- *The buffering principle.* A system which has a capacity in excess of its needs can draw on this capacity in times of need, and so is more resilient.
- *The redundancy principle.* A degree of overlapping function in a system permits the system to change by allowing vital functions to continue while formerly redundant elements take on new functions.

Source: Barnett (2001)

Resiliency principles can be described as follows:

- Adopt a precautionary but positive approach to development that aims not just to avoid further climate change but also to reduce risks and enhance the resiliency of ecosystems and communities.
- Make cities more resilient to climate change, while accommodating higher population densities.
- Require planning to protect, maintain and enhance the multiple and interdependent co-benefits and ecosystem services offered by the total GHG neutrality.
- Develop a strategy which is built around diverse water resources and mixed water infrastructure with smart water grid systems (Kim 2017). It will allow cities the flexibility to access a portfolio of sources at least cost, in terms of environmental impacts and other externalities.
- Develop uniform standards and engineering procedures for designing climate adaptation measures.

2 Flood Hazard and Typology

The natural hazard might lead to disasters. A natural disasters is a major adverse event (damage and loss) resulting from natural processes of the Earth (hazards); examples are floods, hurricanes, tornadoes, volcanic eruptions, earthquakes, tsunamis, storms and other geologic processes (landslides, rock cracks, etc.).

Disasters can be categorized into geologic disasters (earthquakes, volcanic eruptions, etc.), hydrologic disasters (floods, tsunami and limnic eruptions) and meteorological disasters (cyclones, droughts, tornados, etc.). Risk assessment is a process to determine the probability of losses by analyzing potential hazards and evaluating existing conditions of vulnerability that could pose a threat or harm to property, people, livelihoods and the environment on which they depend.

While there are innumerable definitions for the words flood or flooding, after the major Midwest flood of 1993 a fairy complete definition was provided by the U.S. Executive Office of the President (USEOP). The USEOP defined a flood as 'a general and temporary condition of partial or complete inundation of normally dry land areas from the overflow of river and/or tidal waters and/or the unusual accumulation of waters from any source that may be associated with undesirable effects to life and property' (USEOP 1994). Bedient et al. (2013) state that the definition emphasizes certain critical aspects of flooding such as inundation and damage. The following describes some aspects of flooding.

Flash floods can be caused by several things but are most often due to extremely heavy rainfall from thunderstorms. Flash flooding begins within 6 h, and often within 3 h, of the heavy rainfall or other cause. Flash floods can occur due to dam or levee breaks, and/or mudslides.

Urban flooding is the accumulation of floodwaters that results when the inflow of storm water exceeds the capacity of a drainage system to infiltrate water into the soil to carry it away. When a natural landscape is transformed by urban development, its drainage pattern is can be disturbed in ways that impedes the effective drainage of water and causes increased flooding.

Rural flooding is caused by (1) heavy rain over a flat or low-lying area, (2) a river breaching its embankments due to high water flow or heavy rain in the river's catchment area and (3) any drainage obstructions like landslides and land substances after heavy rains.

Riverine or overbank flooding, occurs when water rises over the edges of a river or stream. It can also be caused by heavy snow melt and ice jams.

Although the same physical principles hold as elsewhere in the hydrological cycle, the hydrology of urban areas is dominated by two distinct characteristics: (1) the preponderance of impervious surfaces (e.g. pavement, roofs) and (2) the presence of man-made or hydraulically 'improved' drainage systems (e.g. a sewer system) (Bedient et al. 2013).

3 Flood Disaster-Related Newly Emerging Issues and Solutions

A climate index is a simple diagnostic quantity that is used to characterize an aspect of a geophysical system such as a circulation pattern. It includes temperature, precipitation, sea level pressure and sea surface anomalies. A variety of methods have been used to derive assorted indices which can be used to identify key critical issues.

As most people are well aware, the immediate impacts of flooding include loss of human life, damage to property, destruction of crops and ecosystems, loss of livestock and deterioration of health conditions owing to waterborne diseases. Floods can also traumatize victims and their families for long periods of time.

There have been newly emerging issues for flood risk assessment and management which can be summarized as follows:

3.1 Increase in Frequency and Intensity of Flood Occurrence

Of all natural disasters, river and coastal flooding kills or affects the largest numbers of people and leads to the highest economic losses, each year. On a global level, flood risks are increasing, and flooding events can lead to either temporary or permanent displacement (PBL Netherlands Environmental Assessment Agency 2018). Data shows floods will soon affect millions more (WRI 2022).

Climate change increases frequency and intensity of flood occurrence. We need preparedness plans to prevent and minimize the impacts of frequent and intense flood occurrence owing to warmer temperatures, changing precipitation patterns and rising sea levels.

Table 1 takes into account only the highest estimated death toll for each disaster and lists them accordingly. It does not include epidemics and famines. It does not include several volcanic eruptions with uncertain death tolls resulting from collateral effects such as crop failures. The list also does not include the 1938 Yellow River flood, which was caused by the deliberate destruction of dikes.

Rigby (2021) estimated increase in UK flooding which is shown in Fig. 1.

3.2 Uncertainties: Climate-Related Uncertainties, Climate Impact-Related Uncertainties and Uncertainties Surrounding Adaptation Project Plans and Programmes

Risk has three components to it, probability, consequence and vulnerability. Early warning system needs accurate predictability and probability of flood occurrence. The local and regional models should be used to reduce the local and regional scale uncertainties.

Table 1 Ten deadliest natural disasters by highest estimated death toll excluding epidemics and famines

Death toll (highest estimate)	Event	Location	Date
4,000,000[1][a]	1931 China floods	China	July 1931
2,000,000[2][3][4]	1887 Yellow River flood		September 1887
830,000[5]	1556 Shaanxi earthquake		January 23, 1556
655,000	1976 Tangshan earthquake		July 28, 1976
500,000+[1]	1970 Bhola cyclone	East Pakistan (now Bangladesh)	November 13, 1970
316,000[6]	2010 Haiti earthquake	Haiti	January 12, 2010
300,000+[7]	526 Antioch earthquake	Byzantine Empire (now Turkey)	May 526
≈ 300,000[8]	1839 Coringa cyclone	Andhra Pradesh, India	November 25, 1839
275,000[9][10]	2004 Indian Ocean earthquake and tsunami	Indian Ocean	December 26, 2004
273,400[11]	1920 Haiyuan earthquake	China	

Source Geeks for Geeks (2024)

3.3 *Threats to Poorer Areas of Cities and Regions*

Developing countries are more frequently exposed to flood disasters than developed countries. Therefore, the UNFCCC Adaptation Fund has allocated USD 878 million to adaptation activities and estimates its impact to date in terms of 381,512 ha of natural habitats preserved or restored, 127 concrete, localized adaptation projects, and 31 million beneficiaries in developing countries (Adaptation Finance Bulletin 2021).

3.4 *Limited Access or Transition to Climate Innovation*

Climate innovation has at least three components: integrated, holistic methodological innovation, digital technology innovation, and multicultural and local partnership innovation.

Innovation for climate change adaptation today is less expensive than rebuilding an entire infrastructure system (energy, transportation, water and communication systems) and socio-economic system after a catastrophe.

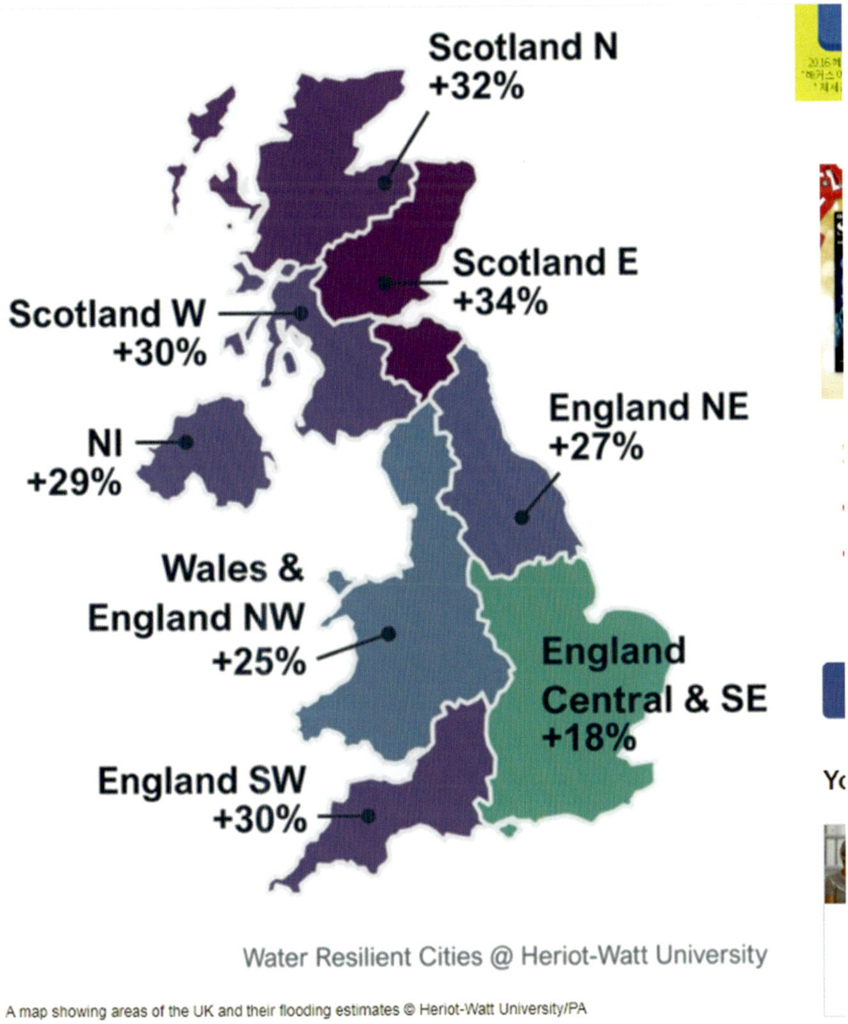

Fig. 1 A map showing areas of the UK and their flooding estimates. *Source* Rigby (2021)

3.5 Fragmentation of Mitigation and Adaptation Measures, and Green Growth for Flood Damage Reduction Actions

Climate change mitigation, adaptation and urban growth patterns are highly connected and thus should be integrated to answer the multidimensional complex questions of flood disasters.

Non-structural and structural measures should be combined as appropriate over future time periods. The interdependencies among and within the two categories of measures should be designed to ensure the synergy effects of them. The interrelation between non-structural and structural methods should be incorporated into project design and performance standards to meet multiple goals.

As an example, waste recovery challenge can help reduce climate disasters and prevent greenhouse gases. Integrated natural flood management can be an adaptive pathway to climate-resilient communities.

3.6 Lack of Flood Risk-Related Big Data

There has been lack of data including observed data, and quantitative and qualitative projection data for flood risk assessment and management.

Data science-climate policy linkages are required to scientifically predict the likelihood or probability of flood occurrence and develop a comprehensive, risk- and science-based adaptation projects and programmes.

3.7 Capacity Gaps to Access to Adaptation Funding

One of the important needs of developing countries is to bridge capacity gaps in accessing adaptation funding.

Following the 20th meeting of the UNFCCC Adaptation Committee (AC), the AC finalized its information note on the topic of capacity gaps in accessing adaptation funding. Based on the submissions from Parties and other stakeholders, the information note highlights capacity gaps in the following areas:

- Raising awareness of climate change impacts and creating enabling environments;
- Making the case for adaptation;
- Navigating and accessing different funding instruments and mechanisms;
- Ensuring capability of the funding seekers; and
- Using and managing funds.

Source UNFCCC Adaptation Committee 2021.

3.8 Lack of Cross-Cultural Practices for Handling Urban Disasters

To solve problems with the capacity gaps described in '3.7', cultural and local context variables should be considered to reflect the method of 'loci' or the 'sense of place' of project areas in adaptation plans.

3.9 Climate Justice Issues at COP27

Use and popularity of climate justice language has increased dramatically in recent years. COP27 has adopted as a slogan 'Action against Hunger' which entails fair consideration of the underlying basis for distributive justice of natural disasters, energy poverty or energy security, food security or agricultural disasters, and water security, and water security. Boxes 2 and 3 below show how a vicious cycle of segregation and disinvestment reinforce the linkage between race and flood risks.

Box 2 Built on a swamp and with aging sewer drains, many neighbourhoods face frequent floods

Flood risk today isn't race blind: Investments in sewers, levees and other infrastructure rescued some neighborhoods from flooding but left others behind. Systemic factors that guided investment exposed Black and Brown households to more severe flood risk. A vicious cycle of segregation and disinvestment reinforced the linkage between race and flood risk.

This pattern makes flooding a matter of environmental justice: Much like asthma, lead exposure, traffic fatalities and other public health threats, flooding disproportionately impacts minority neighborhoods.

Sacramento's disparity is alarming: More than 1 in 5 homes (21.6%) in the city's formerly redlined neighborhoods face a high risk of flooding today. In Sacramento's formerly greenlined areas, just 11.8% of homes face the same risk. That 9.8% disparity in flood risk is the largest among any of the cities analyzed by Redfin. But it isn't the only way to measure this dynamic. In terms of property value, the gap is greater elsewhere.

Chicago
Built on a swamp and with aging sewer drains, many neighborhoods face frequent floods.

Source https://www.bloomberg.com/2021-flood-risk-redlining

Box 3 Historic housing discrimination in Sacramento, USA

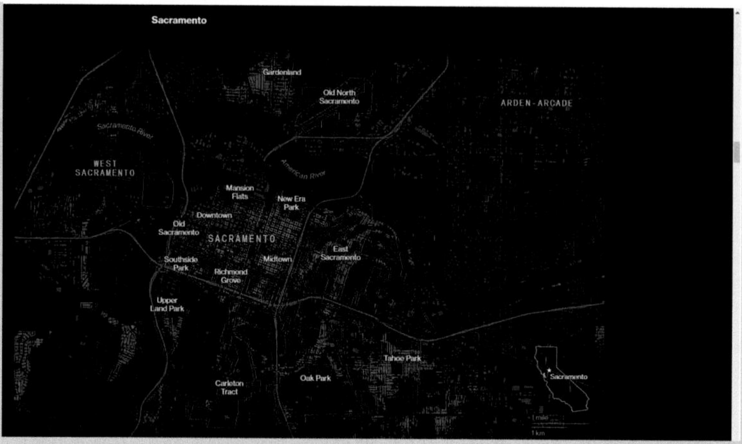

Source Capps and Cannon (2022)

3.10 Climate Change and the Associated Rise in Sea Level

Sea level rise is physically endangering the safety of many coastal cities. The cities need climate change adaptation strategies for the risks posed by the sea level rise to the city's critical infrastructure related to energy, transportation, water and waste, natural resources and communications. However, there are lack of data and information on sea level rise projections derived from global and local climate models.

3.11 Lack of Indicators and Monitoring

Monitoring and reassessment are critical components of any climate change adaptation plan. These should be done taking into account changes in climate science, impacts, technological advancements and adaptation strategies (The New York Academy of Sciences 2010).

The range of flood damage issues and insights is enormous. We need to mobilize our knowledge, politics, strategies and actions in order to provide solutions to the issues.

We need to work holistically, calling on the knowledge of climatologists, engineers, sociologists, economists, public health specialists, designers, architects, community organizers, and more (Plunz and Sutto 1991).

The author has much hope that the rest of this book will inclusively provide integrated framework and methodology for action to close the gap between issues and solutions through a high level of multiplicity of perspectives.

References

Bedient PB et al (2013) Hydrology and floodplain analysis, 5th edn. Pearson, Boston, pp 350, 704
Capps K, Cannon C (2022) Maps of historic housing discrimination show how neighbours that suffered red lining in the 1930s face a far higher risk of flooding today
Geeks for Geeks (2024) Top 10 deadliest floods in history
http://www.bloomberg.com/2021-flood-risk-redlining
Kim KG (2017) Development of an integrated smart water grid model as a portfolio of climate smart cities. J Smart Cities 3(1):23–33. https://doi.org/10.26789/JSC.2017.01.003
New York Academy of Sciences (2010) Climate change adaptation in New York City: building a risk management response, New York City Panel on Climate Change 2010 report. Wiley, Hoboken, p 187
PBL Netherlands Environmental Assessment Agency (2018) The geography of future water challenges. The Hague, p 14
Plunz R, Sutto MP (eds) (1991) Urban climate change crossroads. Ashgate, Surrey
Rigby S (2021) UK flooding could increase by 15–35 per cent by 2080. Science Focus, BBC
UNFCCC Adaptation Committee (2021) Adaptation Finance Bulletin
US Executive Office of President (USEOP) (1994) Flood management executive orders, executive order 11988. February 11, 1994
WORLD BANK BLOGS (2020) Hurricanes don't know borders, neither does a pandemic, 7 Aug 2020
World Resources Institute (WRI) (2022) Securing fresh water for all. Washington DC

International Initiatives and Current Context for Adaptation Planning and Green Recovery

Catherine Atkin

Abstract As international institutions and national and local governments face serial climate impact-driven crises, adaptation has gone from being the focus of a specialized group of planners and disaster risk management experts to one at the center of the global conversation about addressing climate change. Strengthening and accelerating adaptation and green recovery requires an understanding of the global context and on the ground realities in communities. This chapter reviews some of the key global agreements and trends—including the Paris Agreement, UN Sustainable Development Goals, and the movement by vulnerable countries for the funding of "loss and damage"—that define the current context for efforts to move from frameworks and proof points to widespread application and the work to secure the adaptation funding required for impact at scale. It also explores how a range of interrelated institutional, political, economic, socio-cultural, ecological, and technological factors create the enabling environment for change and describes how models for adaptation and climate risk management are evolving to encompass these dynamics.

Keywords Paris Agreement Adaptation Global South · UN Sustainable Development Goals climate adaptation · Sendai Framework for Climate Risk Reduction Global South · Co-benefit solutions adaptation and mitigation · Climate risk management adaptation Global South

1 Introduction

As international institutions and national governments face serial climate impact-driven crises, adaptation has gone from being an area of focus for a specialized group of planners and disaster risk management experts to one at the center of the global conversation about addressing climate change. Governments are not only advancing adaptation and climate risk management agendas in their own countries but also being called on to support adaptation efforts in countries across the Global South that

C. Atkin (✉)
Stanford CodeX Climate Data Policy Initiative, Stanford, USA
e-mail: catherine.atkin@codex.stanford.edu

K. Kim and C. Atkin (eds.), *Cross-Cultural Perspectives on Climate Change Adaptation*, Cities and Nature, https://doi.org/10.1007/978-3-031-50365-8_3

have contributed the least to climate change but are disproportionately experiencing its devastating impacts. This chapter reviews some of the key international agreements and movements that define the current context for efforts to move from frameworks and proof points in climate risk assessment and green recovery to widespread application and the work to secure the adaptation funding required for impact at scale. It also describes how the adaptation to climate change agenda is evolving and identifies the drivers of the strengthened enabling environment that will be required to accelerate our progress.

2 International Agreements and Building Blocks

2.1 UNFCCC COP21 Paris Agreement

While the historic 2015 Paris Agreement is best known for its call for parties to mitigate the impacts of climate change by radically reducing global GHG emissions, the agreement also articulates goals and activities to advance global adaptation efforts to address the impacts of global warming. Article 2 of the agreement calls for "increasing the ability to adapt to the adverse impacts of climate change and foster climate resilience," and Article 7 establishes a global adaptation goal to enhance adaptive capacity, strengthen resilience, and reduce vulnerability to climate change. The agreement also articulates the necessity for climate finance that makes "financial flows consistent with a pathway toward low GHG emissions and climate resilient development" and encourages the ratified countries to prepare Low Emission Development Strategies (LEDS) and Nationally Determined Contributions (NDCs) (Paris Agreement to the United Nations Framework Convention on Climate Change 2015). While international bodies and countries have been slow to advance the adaptation-focused elements of the agreement, including the development of comprehensive adaptation plans, the Paris Agreement has been important in creating an international accountability architecture that includes adapation.

2.2 The UN Sustainable Development Goals

The UN Sustainable Development Goals (SDG) were created and adopted by the United Nations General Assembly in 2015 as a successor construct to the Millennium Development Goals. The development of the SDGs represents a significant advancement of a holistic and integrated approach to people and the environment, eradicating poverty and hunger and promoting sustained and inclusive economic growth. SDG 13 is focused on climate action and the need to "[T]ake urgent action to combat climate change and its impacts." (UN 2015). Of the four targets set out in SDG 13, the first is to "[S]trengthen resilience and adaptive capacity to climate-related hazards and natural disasters in all countries" and includes as indicators of progress the number of countries that adopt and implement the national disaster risk reduction strategies in line with the Sendai Framework for Disaster Risk Reduction as well as the number that adopt and implement local disaster risk reduction strategies in line with national disaster risk reduction strategies (2017 Indicators for UN 2015) (Table 1).

2.3 Sendai Framework for Disaster Risk Reduction

In addition to the adoption of the Paris Agreement and the UN SDGs in 2015, the Sendai Framework for Disaster Risk Reduction Framework (2015–2030), was adopted by the United Nations member states at the World Conference on Disaster Risk Reduction in Sendai, Japan, and endorsed by the UN General Assembly in June 2015. The United Nations Office for Disaster Risk Reduction oversees the implementation of the Sendai Framework. The framework not only set seven global targets but also four priorities for action to prevent new and reduce existing disaster risks: (i) Understanding disaster risk; (ii) Strengthening disaster risk governance to manage disaster risk; (iii) Investing in disaster reduction for resilience; and (iv) Enhancing disaster preparedness for effective response, and to "Build Back Better" in recovery, rehabilitation, and reconstruction (United Nations Office for Disaster Risk Reduction 2015) (Table 2).

2.4 Adaptation Finance

While there has been increased attention to adaptation policy at the national and international levels, there has been minimal progress in securing the estimated $218–$387 billion per year in adaptation finance needed to achieve the goals articulated in the Paris Agreement (UNEP Adaptation and Gap Report 2023). Even though Multilateral Development Banks and Development Finance Institutions have set targets to increase the share of climate finance that goes to fund adaptation projects, current adaptation finance has declined since 2020 to $21.3 billion USD in 2021. While adaptation has risen in prominence at the yearly COPs—including the announcement of the COP26 Glasgow Climate Pact in 2021 which set an ambitious goal to double adaptation finance goals from 2019 levels by 2025—progress toward this goal has been slow and the adaptation funding gap has continued to widen. On a brighter note, concerted pressure by vulnerable countries at COP27 in Sharm el-Sheikh, Egypt, led to the creation of a fund to address "loss and damages" from the severe impacts of climate change.

Table 1 *Source* UN Sustainable Development Goal 13, targets and indicators, https://unstats.un.org/sdgs/metadata/?Text=&Goal=13&Target=

SDG 13 climate action: Take urgent action combat climate change and its impacts	
Target 13.1 Strengthen resilience and adaptive capacity to climate-related hazards and natural disasters in all countries	**Indicator 13.1.1**: Number of deaths, missing persons, and directly affected persons attributed to disasters per 100,000 population **Indicator 13.1.2**: Number of countries that adopt and implement national disaster risk reduction strategies in line with the Sendai Framework for Disaster Risk Reduction 2015–2030 **Indicator 13.1.3**: Proportion of local governments that adopt and implement local disaster risk reduction strategies in line with national disaster risk reduction strategies
Target 13.2 Integrate climate change measures into national policies, strategies, and planning	**Indicator 13.2.1**: Number of countries with nationally determined contributions, long-term strategies, national adaptation plans, and adaptation communications, as reported to the secretariat of the United Nations Framework Convention on Climate Change **Indicator 13.2.2**: Total greenhouse gas emissions per year
Target 13.3 Improve education, awareness-raising, and human and institutional capacity on climate change mitigation, adaptation, impact reduction, and early warning	**Indicator 13.3.1**: Extent to which (i) global citizenship education and (ii) education for sustainable development are mainstreamed in (a) national education policies; (b) curricula; (c) teacher education; and (d) student assessment
Target 13.4 Implement the commitment undertaken by developed-country parties to the United Nations Framework Convention on Climate Change to a goal of mobilizing jointly $100 billion annually by 2020 from all sources to address the needs of developing countries in the context of meaningful mitigation actions and transparency on implementation and fully operationalize the Green Climate Fund through its capitalization as soon as possible	**Indicator 13.a.1**: Amounts provided and mobilized in US dollars per year in relation to the continued existing collective mobilization goal of the $100 billion commitment through 2025
Target 13.5 Promote mechanisms for raising capacity for effective climate change-related planning and management in least developed countries and small island developing States, including focusing on women, youth, and local and marginalized communities	**Indicator 13.b.1**: Number of least developed countries and small island developing States with nationally determined contributions, long-term strategies, national adaptation plans, and adaptation communications, as reported to the secretariat of the United Nations Framework Convention on Climate Change

Table 2 7 global targets articulated in Sendai Framework for Disaster Risk Reduction Framework

1. Substantially reduce global disaster mortality by 2030, aiming to lower average per 100,000 global mortality between 2020 and 2030 compared to 2005–2015
2. Substantially reduce the number of affected people globally by 2030, aiming to lower the average global figure per 100,000 between 2020 and 2030 compared to 2005–2015
3. Reduce direct disaster economic loss in relation to global gross domestic product by 2030
4. Substantially reduce disaster damage to critical infrastructure and disruption of basic services, among them health and educational facilities, including through developing their resilience by 2030
5. Substantially increase the number of countries with national and local disaster risk reduction strategies by 2020
6. Substantially enhance international cooperation to developing countries through adequate and sustainable support to complement their national actions for implementation of the framework by 2030
7. Substantially increase the availability of and access to multi-hazard early warning systems and disaster risk information and assessments to the people by 2030

Source Sendai Framework for Disaster Risk Reduction Framework https://www.undrr.org/publication/sendai-framework-disaster-risk-reduction-2015-2030

3 Global Trends in Adaptation Theory and Practice

3.1 Integrated Approaches to Adaptation, Mitigation, and Sustainable Development

With siloed approaches by international, national and local governments and institutions to advancing adaptation, mitigation, and sustainability goals often yielding disappointing results, new approaches must address the interrelated needs of communities by advancing integrated, co-benefit strategies and solutions. The IPCC AR 6 Adaptation Report promotes the concept of climate resilient development as a way to move us beyond siloed efforts and towards powerful and integrated approaches that advance sustainable development in communities: "Adaptation and mitigation can reduce climate-related risks. Implementing these two types of climate action together increases their effectiveness by exploiting synergies and reducing trade-offs among them. In addition, implementing adaptation and mitigation as an integral part of development can similarly make all three more effective (Section 18.2.3)". (Ara Begum et al. 2022). The foundation for this change begins with comprehensive policymaking and integrated funding at all levels of government. In addition, at the planning and implementation levels, it is necessary to develop integrated frameworks and tools that "bake in" a more comprehensive approach, as well as put in place new structures in government that bring together planners with those focused on mitigation and sustainable development.

3.2 *Equity and Climate Justice*

Centering social justice issues and ensuring equitable outcomes as part of climate risk management theory and practice is moving to center stage. Climate risk management, done well, identifies the differential risks and burdens experienced by people in nations, regions, and communities and allows for the targeting of resources to serve the most vulnerable first. Communities contending with high unemployment, food instability, and poor shelter options are more vulnerable and less capable of responding to climate hazards and addressing these core human needs must be part of the design of the set of strategies and solutions to address climate-related hazards. The IPCC report states with high confidence that "integrated, multi-sectoral solutions that address social inequities, differentiate responses based on climate risk and cut across systems, increase the feasibility and effectiveness of adaptation in multiple sectors." (Johnston et al. 2022).

There is an emerging body of research on the increased efficacy of adaptation initiatives where communities are given more decision-making authority. As a result, local governments are reexamining legacy approaches to community engagement where the focus has been to inform and consult with affected communities and instead implement more collaborative and community-led approaches (Castellanos et al. 2022). This means not only including robust community voice in the risk assessment process but also enfranchising community members to be part of—and even lead—decision making about which hazard reduction measures should be prioritized and ensuring their continued role in the implementation phases. There is also a recognition of the importance of lived experience and indigenous knowledge in local communities (Johnston et al. 2022).

3.3 *Importance of Strengthening Adaptation Enabling Environment*

The vulnerability of communities to the negative impacts of a changed climate is determined by a range of interrelated institutional, political, economic, sociocultural, ecological, and technological factors that create the enabling environment for change. As the sector grapples with less than expected impacts from adaptation projects across the globe there has been increasing focus on strengthening the enabling conditions "that enhance the feasibility of adaptation and mitigation options… includ[ing] finance, technological innovation, strengthening policy instruments, institutional capacity, multi-level governance and changes in human behavior and lifestyles" (IPCC 2022).

Examples of ways to strengthen the enabling conditions include: building the political will for increased adaptation funding and the development of public, private, and blended finance mechanisms, implementing policies for comprehensive and vertically connected adaptation planning at all levels of government, developing comprehensive land use planning policy and regulations, and investing in technology for the digitalization of climate risk management as well as the leveraging Artificial Intelligence (AI), Internet of Things (IoT) and Distributed Ledget Technology (DLT) to allow for the rapid scale and deployment of climate tech solutions.

3.4 Adaptation and Resilience at the City Level

Although a bias towards centralized adaptation planning at the country and regional levels has predominated, there is increasing awareness that what happens on the ground in communities is the ultimate measure of success. Given that rapid urbanization means that by 2050 it is estimated that 70 out of 100 people will live in cities by 2050, urban areas must be the focus of global efforts to support adaptation and resilience (World Bank 2024). ICLEI - Local Governments for Sustainability - a global network of over 2500 cities explains that "[L]ocal governments are closest to the people and their needs, enabling them to drive bottom-up approaches and address global challenges locally, where they have the greatest impact. Policy integration and coherence with the global agendas has progressively happened at the local level. Additionally, major international networks of local governments have undertaken joint advocacy in international policy processes and have supported their constituencies to implement programs and solutions to promote resilient development." While cities across the globe are making adaptation and resilience building top priorities, the reality is that progress in advancing these agendas will be limited until new and robust adaptation and resilience financing at the city level is made available.

References

2017 Indicators for UN (2015) Transforming our world: the 2030 agenda for sustainable development. Resolution Adopted by the General Assembly on 25 September 2015, 42809, 1–13

Ara Begum R, Lempert R, Ali E, Benjaminsen TA, Bernauer T, Cramer W, Cui X, Mach K, Nagy G, Stenseth NC, Sukumar R, Wester P (2022) Point of departure and key concepts. In: Climate change 2022: impacts, adaptation, and vulnerability. Contribution of Working Group II to the Sixth Assessment Report of the Intergovernmental Panel on Climate Change [Pörtner H-O, Roberts DC, Tignor M, Poloczanska ES, Mintenbeck K, Alegría A, Craig M, Langsdorf S, Löschke S, Möller V, Okem A, Rama B (eds)]. Cambridge University Press, Cambridge, UK and New York, NY, USA, pp 121–196. https://doi.org/10.1017/9781009325844.003

Castellanos E, Lemos MF, Astigarraga L, Chacón N, Cuvi N, Huggel C, Miranda L, Moncassim Vale M, Ometto JP, Peri PL, Postigo JC, Ramajo L, Roco L, Rusticucci M (2022) Central and South America. In: Climate change 2022: impacts, adaptation, and vulnerability. Contribution of Working Group II to the Sixth Assessment Report of the Intergovernmental Panel on Climate Change [Pörtner H-O, Roberts DC, Tignor M, Poloczanska ES, Mintenbeck K, Alegría A, Craig

M, Langsdorf S, Löschke S, Möller V, Okem A, Rama B (eds)] Cambridge University Press, Cambridge, UK and New York, NY, USA, pp 1689–1816. https://doi.org/10.1017/9781009325844.014

https://e-lib.iclei.org/publications/Resilient-Cities-Thriving-Cities_The-Evolution-of-Urban-Resilience.pdf

IPCC (2022) Summary for policymakers [Pörtner H-O, Roberts DC, Poloczanska ES, Mintenbeck K, Tignor M, Alegría A, Craig M, Langsdorf S, Löschke S, Möller V, Okem A (eds)]. In: Climate change 2022: impacts, adaptation, and vulnerability. Contribution of Working Group II to the Sixth Assessment Report of the Intergovernmental Panel on Climate Change [Pörtner H-O, Roberts DC, Tignor M, Poloczanska ES, Mintenbeck K, Alegría A, Craig M, Langsdorf S, Löschke S, Möller V, Okem A, Rama B (eds)]. Cambridge University Press, Cambridge, UK and New York, NY, USA, pp 3–33. https://doi.org/10.1017/9781009325844.001

Johnston KA, Taylor M, Ryan B (2022) Engaging communities to prepare for natural hazards: a conceptual model. Nat Hazards 112:2831–2851. https://doi.org/10.1007/s11069-022-05290-2

Organisation for Economic Co-operation and Development (2016) OECD DAC Rio markers for climate. Handbook. The Development Assistance Committee. https://www.oecd.org/dac/environment-development/Revised%20climate%20marker%20handbook_FINAL.pdf; Aggregate Trends of climate finance provided and mobilised by developed countries in 2016–2020: insights from disaggregated analysis. Paris, France. https://doi.org/10.1787/d28f963c-en

Paris Agreement to the United Nations Framework Convention on Climate Change, Dec. 12, 2015, T.I.A.S. No. 16-1104

UN (2015) Transforming our world: the 2030 agenda for sustainable development. Resolution adopted by the General Assembly on 25 September 2015, 42809, 1–13

UN world conference on disaster risk reduction, 2015 March 14–18, Sendai, Japan. United Nations Office for Disaster Risk Reduction, Geneva

UNEP Adaptation and Gap Report (2023) file:///Users/catherineatkin/Downloads/adaptation_gap_report_2023.pdf

World Bank (2024) https://www.worldbank.org/en/topic/urbandevelopment/overview#:~:text=Today%2C%20some%2056%25%20of%20the,people%20will%20live%20in%20cities

Climate Risk Management Frameworks and Overview to Case Studies

Nine-Step Conceptual Framework for Climate Disaster Risk Assessment and Green Recovery Planning

Kwi-Gon Kim

Abstract The objective of this chapter is to present a new and integrated holistic approach to climate disaster assessment and green recovery developed by professor Kwi-Gon Kim which embraces digital technology, and cultural and local context variables within a broad frame of planning and innovation in line with international movements described in Chap. "Climate Disaster Risk Management—Risk Assessment and Green Recovery Planning: Core Concepts and Identification of Current Issues". The nine steps of Professor Kim's conceptual framework are detailed and lay the foundation for its application to flood risk assessment and green recovery and used by a number of case authors of the case studies in Part 4. The climate disaster and green recovery process consist of nine steps to help practitioners started with adaptation, (Steps 1–2)) delineate boundary of the project, (Steps 3–6)) understand and assess the impacts of current and future climate change, (Step 7) identify significant flood risks, (Step 8) prioritize flood damage reduction options, and (Step 9) monitor and evaluate periodically.

Keywords Integrated holistic approach · Green recovery · Three pillars · Digital technology solutions · Cultural and local context · Catchment area · Inventory map · Composite maps · Watersheds · Hydrologic cycle · Multi-level methods · Overland flow · Urban hydrological system

1 Three Pillars of Climate Disaster and Green Recovery Planning Framework

Innovative climate solutions and sustainable development outcomes are a framework that incorporates comprehensive planning and implementation, and available digital technological solutions in addition to cultural and local context considerations for climate disaster and green recovery planning. Figure 1 illustrates the concept of three pillars of climate disaster and green recovery planning framework.

K.-G. Kim (✉)
Seoul National University, Seoul, South Korea
e-mail: kwigon@snu.ac.kr

K. Kim and C. Atkin (eds.), *Cross-Cultural Perspectives on Climate Change Adaptation*, Cities and Nature, https://doi.org/10.1007/978-3-031-50365-8_4

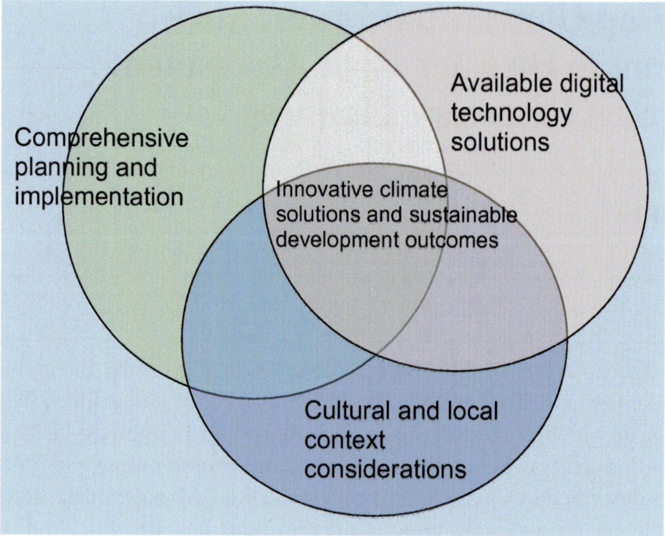

Fig. 1 Three pillars of climate disaster and green recovery planning

2 A New Nine-Step Flood Risk Assessment and Green Recovery Planning Framework

Figure 2 shows the traditional IPCC "propeller" images of hazard, vulnerability, expose, and response. In contrast with this traditional framework, this section explores alternative opportunities to assess flood risks and to plan green recovery and proposes a nine-step approach.

The nine-step framework integrates three pillars in Fig. 1 into risk assessment and management process and employs risk mapping technology for flood risk including an innovative map-based indicator approach. One of the main features of the steps is to combine land suitability and risk analysis which is a tool for urban green growth and sustainable development at the site level in a way of harmonizing negative and positive thinking (WCED, 2009), as shown in Fig. 3.

The nine steps in Figure 3 provide a comprehensive process of risk assessment and green recovery planning. The basic steps are as follows:

Step 1: The delineation of the boundary of landscape district and place unit for project areas should be based on the bioregion with understanding of a sense of place.

Step 2: Catchment area boundary should also be defined with respect to the place units because we are continually interacting with places, both natural and man-made.

The watershed divide is the loci of points (the ridge line) that separate two adjacent watersheds, which then drain into different outlets. Figure 4 depicts several watershed areas that have been defined based on topographic or elevation data.

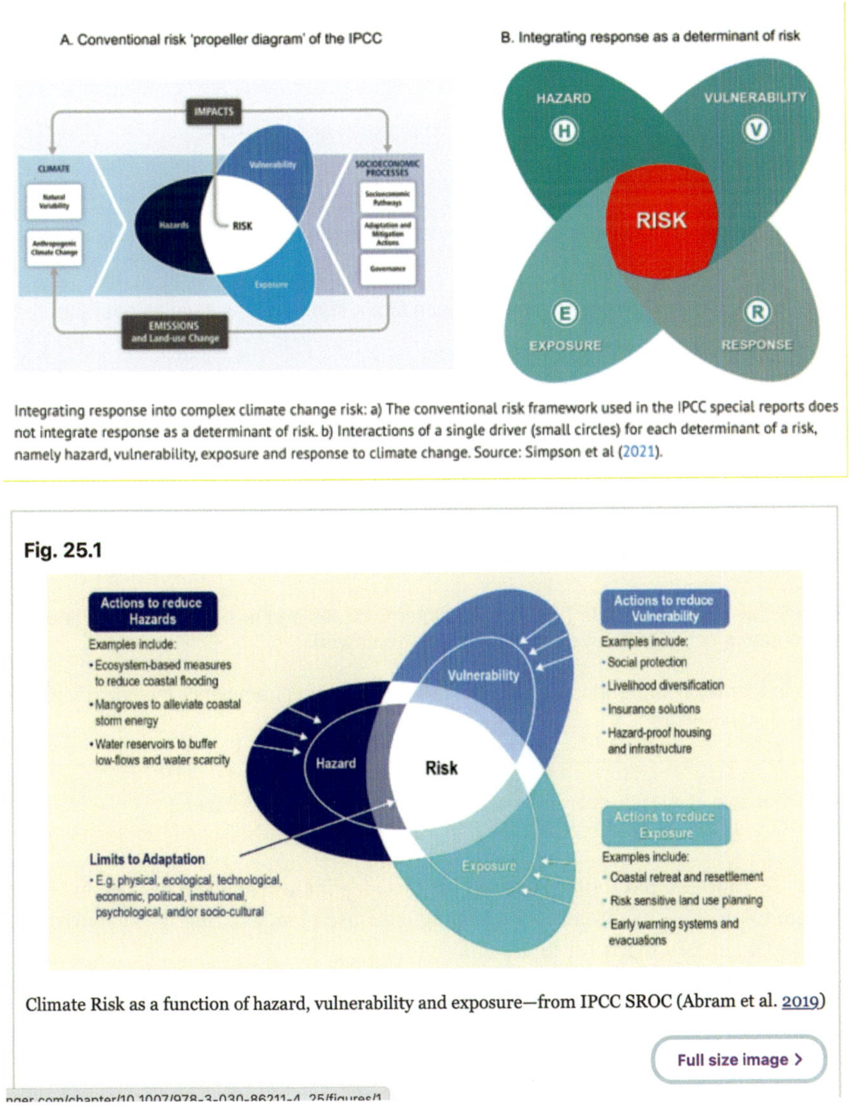

Fig. 2 Traditional IPCC "propeller" images of hazard, vulnerability, exposure, and response. *Source* IPCC WGIl AR5 (2014)

Step 3: Factors should be identified to be included in risk assessment exercise. In this step, it is recommended that factors not only for risk assessment but also for land suitability analysis be included in data collection and analysis for the map overlay technique at the various levels.

9-Step Climate Disaster Risk Assessment and Green Recovery Planning (Dr. Kwi-Gon Kim)

Step 1: Delineate the boundary of landscape district and place unit

Step 2: Delineate boundary of catchment area

Step 3: Identify factors (indicators) to be included in the flood risk assessment exercise

Step 4: Prepare an "inventory map" for each factor showing how it varies over the study area

Step 5: Create composite maps by overlaying two or more inventory maps

Step 6: Analyze the composite maps to make influences relevant to flooding risk assessment

Step 7: Use spatial multi-criteria evaluation - including hazard mapping and vulnerability assessment - to assess overall risk including the probability of damage and loss

Step 8: Identify appropriate hazard reduction measures and remediation strategies and create a comprehensive implementation framework

Step 9: Identify comprehensive monitoring and evaluation framework including MRV and ICT applications

Fig. 3 Nine-step climate disaster risk assessment and green recovery planning

Step 4: Techniques and tools proposed for this step include the use of GIS and other software to produce location maps, existing land-use plan, existing flood zone maps, and other inventory maps for land suitability.

Step 5: Multi-layered maps are produced to overlay different types of inventory maps including maps described in step 4 and layers of physical, digital, and socio-financial world.

Step 6: This step includes an analysis of existing situation, SWOT analysis, land-use plan, and identification of hazardous zones. Details of each step are described in Chap. Methods and Technologies for Developing the Indicators of Climate Disaster Risk Assessment and Green Recovery Planning: The Standard Techniques for a Whole System.

As an example, potential 1-in-100 year flood zones for case study areas can be developed by using a series of maps under different climate change scenarios (Fig. 5).

Step 7: This step is a multi-layered process. Even cyber-layers can be brought into the process (Kim and Thioye 2022, p. 21). The likelihood or probability categories defined in the IPCC Working Group/Technical Summary (2007) can be used here.

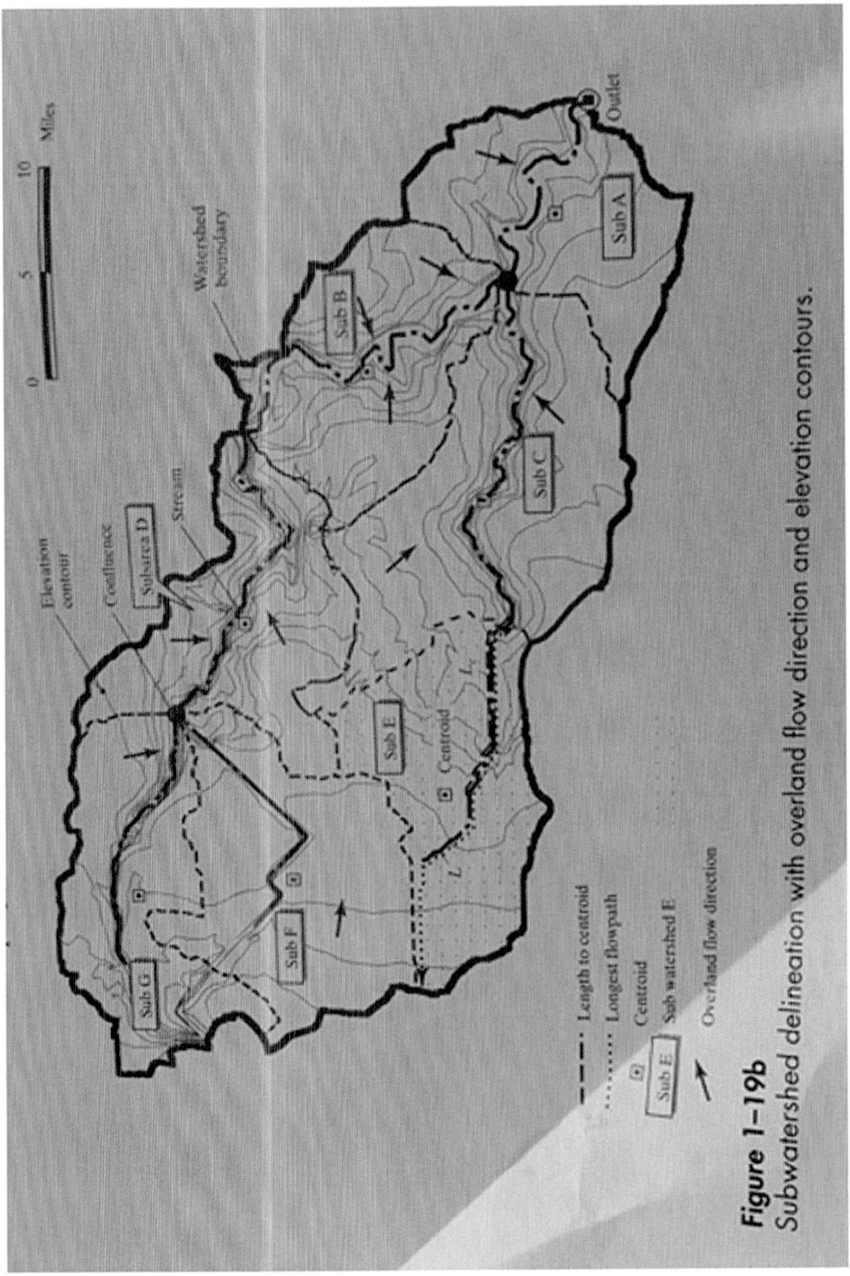

Fig. 4 Subwatershed delineation with overland flow direction and elevation contours. *Source* (Bedient et al. 2013)

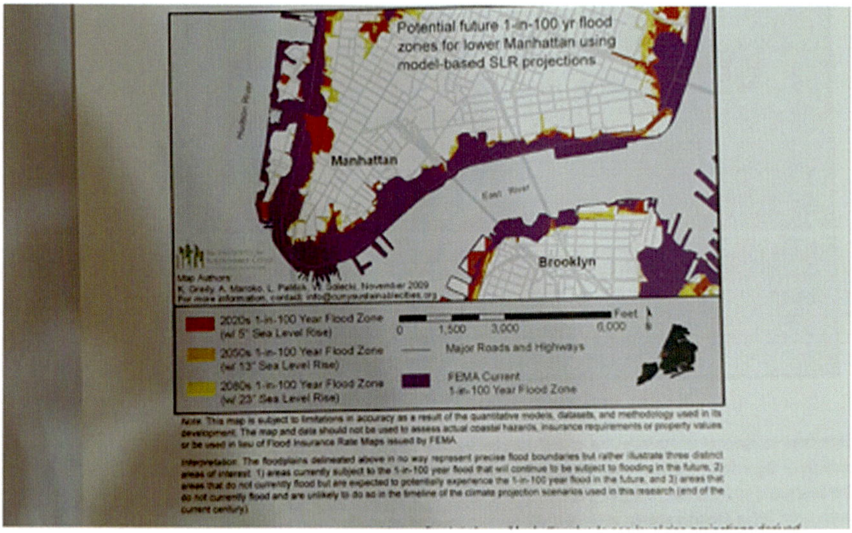

Fig. 5 Areas potentially at-risk to the 1-in-100 year floods: The case of Lower Manhattan due to sea level rise projections derived from global climate models. *Source* The New York Academy of Sciences (2010, p. 336)

It is important to note that the assignment of climate hazards and vulnerability to these categories should be based on global and local climate simulations, published literature, and expert judgment when it is applied to cities, communities, and sites.

Step 8: If it is determined what areas would be vulnerable to climate hazards, then mitigation measures or control measures should be presented. Excellent references are available describing various structural and non-structural measures that can be utilized for various areas.

Step 9: The final step in Fig. 3 is a discussion of monitoring and evaluating changes resulting from implementation of adaptation plans. This step recommends the agencies concerned to establish an annual State and Trends in Adaptation Report to show progress to date toward further resilience to climate impacts. "The State and Trends in Adaptation Report 2021: How Adaptation Can Make Africa Safer, Greener and More Prosperous in a Warming World" is an excellent example in the field (Global Center on Adaptation, 2021).

3 Multilevel Approach to Flood Risk Assessment and Management Framework at the Different Scales

The basic components of hydrologic cycle include precipitation, evaporation, evapotranspiration, infiltration, overland flow, streamflow, and groundwater flow (Bedient et al. 2013). The movement of water (rainfall and runoff) through various phases of the hydrological cycle varies greatly in time and space, giving rise to extremes of floods. The magnitude and the frequency of occurrence of these extremes are of great interest to natural disaster planner from a design and operation standpoint.

Figure 6 shows urban hydrological system and cycle which combine runoff from natural surroundings and man-made drainage systems. Different design and operation methods are needed for different hydrological systems.

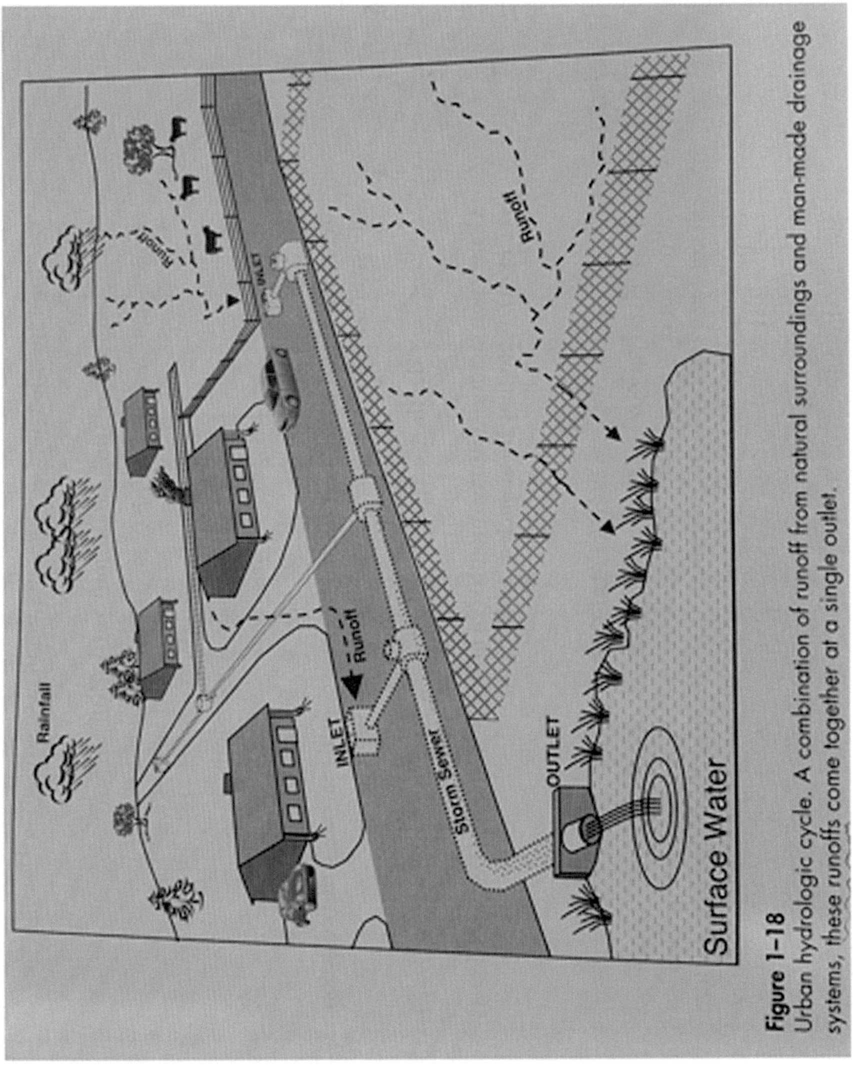

Fig. 6 Urban hydrologic cycle. A combination of runoff from natural surroundings and man-made drainage systems, these runoffs come together at a single outlet. *Source* Bedient et al. (2013, p. 53)

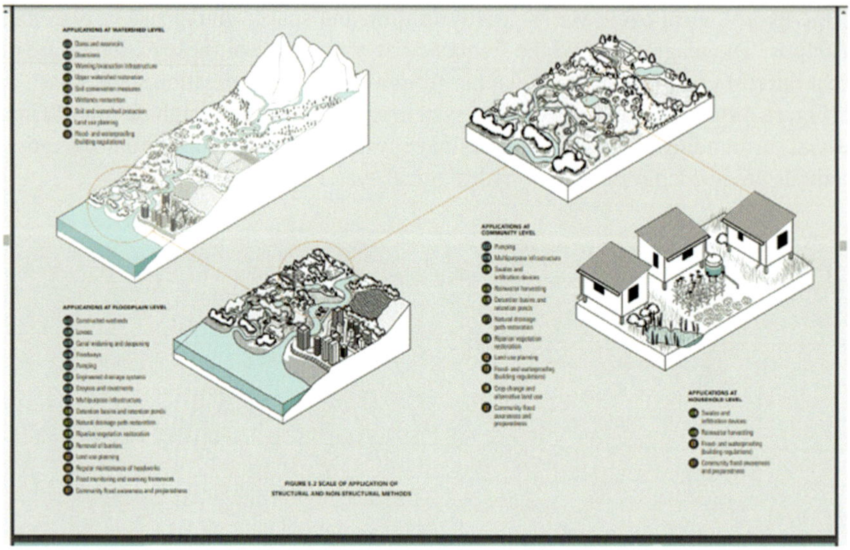

Fig. 7 Scale of application of structural and non-structural methods. *Source* USAID, WWF (2017, p. 82)

Floods have consequences at multiple levels, including national/regional, floodplain, community, and household. USAID and WWF (2017) suggested multiple-level methods which should be selected based on the specific requirements at different scales.

In order to be able to include the damage reduction measures in the framework described in the previous chapter at different scales, it is necessary to look at different levels of framework in which the different actions are to take place (Fig. 7).

World Bank (2012) has further elaborated the multi-level methods, as shown in Fig. 8.

4 Establishment of New Urban Agenda: A Practical Application to Climate Innovation Land-Use Planning

We need to establish a new urban agenda for the nine-step conceptual framework at the different scales to make cities and human settlements inclusive, safe, resilient, and sustainable.

Forward-looking advanced climate innovation planning, in particular land-use planning, can influence both natural disaster and GHG emissions by:

(a) locating less vulnerable cities and communities; and
(b) reducing consumption demand and hence emissions.

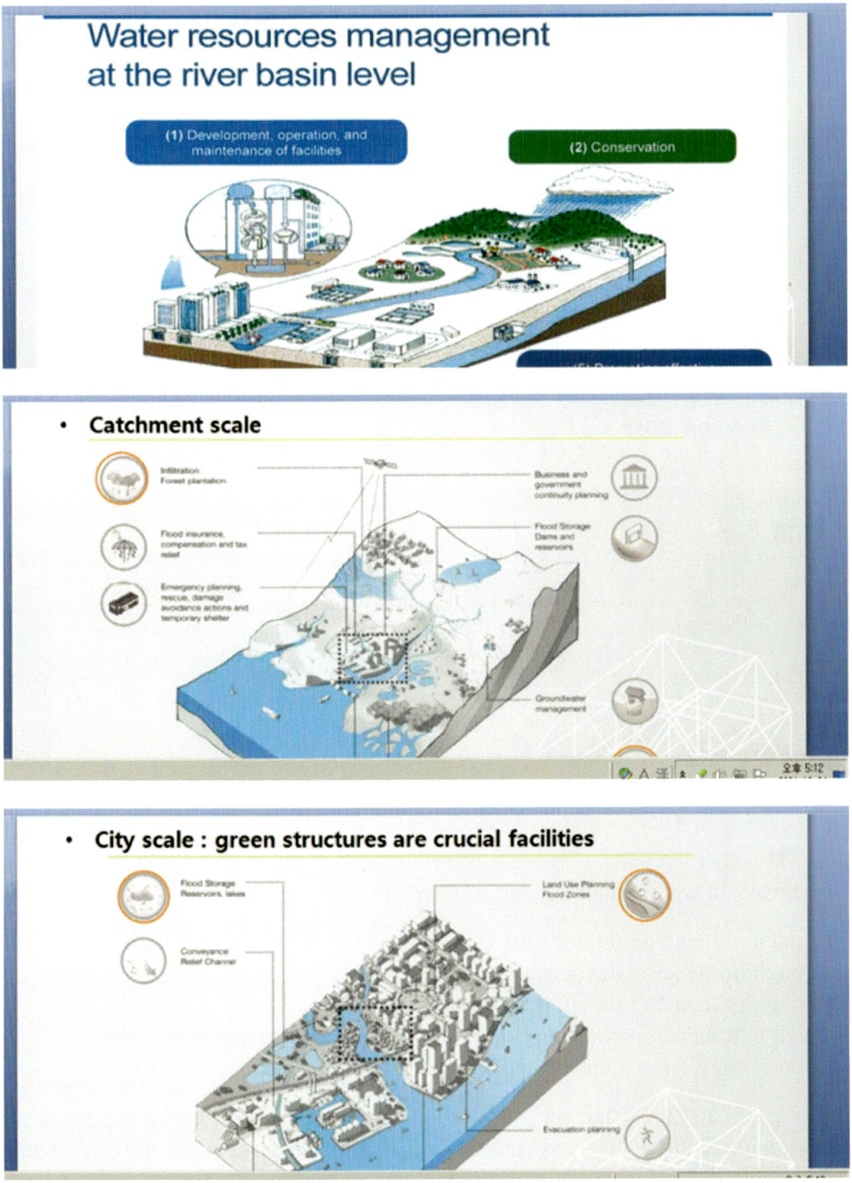

Fig. 8 Integrated approach to flood risk management framework at the different scales. *Source* World Bank (2012)

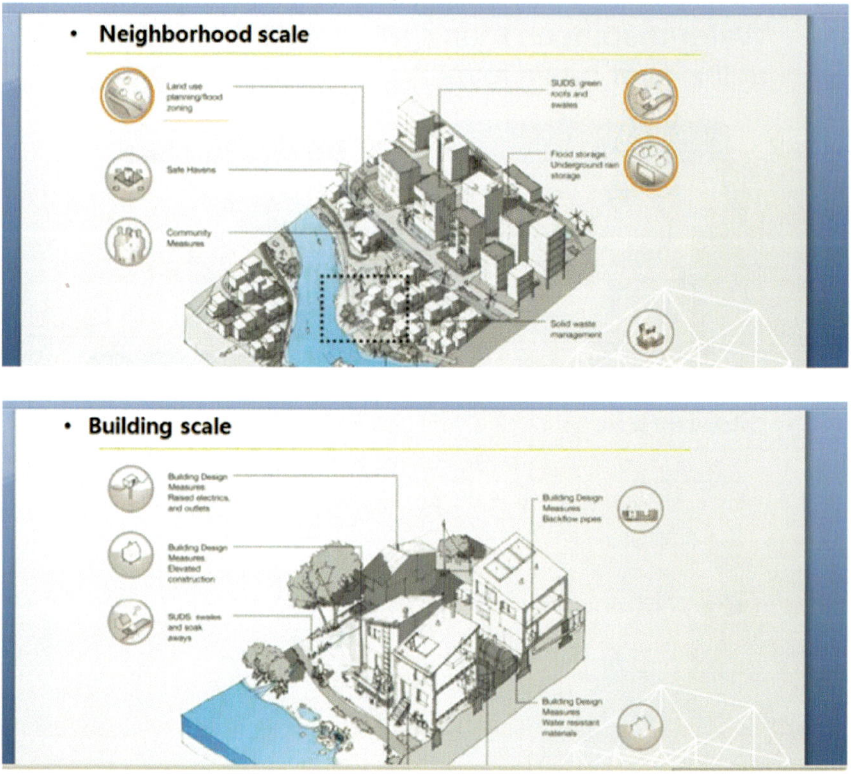

Fig. 8 (continued)

The ways in which planning can achieve this include:

- through influencing the location of development and settlement patterns;
- through influencing the density of development;
- through influencing the mixture of land uses;
- through influencing consumption behavior and land-use patterns; and
- through simulating alternative development patterns

Research can be carried out to improve estimates of GHG emissions, improve adaptation indicators and risk index, and identify measures of mitigation and adaptation from land-use and land management in space and time.

Table 1 shows an interaction matrix between adaptation and mitigation associated with land-use planning and data sources.

Although this chapter suggests an integrated framework for the nine steps, many analyses are not fully synthesized in a formal way in practice. It is still a developing framework which may require substantial research. A reflection of the usefulness of the framework will be given in the cross-cultural case studies discussed later in this book.

Table 1 Land-use planning matrix for shelter sector (buildings)

Land-use planning	Adaptation	Mitigation	Leading data sources related to human activities
Location of development and settlement patterns	• Location of development in compliance with policies set out in adaptation action plans • Index of land suitability and adaptation (risk) indicators • Flood-prone zoning • High-lying areas • Unified watershed assessments and policy • Watershed restoration and protection • Digital twin water management platform • Flood defense standards and criteria: Locational criteria for residential care homes and nursing homes • Bridges between natural disaster and natural resources programs • Response to growth pressure on sensitive coastal areas • Enhancement of the condition of riparian areas and stream corridors • Geographic-based planning to protect and restore wetlands	• Sustainable forms of urbanization • Energy-efficient spatial structures • Location of development in compliance with policies set out mitigation action plans • Index of carbon neutrality indicators • Low-carbon emission zoning • Compact development • Carbon emission standards and criteria	• Land-use maps • Planned unit development maps • Watershed and catchment maps • Unitary development plans – Housing prices – Residential stability • Flood risk maps for people and buildings • River system maps • Climate action plans • Watershed restoration action strategy maps • Flood maps • CO_2 monitoring maps

(continued)

Table 1 (continued)

Land-use planning	Adaptation	Mitigation	Leading data sources related to human activities
Density of development	• Density control • Green remodeling • NBS • Household-associated infrastructure	• Control of urban sprawl • Higher residential densities within settlements • Green remodeling • NBS • Household-associated transport	
Mixture of land uses	• Dwelling mix – Family units – smaller units – Studio flats • Designation of flood shelter • Delineation of riparian areas: and establishing conservation buffers	• Dwelling mix: Mixture of building size • Designation of carbon forests	
Community-based behavior and land-use patterns	• A community-based approach to riverine ecosystem restoration and flood hazard mitigation projects • Prior negotiation with stakeholders • Purchase of easements and land acquisition • Building community partnership – Building strong partnerships to speed restoration and protection	• Improving information and citizens' right to know – Climate information network • Prior negotiation with stakeholders	
Simulating alternative development patterns	Appropriate attenuation measures when development is likely to increase the risk of flooding downstream due to the generation of additional surface water run-off from roofs and paved surfaces	• The worst scenario of development patterns and CO_2 emissions • The best scenario of development patterns and CO_2 emissions	

References

Bedient PB, Huber WC, Vieux BE (2013) Hydrology and floodplain analysis, 5th edn. Pearson, Boston, p 53

Global Center on Adaptation (2021) State and trends in adaptation report 2021: how adaptation can make Africa safer, greener and more prosperous in a warming world, Rotterdam, The Netherlands

IPCC WGll AR5 (2014) Climate change 2014: impacts, adaptation and vulnerability

Kim KG, Thioye M (eds) (2022) Planning climate smart and wise cities: a multidisciplinary approach. Springer, Heidelberg, p 21

The New York Academy of Sciences (2010) Climate change adaptation in New York City: building a risk management response, New York City Panel on climate change 2010 report, Blackwell, Boston, Massachusetts, p 336

United Nations, World Commission on Environment and Development (WCED) (2009) Our common future. Oxford University Press, Oxford

USAID, WWF (2017) Natural and nature-based flood management: a green guide, p 82

World Bank (2012) Unpublished

Methods and Technologies for Developing the Indicators of Climate Disaster Risk Assessment and Green Recovery Planning: The Standard Techniques for a Whole System

Kwi-Gon Kim

Abstract Those using the nine-step risk assessment and green recovery framework outlined in Chap. "International Initiatives and Current Context for Adaptation Planning and Green Recovery" need access to a broad set of evidence-based scientific integrated methods and technologies—including smart digital technologies—for use with each step as they seek to assess and manage the risks posed by flooding and other climate-related events in cities and communities across the globe. Although floods are one of the most frequently occurring natural hazards worldwide, successful adaptation can be achieved through avoidance, planning control, and smart and wise measures. In this book, a risk-based indicator approach to flood risk assessment is used as an example of climate-related hazards and natural disaster platform project models. Indicator-based approaches are semi-quantitative methods for evaluating climate-based land use suitability mapping and risk mapping—including the components of community development, hazard frequency and intensity, physical vulnerability, and risk probability—to compare different areas or communities. As part of this process, data for each of these indicators is collected at a particular spatial level, by administrative units. These indicators are then standardized (e.g., by reclassifying them between 1 and 0), weighted internally within objectives, and then the various objectives are also weighted among themselves. Although the individual indicators normally consist of quantitative data (e.g., population and property statistics), the resulting hazard, vulnerability, and risk probability results are scaled between 0 and 1. These relative data allow the user to compare the indicators for the various administrative units. These methods can be carried out at different levels, ranging from specific sites, local communities, cities to countries.

Keywords Avoidance · Planning control · Smart and wise measures · Non-structural measures · Structural measures · Smart digital technologies · Guiding principles · Domain · Early warning systems · SWOT analysis · Overlay

K.-G. Kim (✉)
Seoul National University, Seoul, Korea
e-mail: kwigon@snu.ac.kr

K. Kim and C. Atkin (eds.), *Cross-Cultural Perspectives on Climate Change Adaptation*, Cities and Nature, https://doi.org/10.1007/978-3-031-50365-8_5

analysis · Adaptation indicators · Land use planning · Flood proofing · Flood insurance programs · Low Impact Development (LID)

1 The Methodology for Flood Disaster and Green Recovery Planning

The methodology for nine steps described in Chap. "International Initiatives and Current Context for Adaptation Planning and Green Recovery" is developed for data sources, methods, and analysis and assessment in Table 1. It includes the whole flow of data sources, methods, and analysis and assessment for each procedure which is described below.

The low level of technical procedures in the broader framework discussed in the previous chapter is suggested for the considerations of technical process. The detailed description of each method is beyond this book. Many references are available from various sources.

While the components of community development, hazard frequency and intensity, physical vulnerability, risk probability, and risk management response might be desirable to achieve full characterization of flood disaster and green recovery plans, limited time and funds for surveys and analysis, scenario development, assessment and management proposals with available appropriate methods and techniques listed in Table 1 invariably make it necessary to restrict studies to a more limited range of components and parameters.

For these reasons, Table 1 has resulted in the development of 'focusing procedures,' which can be used to identify key components of processes to rationalize the subsequent methods and techniques for applications. More elaborated criteria for selecting methods and techniques should be based on basic requirements for each method and technique, and local context variables which will be discussed later in this book. Tools must be applied with professional judgment, and their results must also be interpreted using professional judgment.

The following sections provide the scientific basis for techniques of hazard reduction measures and action strategies, risk management methods, land suitability and resilience analyses for indicator approach, and practical examples of their applications.

Table 1 Methodology for flood disaster and green recovery: a suitability and risk-based indicator approach to a whole system of planning

Technical procedure	Data source	Collection method	Analysis and assessment
1. Select a project site	• Published literature • The press • Key informants	• Literature review in relevant city, community, and site • Survey in the field • Expert interview	• Location analysis • Regional context analysis • Situation analysis
2. Delineate boundary of the site	• Biogeographic map • Watershed map • Catchment map • Rainfall data • Soil map • Topographic map (contour line)	• Hydrological cycle survey • Inventory for selected indicators or factors	• Hydrological analysis(surfaces runoff analysis) • Ecological analysis • Economic analysis • Socio-cultural analysis
3. Describe site profile	• Existing and planned land map • Ecological map • Satellite image	• Literature review • Field survey • Public participation	• Criteria for selecting indicators • Indicator analysis • SWOT analysis
4. Describe historical event of flood hazard and develop future scenarios: Identification of current and future climate hazards	• Historic weather events with record rainfall • Past episodes and events • Historical data (observed data) or hypothetical data • Future projection data	• Literature review • Report review • Expert contact	• Layer system analysis: spatial distribution of flood hazards • Predictive mathematical analysis • In-100-year flood zones (potential future) • Hydrometeorological mapping
Technical procedure	Data source	Collection method	Analysis and assessment
5. Analyze the vulnerability of the site	• GIS map (distribution of people, property, ecosystem, and livelihood) • Topographic map	• Semi-structured interviews • Government publications and documents	• Model-based vulnerability analysis • Vulnerable area and group mapping • Map-based vulnerability analysis
6. Assess flooding risk	• Risk perception study • Research reports	• Households and community leaders	• Composite risk mapping • Mathematical modeling: Model-based probability • Flood index mapping and analysis • I PCC definition of probability

(continued)

Table 1 (continued)

Technical procedure	Data source	Collection method	Analysis and assessment
7. Prepare restoration and green recovery with project activities: Flood hazard mitigation	• Key informants for disaster policy and governance • Institutional framework	• Questionnaire survey • Semi-structured interviews with community leaders and residents • Agency documents	• Effectiveness of potential flood risk mitigation measures
8. Monitoring with ICT application and periodic evaluation	• Climate indicators • Climate-related impact indicators	• Various monitoring reports	• Understanding of advances in climate science, technology and reduction strategies • Trend analysis • Reassessment of non-climatic impact factors that are based on uncertain projection

2 Hazard Reduction Measures and Action Strategies for City Climate Resilience and Adaptation for Indicator Approach

Even with global action to reduce emissions, the impacts of climate change will continue to increase over the coming decades. Strategies should be designed to support governments, communities, and businesses to better carry out the framework described in Chap. "International Initiatives and Current Context for Adaptation Planning and Green Recovery" in order to drive forward adaptation. Guide principles for the strategies are as follows:

- Implemented hand-in-hand with action to reduce emissions;
- Informed by the best available science and information;
- Delivered through partnerships and investments;
- Guided by effective governance and coordination;
- Ensuring that climate change projections are reflected in the design, location, and investment decisions of long-life assets such as hospitals, factories, schools, and other critical infrastructure;
- Evidence-based, risk management approach;
- Flood-resilient development; and
- Collaborative, value-based choices.

Climate change will impact each domain differently. It is important to recognize four domains, natural, economic, social, and built domains (Fig. 1) which are

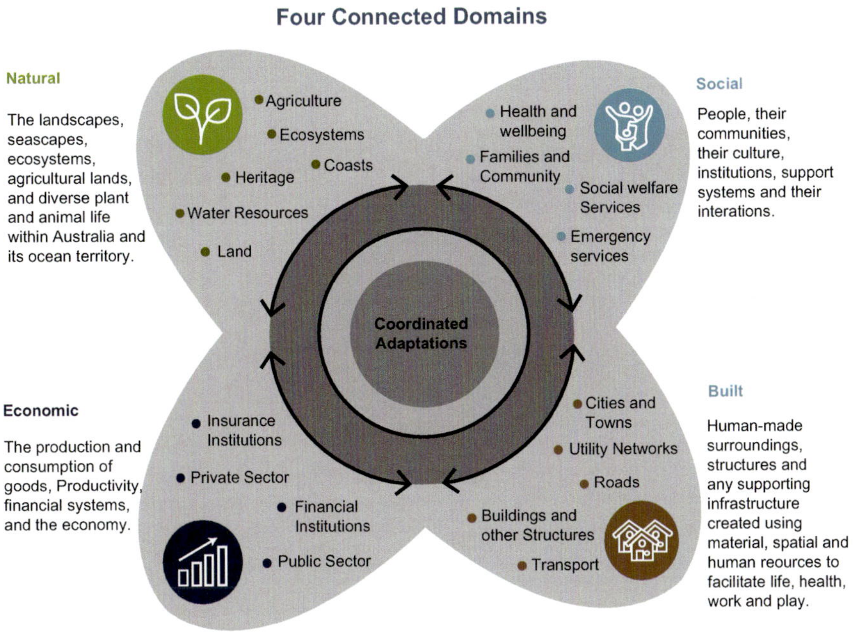

Fig. 1 Four connected domains impacted by climate change and operated for adaptation action strategies *Source* Australian Government, National Climate Resilience and Adaptation Strategy 2021–2025: positioning Australia to better anticipate, manage, and adapt to our changing climate, p. 11

facing individuals, businesses, communities, organizations, and governments, all facing unique challenges and barriers. To respond these challenges, we need to work together on tailored, co-designed solutions. Figure 2 shows examples of climate change strategies at different levels of government and business in Australia.

Australian Government expects that implementation of the strategies will continue to support the integration of climate information in new plans and strategies as they are developed.

3 Integrated Flood Risk Management Methods

Methods and technologies for developing the indicators of climate disasters and green recovery include:

- Non-structural methods

  Watershed city planning approach;
  Flood disaster preparedness activities;
  Flood proofing: Resilience strategies and practices;

Fig. 2 Examples of climate change adaptation strategies at different levels of government and business in Australia. *Source* Australian Government, ibid., p. 15

Early warning systems and mechanisms;
Land-use controls such as zoning and development ordinances;
Flood insurance programs;
Public awareness and education programs;
Existing primary and secondary drainage system maintenance;
Acquisition of floodplain land or voluntary flood-prone property buyouts;
Rescue plan;
Evacuation orders; and
Monitoring, evaluation, and mainstreaming of flood disaster management.

- Structural methods

Channelization scheme: Channel modifications;
Building and maintaining dams;
Rectification projects of all navigable waters;
Flood control reservoirs (insurability and insurance affordability);
Retention and detention pond, and riparian plain;
Diversion channels to redirect flood flows, levees, and by-pass channels (or off-channel floodway);
Low impact development (LID);
Regenerative agriculture.

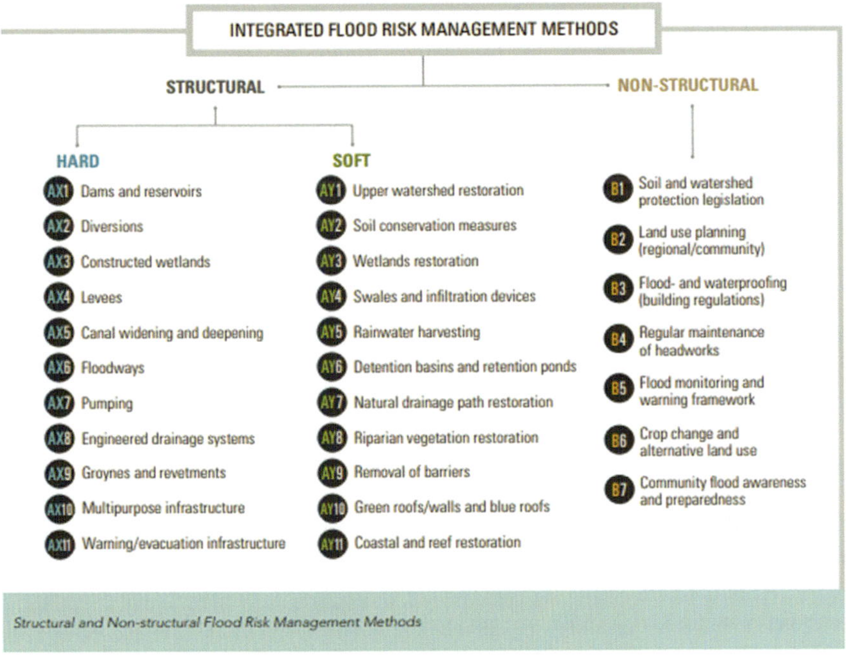

Fig. 3 Structural and non-structural flood risk management methods. *Source* USAID, WWF (2017) Natural and Nature-based Flood Management: A Green Guide

- Smart digital technologies: Status and future task of advanced ICT technology for climate change adaptation

  Discover new technology and innovation trends;
  Remote sensing and geographic information systems in disaster management;
  Internet of Things (IoT), Big Data, Cloud Computing, and AI in flood hazard and disaster management, and green recovery.

USAID and WWF (2017) identified integrated flood risk management methods in Fig. 3.

4 Land Suitability and Resiliency Analyses: An Indicator Approach

Indicator-based approaches are semi-quantitative methods for evaluating climate-based land use suitability mapping and risk mapping—including the components of community development, hazard frequency and intensity, physical vulnerability, and risk probability—to compare different areas or communities. As part of this process, data for each of these indicators is collected at a particular spatial level, by administrative units. These indicators are then standardized (e.g., by reclassifying them between 1 and 0), weighted internally within objectives, and then the various objectives are also weighted among themselves. Although the individual indicators normally consist of quantitative data (e.g., population and property statistics), the resulting hazard, vulnerability, and risk probability results are scaled between 0 and 1. These relative data allow the user to compare the indicators for the various administrative units. These methods can be carried out at different levels, ranging from specific sites, local communities, and cities to countries.

Adaptation indicators include climate parameters, climate change impacts, adaptation actions, and adaptation results described above in an integrated manner. Innovative applications of the risk assessment approach involve its use in conjunction with land suitability analysis. The tandem use of both analytical procedures is demonstrated in Fig. 4.

Figure 4 contains land suitability model and comparative indicator-based flood risk assessment for land-climate suitability and resiliency analysis.

In Fig. 4, the five-step spatial distribution analysis is carried out to get the suitability maps for land use (Fig. 5). The next step is to analyze seven-step spatial distribution which is conducted to develop so-called flood risk indices based on the use of spatial multi-criteria evaluation (Fig. 6). Then, land suitability indices are combined with flood risk indices to get land-climate suitability and resiliency scoring

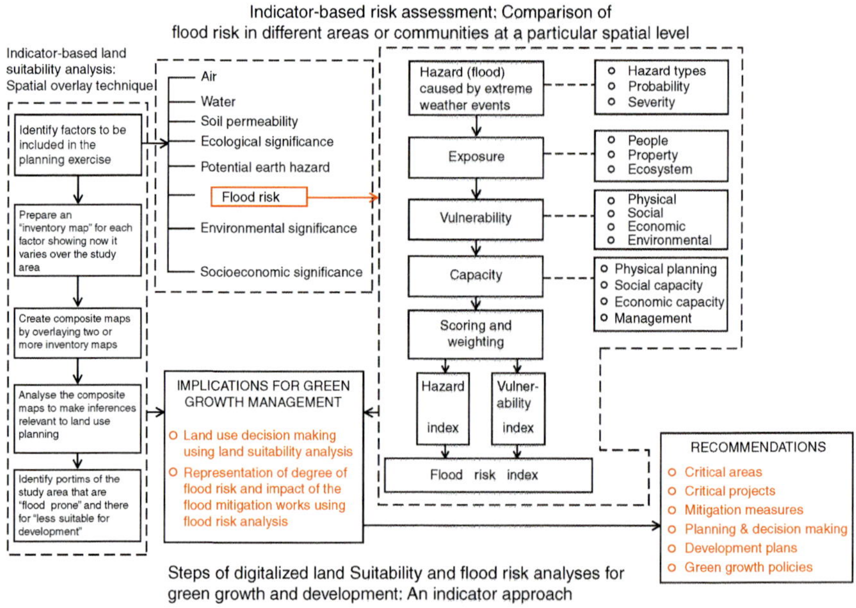

Fig. 4 Steps of digitalized land suitability and flood risk analyses for green growth and development: Tandem use of land suitability and flood risk analyses

maps. These maps provide policymakers and planners with a visible option for flood damage reduction and green urban development.

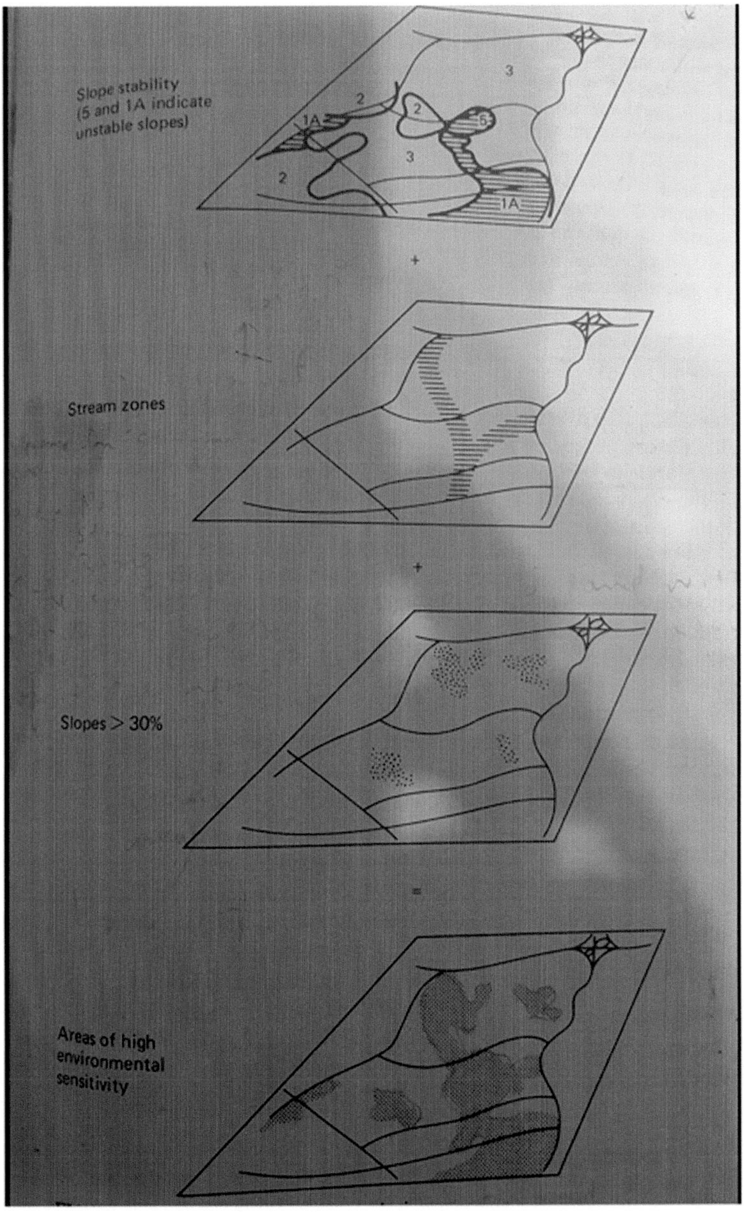

Fig. 5 Map overlay technique for land suitability analysis. *Source* Ortolano (1984)

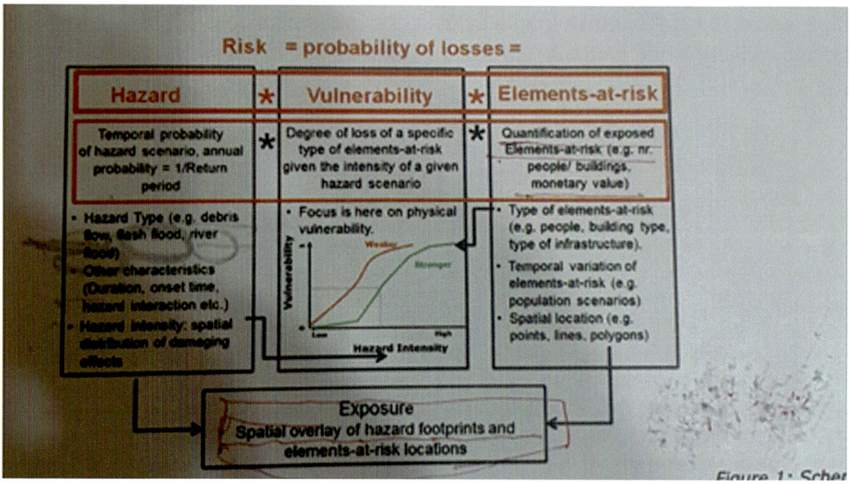

Fig. 6 Schematic representation of risk as the multiplication of hazard, vulnerability, and quantification of the exposed elements-at-risk. *Source* Weston C J v (http://www.charim.net/printpdf/88)

The impact of climate change in the form of increased frequency and intensity of floods can be used to re-adjust land management. In view of this, all land managers and stakeholders involved may consider making low flood risk land use practice as an important component of their commitments to tackle climate change and enhance the resiliency and suitability of the land use sector.

Sensitivity analysis can be undertaken with different sets of weighting value of indicators for scoring land-climate suitability and resiliency. The analysis results can be used to calculate the potential of flood mitigation measures by assuming that all study areas are continuously flooded (baseline scenario) and land development is applied to areas of moderate and least suitability (new land development adoption scenario). Under the land development adoption scenario, flood disaster damages can be compared with those in the baseline scenario.

The city' watersheds, ridges, rivers, floodplains, and wetland soils are critical features the shaping quality of life for current and future generations. The optimal pattern of future growth must complement such natural systems. New growth areas should preserve natural assets and minimize flood hazard risk. Land suitability and flood risk analyses help smart growth management.

The integration platform of land suitability and flood risk assessment analyses includes frequency and intensity of floods in the study area. After considering the flood factors and risk index, the land in the area is divided into several 'risk levels.' These study results provide the basis for smart growth patterns and allowable percentage of the land that could be covered with buildings, houses, physical facilities, and other types of developments.

However, many previous studies and projects did not integrate risk management goals into growth management for minimizing flood hazard and disasters.

Through the identification of hazards and risks, and disaster risk reduction programs for resilience to disasters and climate change, we can increase local communities' capacity to respond and to build back better after flood disasters, as shown in the next section.

5 Climate Resilient Lands Strategy: The Case of Sonoma County, California, USA, for an Application of the Indicator Approach

This particular case study uses land-based system approach to climate resilience and carbon removal. The study is based on the idea that some environmental solutions can be addressed through natural processes if we work toward resilient systems and increase carbon sequestration.

The climate-resilient lands strategy was prepared for the County of Sonoma Eastern Research Group (ERG). This strategy was directed and funded by the County of Sonoma Climate Action and Resilience Division and the Sonoma County (2022) Agricultural Preservation and Open Space District.

This strategy aims to address the following objectives:

- Conserve, manage, and restore as much of the county as possible across public, private, natural, developed, and agricultural lands.
- Focus early actions on areas with the greatest potential for climate risk reduction and biodiversity enhancement, and, where possible, promote carbon sequestration opportunities.
- Provide a forum for coordinated action on climate resilience in Sonoma County.
- Reduce fragmentation of the natural lands system by adding to conserved spaces, increasing connections and corridors, and working with private land owners to develop shared management strategies.
- Prioritize equity and climate justice approaches that are measurable and clear.

The strategy uses landscape- and watershed-scale approaches based on some of the vulnerability characteristics, which is discussed in Chap. "International Initiatives and Current Context for Adaptation Planning and Green Recovery".

In this strategy, two types of indicators were identified from different scales: at the eco-regional scale (Fig. 7) and at land type scale or sub-geographic or sub-eco-regional scale. Indicators are designed to provide a way to identify desired outcomes and measure progress toward the resilience of natural and working lands. On the other hand, performance criteria are designed for use toward the end of project designed to assess how well each project addresses a more limited and measurable set of key criteria. The indicators of resilience adopted for the strategy are listed in Appendix F of the strategy which is related to three major categories: physical climate resilience, social climate resilience, and equity and community benefits.

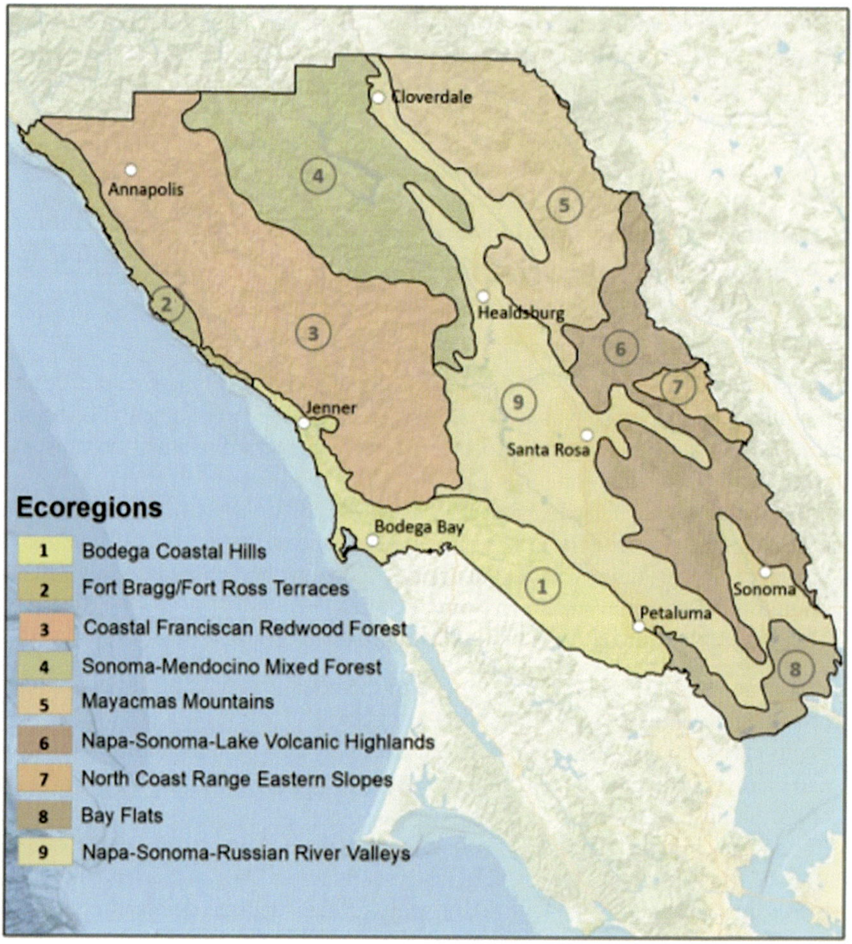

Fig. 7 Ecoregions in Sonoma County. Eastern Research Group (ERG) and Sonoma County (2022)

The recommendations suggested below are non-regulatory actions the County should consider prioritizing at the county-wide scale.

- Bring climate resilience to people most at risk;
- Develop partnerships to increase climate resilience;
- Increase and connect the amount of conserved lands;
- Make Sonoma County a sponge;
- Support and increase regenerative agricultural practices.

In summary, it is important to note that reducing the probability and impacts of flooding requires an integrated flood risk strategy, encompassing warning and evacuation systems, improved protection measures (levees, dykes, and storm surge barriers) and disaster and recovery plans. Ecosystem-based approaches or nature-based solutions, spatial planning (where to build), and the design of the built-up area (how to build) are powerful instruments for reducing flood risks and future vulnerability (PBL Netherlands Environmental Assessment Agency 2018).

One of the attractive features of the strategy is that nature-based approaches were utilized to protect and conserve shoreline and increase carbon sequestration. The approaches include:

(1) Green/nature-based techniques to stabilize highly erodible shoreline using vegetation (low initial cost, high management effort);
(2) Edging (low initial cost; higher management effort);
(3) Living reefs/submerged aquatic vegetation (higher initial cost, lower management effort).

6 An Example of Software User Guide for Local Flood Risk Assessment and Management

UK and Northern Ireland Department for Environment, Food & Rural Affairs and Environment Agency have developed 'Framework and tools for local flood risk assessment: software user guide.' This project aimed to produce a framework, methods, and prototype software tools to help local authorities better understand and manage flood risk in their area.

Basic steps for the soft user guide are as follows:

i. Step 1: Set project name
ii. Step 2: Specify location and name of output geo-database
iii. Step 3: Define study area
iv. Step 4: Select reporting unit polygons
v. Step 5: Load flooding datasets
vi. Step 6: Load receptor datasets
vii. Step 7: Define basic flood risk metrics
viii. Step 8: Define extended flood risk metrics (optional)
ix. Step 9: Run the LFRA Calculator.

This process has hypothetically been tested with Gloucestershire data which was available. The guide is very useful as a technique for integrating the steps described in Chap. "International Initiatives and Current Context for Adaptation Planning and Green Recovery" and technical procedures in Chap. "Nine-Step Conceptual Framework for Climate Disaster Risk Assessment and Green Recovery Planning" of this book with digital technologies.

Official guidance on appropriate methods and techniques for flood hazard and risk assessment has been limited, and in many cases, the lack of formalized procedures has resulted in neglect of some key issues and problems. Hopefully, this chapter will help to ensure that methodological and technological issues are given full and careful consideration in all areas of risk assessment and green recovery planning.

References

Australian Government, National Climate Resilience and Adaptation Strategy 2021–2025: positioning Australia to better anticipate, manage and adapt to our changing climate, p 11, p 15

Bedient PB, Huber WC, Vieux BE (2013) Hydrology and floodplain analysis, 5th edn. Pearson, Boston

Danumah JH et al (2016) 2016) Flood risk assessment and mapping in Abidjan district Using multi-criteria analysis (AHP) model and geoinformation techniques, (cote divoire. Geoenvironmental Disasters 3:10. https://doi.org/10.1186/s4067-016-0044-y

Department for Environment, Food & Rural Affairs and Environment Agency (2021) Framework and tools for local flood risk assessment: software user guide, SC070059/R4, Flood and Coastal Erosion Risk Management Research and Development Programme

ERG (2022) Sonoma County climate resilient lands strategy, Sonoma County, 225 pp

GLA, Environment Agency (2018) London regional flood risk appraisal 2018

Ortolano L (1984) Environmental planning and decision making. Wiley, New York, p 235

PBL Netherlands Environmental Assessment Agency (2018) The geography of future water challenges, The Hague, p 53

World Bank and CAPRA (2016) Methods in flood hazard and risk assessment, Washington DC

Westen C.J.v. http://www.charim.net/printpdf/88

Overview and Exploration of Nature-Based Flooding Solutions: The Green Transition

Kwi-Gon Kim

Abstract Nature protection and restoration are necessary for climate change adaptation and mitigation. Nature markets are valued at $7 trillion (Climate Action, 28 September 2022). NBSs reveal multiple synergies and few trade-offs in urban green space management. Nature provides a blueprint for urban resiliency and sustainability. There is a need for protecting and restoring our forests, grasslands, wetlands, wildlife, soils, water and land against exploitation (SDG 15: Life on Land). Nature-based solutions (NBSs) are used to combat climate change by conserving and restoring vital and vulnerable ecosystems. Investing in nature helps both adaptation and mitigation. They provide multiple benefits including enhancing biodiversity, sequestering carbon and making cities and communities safer and more resilient. The nature-based solutions increase the resilience and effectiveness of flood mitigation. The approaches of NBS that blends grey and green infrastructure can generate net benefits for residents. This chapter describes the steps of nature-based solutions framework and examples of NBS. The chapter gives an overview of various NBS techniques which include levee setback, floodplain restoration, wetland restoration and vegetation management. Hybrid nature-based flood-mitigation solutions including innovative digital blue-green approaches are discussed as cost-effective solutions to build back better. The chapter also describes sustainable use of natural capital through NBS.

Keywords Nature-based solutions · Urban resilience · Vulnerable ecosystems · Carbon sequestration · Cost-effective solutions · Multiple benefits · Ecological coherence · Nature restoration · CO_2 balance · Monitoring solutions · Soil holding capacity · Net ecosystem CO_2 exchange (NEE) · Hybrid solutions

K.-G. Kim (✉)
Seoul National University, Seoul, South Korea
e-mail: kwigon@snu.ac.kr

K. Kim and C. Atkin (eds.), *Cross-Cultural Perspectives on Climate Change Adaptation*, Cities and Nature, https://doi.org/10.1007/978-3-031-50365-8_6

1 Steps of Nature-Based Flood Solutions Framework

World Wildlife Fund (WWF) (2017) developed 'Natural and Nature-based Flood Management: A Green Guide' to support local communities around the world in using natural and nature-based methods for flood risk management. It describes a basic nature-based flood management project cycle consisting of five steps (Fig. 1).

The guide instructs users to do the following:

- Identify the types of information and analyses required to characterize the flood management context and risk.
- Identify and select a potential combination of flood management methods (hard, soft and non-structural) applicable within the given context and watershed.
- Develop maintenance monitoring plans for the project.
- Evaluate chosen methods and the project as a whole.

Figure 2 depicts details of nature-based flood solutions framework which contains flood green guide framework stages and steps. It is important to note that decisions for each step should be context-specific, and subject to social, political and economic factors.

As an example, step 1 of stage 1 in Fig. 2 suggests to understand key crosscutting issues which will be discussed in the cross-cultural comparative case studies. The crosscutting issues for NBS include institutions, regulation, community engagement, gender and finance.

Fig. 1 Steps of nature-based flood solutions framework. *Source* WWF (2017) Natural and Nature-based Flood Management: A Green Guide, p. 6

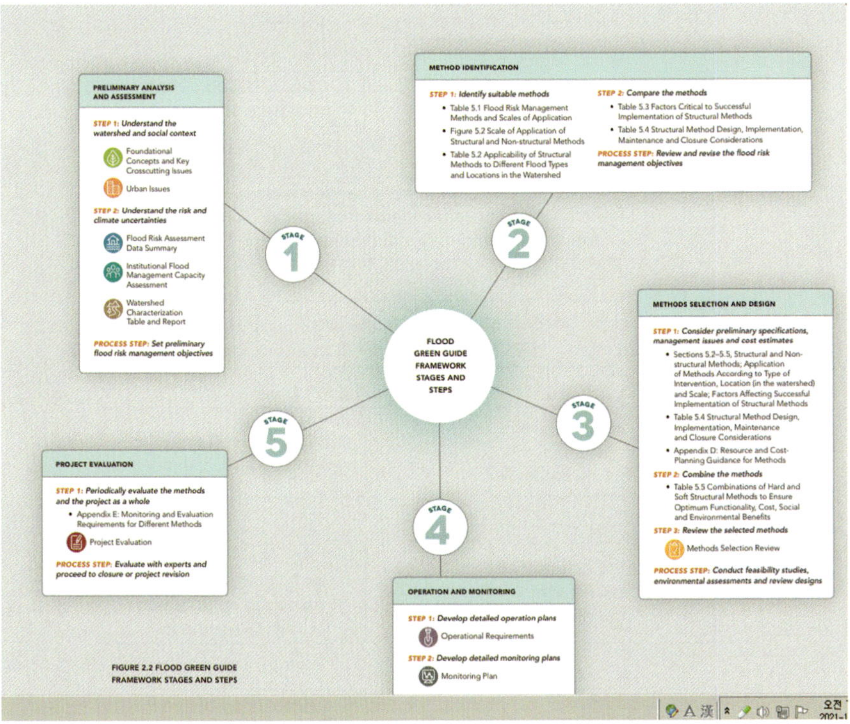

Fig. 2 Details of nature-based flood solutions framework. *Source* USAID, WWF (2017) ibid., p.14

2 Integration of Nature-Based Adaptation and Mitigation Through Land Use Practices

Nature-based solutions are sometimes referred to as green methods, such as wetland protection and restoration, upper watershed restoration or rain gardens (LID). These methods are also used for climate change mitigation. We are in need of including the consideration of multiple values of nature in policy decision.

Here a question is how to use nature-based solutions responsible for both adaptation and mitigation.

There is likely to be more focus on mitigation through land use practices in future NBS, particularly preventing loss of carbon and where possible increasing its storage. Most of the terrestrial carbon is in the soil and measures to reduce modification of soils—for example remedying the adverse effects of drainage on peat soils, or managing vegetation to minimize organic-rich soil disturbance—should be incorporated into climate action plans for adaptation and mitigation.

Figure 3 shows multistages of a typical adaptation and mitigation project cycle which goes hand in hand action to reduce emissions and handle the impacts of floods through meeting of adaptation and mitigation projects.

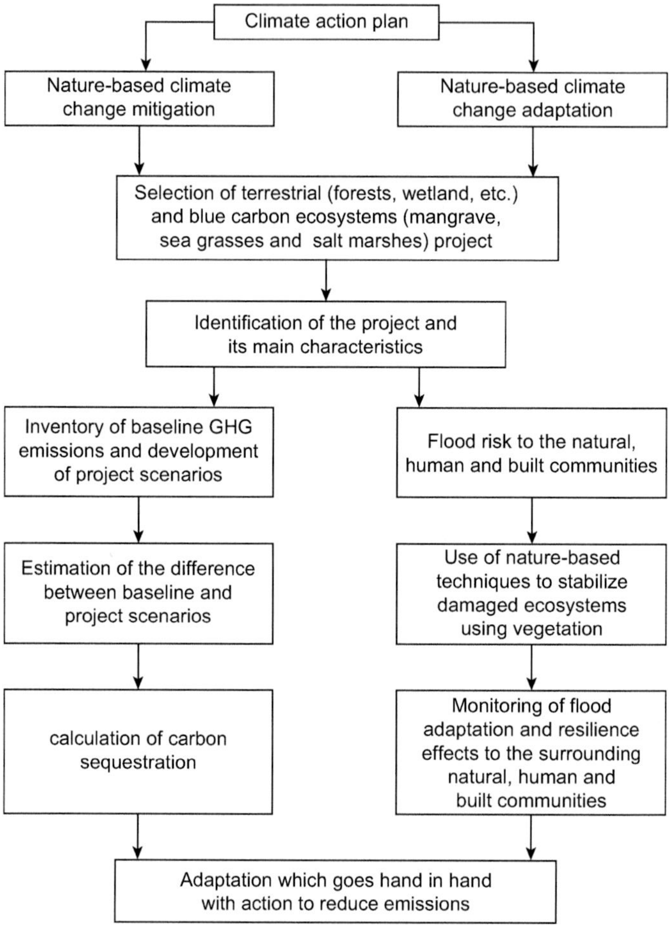

Fig. 3 Integration of nature-based adaptation and mitigation: A convergence approach

Thinking about NBS as a form of adaptive management has some advantages because such an approach explicitly acknowledges that we understand the near future is going to be a time of rapid change and great uncertainty, and thus, any ecosystem management we undertake has to be flexible (Allison 2012).

3 Examples of Nature-Based Solutions for Flood Risk Management

NBSs include protection and wise use of ecosystem with multiple co-benefits and IPCC guidelines for measuring mitigation value.

It is important to understand the impact of climate change on natural systems, with particular emphasis on understanding the future impacts of human-induced climate change (Figs. 4, 5 and 6).

This section describes successful examples of nature-based solutions for flood risk management at the different scales. It also addresses best practices of technology applications for NBS which could increase its environmental performance and social impact. The technology can include:

(1) Technology in Green: Embedding technologies inside NBS could include environmental sensors measuring the health or integrity of nature and biodiversity; green as urban displays and communication interfaces to people for

Fig. 4 Death of mountain bamboo community due to climate change in Odaesan National Park, Korea (*Photo* by Kwi-Gon Kim)

Fig. 5 Seedlings of fir trees in devastated slope after the disappearance of mountain bamboo (*Photo* by Kwi-Gon Kim)

Fig. 6 Natural regeneration of mountain bamboo at the bottom of the mountain (*Photo* by Kwi-Gon Kim)

communicating relevant messages; sensors for co-monitoring campaigns; visual rendering of de-carbonisation (digital twins, metaverse cyberspace, etc.); early warning system, as part of citizen-science activities around urban greening.

(2) Technology for Green: Outside and beyond NBS per se, technologies can be applied to enhance the impact of NBS in cities; ICT can be used to engage citizens in taking care of green (co-maintenance, co-monitoring), thus generating a stronger sense of belongings and social bonds within communities, with emphasis on the inclusion of marginalized and vulnerable groups to floods; moreover, mapping technologies such as remote sensing, LiDAR, digital twins to urban green, can support monitoring the overall progress towards greener cities, contributing to assessing the global targets of the 2030 Agenda at the local scale (e.g. localizing SDGs in cities) and reporting on ESG criteria.

3.1 'Nature-Based Flood Mitigation Can Help Mississippi River Farmers', World Resources Institute, Washington, DC (2021)

Gonzalez and Kuzma (2020) suggested hybrid flood-mitigation solutions which can generate net benefits for farmers in the Mississippi Basin. They state that hybrid solutions are not the panacea, but rather, a form of eco-friendly insurance that minimizes the risk of a total loss in production for prolonged periods of time (Fig. 7). If

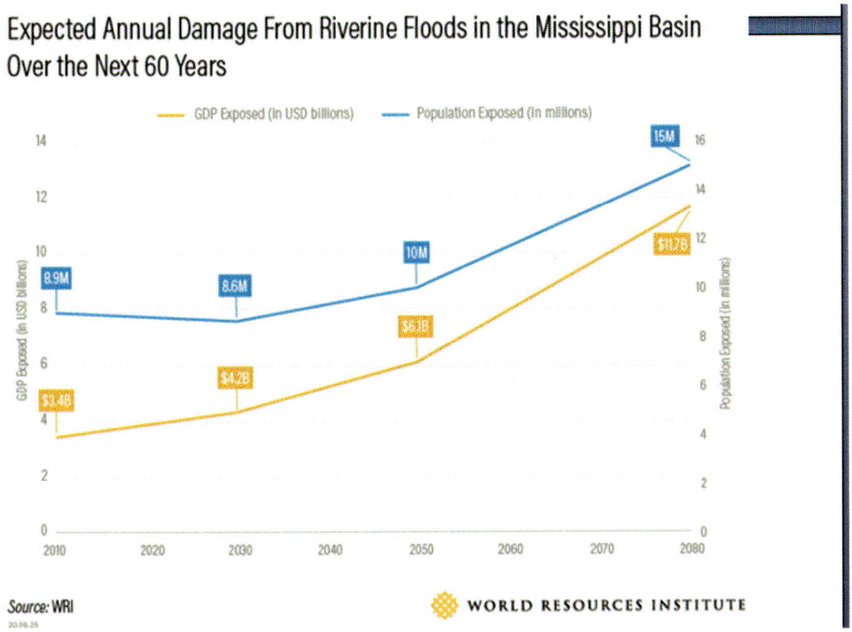

Fig. 7 Expected annual damage from riverine floods in the Mississippi Basin over the next 60 years. *Source* WRI (2022)

implemented on entire landscapes and done through adequate incentives and compensation, these solutions will generate long-term net benefits for farmers. Engineers, scientists and mayors have been calling for a new system of flood management in the Mississippi region.

The suggested measures and techniques for hybrid solutions include the following:

- Levy setbacks

Levy setbacks are an optimal and cost-effective first step. They consist of relocating a levee farther away from the river. This option retains the added protection of the levee while creating space for nature-based solutions that increase the resilience and effectiveness of flood mitigation.

- Wetland restoration

The World Resources Institute (WRI) has suggested that restoring wetlands is a critical defence against flood damage in the Mississippi region. Wetlands function as sponges that retain water during heavy rainfall and then slowly release over time. Restoring wetlands on just 1.5% of a landscape is shown to reduce flood peaks (the maximum height and volume of water flow) by as much as 29%.

- Growing cover crops

Growing cover crops—plants that cover the soil but are not harvested—can protect soil from erosion, improve fertility and control pests. This technique increases the soils' capacity to recover from extreme weather and contributes to flood mitigation by storing nearly 3 times more water than bare soil. Cover crops have a low cost of $37 dollars per acre and reduce peak water levels and economic losses.

- Combination of policy and market incentives

Authors suggest that a great effort needs to be made through policies and funding that ensure large-scale implementation, appropriate compensation, and a high confidence in potential avoided costs of flood damage.

For instance, through the right combination of policy and market incentives, farmers in corn belt have more than doubled crop cover from 1.9 million to 4.2 million acres over the last two decades.

3.2 Ecological Coherence: A Practitioners' Guide, EcoCo, Scotland, UK (2021)

A large proportion of the carbon stored in the UK soils and vegetation is within Protected Areas and other high-quality wildlife sites, especially in peatlands, and their management is a significant responsibility for land managers. There has, however, been very little quantification of the way in which conservation management practices impact upon carbon in the soil and vegetation. Such knowledge is urgently needed in order to ensure that they do not result in the increased release of GHG gases and, where possible, do result in carbon being removed from the atmosphere (UK Department of Environment, Food and Rural Affairs 2007).

The Scottish EcoCo Project (2014–2019) developed an 'ecological coherence approach protocol' to identify the best places across central Scotland for carrying out management interventions to maximize ecological, ecosystem services and socio-economic benefits.

This guide introduces practitioners to the streamlined ecological coherence approach and provides guidance to those looking to undertake similar work in their landscape, land-holding or area (EcoCo 2021). The guide can be used to build connectivity to existing protection with new protection for climate change connectivity (Hannah 2011).

3.3 *Co-Creation of NBS for Climate Change Adaptation: The Case of Rotterdam, Netherlands*

There is an additional methodology for NBS, which has been presented by Tillie (2022) at European Union side events, COP27. It is called 'Co-creation of nature-based solutions for climate change adaptation'. The methodology has been applied to Rotterdam, Netherlands, which drops 233 million euros on green spaces. The characteristics of the co-creation are as follows:

- 'City as ecosystem; Urban Ecosystem Design in three steps,' system, habitat and species.
- Co-creating initiatives from bottom up and top down meet.
- Invest in what people understand and like; parks, playgrounds, beaches but make it function as NBS.
- Design and plan for synergies in infrastructure, energy, adaptation, circularity, housing, recreation, biodiversity and NBS-too many are left behind, get everyone on board and co-create synergies for them for society! (Fig. 8).

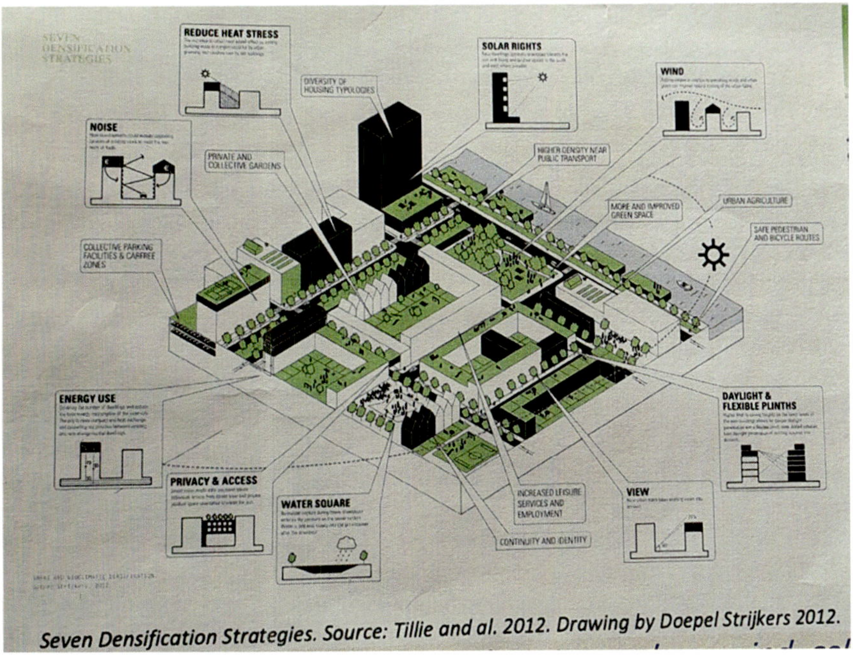

Fig. 8 Seven densification strategies. *Source* Tillie and et al. (2012) cited by tillie, N (2022) NBS education to practice, Paper Presented at European Union side events, COP27, Sharm El-Sheikh, Egypt

3.4 *Blue-Green Infrastructure: The Case of Malmo, Sweden*

One of the many important features of blue-green infrastructure is the ability to lower flood risks by detention of stormwater. This ability of flood risk reduction has for the first time been evaluated with empirical data in this study. In 2014, Malmo, Sweden, was hit with extreme precipitation corresponding to a return period of 50–100 years that led to severe pluvial flooding. This and other large events presented the opportunity to evaluate the efficiency of the 15-year-old stormwater system retrofit in the Augustenborg area (30 ha).

Blue-green infrastructure is widely promoted for climate change adaptation, making this unique case important in the discussion of the capabilities of the related technical solutions. The long-term trends showed less flood damage in Augustenborg than in similar neighbourhoods that have conventional sewer systems (combined or separate), indicating a direct effect of the retrofit with stormwater control measures. Even though a few properties were flooded in Augustenborg, it was shown that the retrofitted stormwater system performed successfully during the extreme event in 2014.

3.5 *Restoration of Nature: Transformative Actions for the Wise Use of Wetland: The Case of Seoul, Korea*

The extent of wetlands, the magnitude of loss, and the related socio-economic ramification of the destruction of wetlands are of global significance. The carbon density and rates of land-cover change in these ecosystems are amongst the highest of any wetlands on earth.

At a time where the world needs every tool available to mitigate climate change and advance climate adaptation, NBS remain underutilized (World Resource Institute 2022).

Nature restoration is one of the important measures for flood damage reduction. It is also good for GHG balances. Nature contains the elements that have been transformed through the advancement of civilization since industrialization. Climate change accompanying global warming due to the industrialization creates greater risks of macro-scale changes to the ecosystems as a result of frequent strikes of abnormal weather conditions.

We need not only to repair damaged nature but also to create near-natural environment or to mimic nature. At the same time, we must not forget the important task that we make reliable estimation of accompanying risks and establishing the technological foundation necessary to manage the existing and future risks.

The followings are the thematic issues related to the direction, methods and techniques towards the restoration of nature as nature-based solutions to flood risks.

- Criteria questions for wetland assessment, planning, restoration design, construction oversight, environmental monitoring, and community outreach. Table 1 presents questions that represent criteria associated with urban wetlands restoration projects.

- Restoration of riverine wetlands

Healthy flood plains provide numerous benefits to ecosystems and communities and are vital resources for NBS to protect, manage and restore wetlands.

In Seoul, Korea, according to the restoration criteria and principles described above, the city planted trees and herbaceous plants to increase water absorption, catch rainfall, slow down surface water run-off, improve soil cover, reduce water pollution, create areas to store water, enhance biodiversity and increase carbon sinks along rivers. Figures 9, 10 and 11 show restored riverine wetlands in the heart of Seoul, with walkways and reedbeds.

Table 1 Criteria questions for wetland assessment, planning, restoration design, construction oversight, environmental monitoring and community outreach. *Source* Kim K-G (2011)

Criteria	Question
Wetland assessment	Whether a proper survey of wetland characteristics was conducted on a selected site Whether a selected site had wildlife habitat conditions
Wetland planning	Whether division for wetland management (core, buffer, transitional area) was assessed Whether a plan was established to facilitate the diverse functions of a wetland (biodiversity, water purification, etc.)
Wetland restoration design	Whether the restored wetland had a similar structure as the original ecosystem Whether the restored wetland had similar productivity as the original ecosystem
Construction oversight	Whether the construction was done correctly and safely within the allotted budget and time period Whether the construction was done considering the impact on the surrounding ecosystem Whether materials from the local area were used for construction
Environmental monitoring	Whether regular monitoring was conducted on wetland components and species Whether early warning systems were created to predict and evaluate ecological changes in the wetland
Community Outreach	Whether educational or recreational activities on the wetland were implemented Whether local citizens participated in wetland management

Fig. 9 Yangjaecheon riverine wetland: Vegetation, Seoul, Korea

4 Newly Emerging Research Topics

In this chapter, the review of NBS has led to the delineation of research needs associated with the solutions in project decision-making. Newly emerging research topics are as follows:

- Climate regulation contributing to natural and managed ecosystems.
- Tropical peatlands: Carbon stores, carbon gas emissions and contribution to climate change processes.
- Short-term and long-term carbon dynamics in wetlands-stream-lake continuum (connectivity): A catchment approach for applications of digital technology including blockchain.

Fig. 10 Yangjaecheon riverine wetland after it was flooded, Seoul, Korea

Fig. 11 Tancheon riverine wetland: Vegetation, Seoul, Korea

Figure 12 shows a conceptual sketch of a wetlandscape, while Fig. 13 is a system of wetlands and their connectivity.

- Integrating coastal blue carbon in climate mitigation and adaptation policy strategies.
- The interrelations between soil and climate change.
- The use of rain gardens for flood control as an adaptation to climate change.

Shifting climate patterns due to global warming increases the frequency of severe and destructive storms and associated widespread flooding.

Rain garden technology introduces urban design elements with nature that will deal with periods of heavy rainfall that might normally give rise to flash flooding.

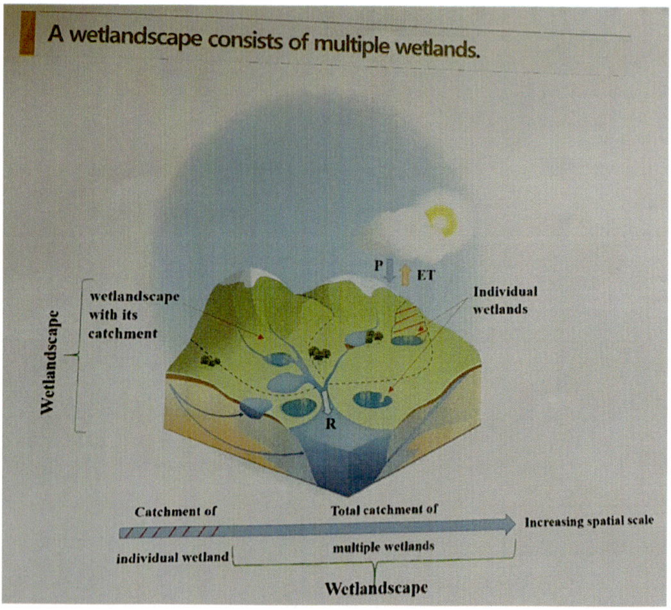

Fig. 12 A conceptual sketch of a wetlandscape. *Source* WWF (2017)

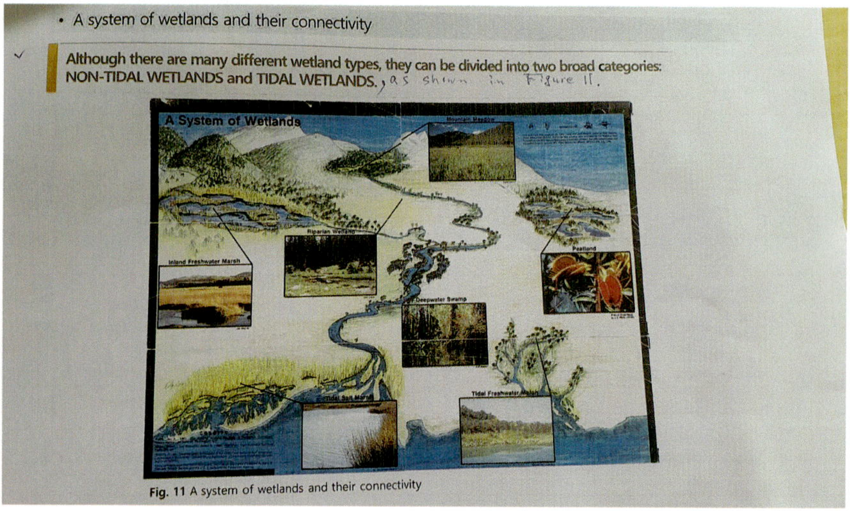

Fig. 13 A system of wetlands and their connectivity. *Source* US National Park Service (1998)

					Some distance from buildings			
				In proximity to buildings/built development				
	Immediate curtilage of the building							
Component	Green roofs	Green facades/ vertical green	Rain barrels and water butts	Storm-water planters	Porous pavements	Rain gardens, infiltration strips	Landscape swales	Vegetation filters, constructed wetlands
Location/ application	Planted and vegetated roof surfaces	Climbers, green wall systems, vertical swamps	Rainwater collection directly from roof surfaces	Raised or at grade planters for rainwater storage and interception immediately at base of building	Hard surfaces that allow infiltration of rainwater	Planted rainwater collection areas	Parks, housing and commercial landscapes, urban infra-structure	Parks, housing and commercial landscapes, urban infra-structure
Run-off prevention	●		●	●	●	●		
Retention	●	●		●	●	●	●	●
Detention	●		●	●			●	●
Conveyance							●	
Filtration				●		●	●	●
Habitat	●	●		●	●	●	●	●
Amenity	●	●		●	●	●	●	●

Fig. 14 Elements of a stormwater chain and how they can be integrated into typical residential and public or commercial areas. *Source* Dunnett N, Clayden A (2008) Rain Gardens: Managing water sustainability in the garden and designed landscape, Timber Press, Portland, p. 48

Figure 14 provides elements of a stormwater chain, and how they can be integrated into typical residential and public or commercial areas.

- Google 2030 Goal: Reforesting cities using AI.
- Nature-based ESG.
- Soil holding capacity to respond to sea level rise.

Mangrove root systems and other plants in salt water can be used to increase soil holding capacity to respond to sea level rise. Lots of experiment of plant roots have been undertaken by the Royal Netherlands Institute of Sea Research (NIOZ).

"Setting Down Roots: Mangroves interlock with communities on Puerto Rico's North Coast" is a success story to integrate ecosystem restoration with disaster risk reduction. There is a need for monitoring, reporting and verification (MRV) systems for carbon in such soils and vegetation.

- Climate change mitigation: CO_2 balance and net ecosystem CO_2 exchange (NEE)

For instance, the area under the cultivation of perennial bioenergy crops on organic soils in the northern countries is fast increasing. To understand the impact of reed canary grass (RCG, Phalaris arundinaceae L.) cultivation on the carbon dioxide (CO_2) balance of an organic soil, net ecosystem CO_2 exchange (NEE) was measured.

- Diving into the digital coast—Nature-based solutions and coastal inundation risk information with satellite images and sensors

Digital technology can help dive into the digital coast—Nature-based solutions and develop digitalized coastal inundation risk information system, including the use of remote sensing and drones to monitor land use and land-cover change in coastal areas.

- Mitigating climate change through restoration and management of coastal wetlands and near-shore marine ecosystems

5 Investing in Nature-Based Solutions

Very recently, EU adopted the Nature Protection Law (2024). The new law sets a target for the EU to restore at least 20% of the EU's land and sea by 2030 and all ecosystems in need of restoration by 2050.

Australia is working with their partners in the Indo-Pacific to use nature-based solutions to combat climate change by conserving and restoring vital and vulnerable ecosystems. Australia's Development Program (2021) states that ecosystems across the Indo-Pacific are at risk from climate change and human activity.

Investing in nature helps both adaptation and mitigation. It provides multiple benefits including boosting biodiversity, sequestering carbon and making communities more resilient.

Stockholm Environment Institute (SEI) research assessed the scale of funding for nature-based solutions to climate change aimed at developing countries. This institute used a global data set from the Organisation for Economic Co-operation and Development (OECD) for the assessment. It also examined how effectively this funding is being put to use. Figure 15 shows total climate finance commitments and total NBS-like finance commitments, 2016–2020.

The paper concluded that, while the NBS concept may resonate in academic circles and the rhetoric of international organizations, their observation of a declining trend over recent years could suggest that the discourse is not trickling down to practice nor being translated into financial decisions.

The utility of NBS will be characterized by manifold activities. Current and future research efforts, and actual practices on the ground will provide information relevant to many of today's unanswered questions.

It is expected that the NBS resonation in academic circles and the rhetoric of international organizations will actually be translated into financial decisions.

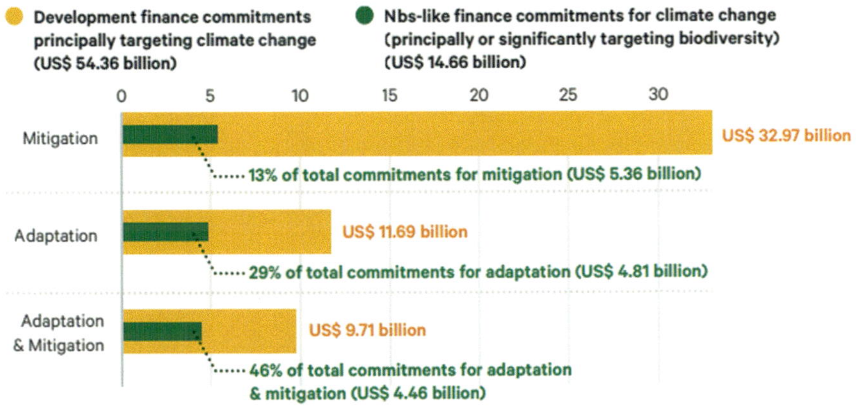

Fig. 15 Total climate finance commitments and total NBS-like finance commitments, 2016–2020, p. 8. *Source* Stockholm Environment Institute (SEI) (2022)

References

Allison SK (2012) Ecological restoration and environmental change: renewing damaged ecosystems. Routledge, New York, p 107

Atteridge A, Bhatpuria D, Macura B, Barquet K, Green J (2022) Assessing finance for nature-based solutions to climate change, SEI working paper. Stockholm Environment Institute, Stockholm, https://doi.org/10.51414/sei 2022.052

Australian Government (2021) Building climate resilience: addressing climate change through Australian's Development Program, pp. 12–13

Drosler M, Restoration of peatlands and greenhouse gas balances

Dunnett N, Clayden A (2008) Rain gardens: managing water sustainability in the garden and designed landscape. Timber Press, Portland, p 48

Food resilience to tackle nature crisis

Hannah L (2011) Climate change biology, Elsevier, Amsterdam, p. 313

Kim K-G (2008) Flood control as an adaptation to climate change, the use of rain gardens for LID, 2nd UCLG ASPAC congress 2008, 16 July 2008

Kim K-G (2011) Wetland restoration to enhance biodiversity in urban areas: a comparative analysis. Landscape Ecol Eng 2011(7):27–32. https://doi.org/10.1007/s1135-010-0144-x

Kwi-Gon Kim, A study on the establishment of water circulation system for the eastern Pangyo New Town

The Land Sector in the 2030 EU climate change policy framework: Look at future

Tillie A (2022) NBS education to practice, Paper Presented at the European Union side events at COP27, Sharm El-Sheikh, Egypt

U.K Department of Environment, Food and Rural Affairs (2007) Conserving biodiversity in a changing climate: guidance on building capacity to adapt, London, p. 24

World Resources Institute (2022) Nature-based flood mitigation can help mississippi river farmers, Washington DC

WWF (2017) Natural and nature-based flood management: agreen guide, p.6

www.ecocolife.scot

Overview of Case Studies: Common Elements and Framework for Analysis

Catherine Atkin and Kwi-Gon Kim

Abstract This chapter provides an overview and explanation of the elements of the thirteen case studies from Africa, South and Southeast Asia, South America and the Middle East. It introduces the reader to the elements of comprehensive climate risk management and green recovery planning covered by the case studies including flood risk assessment, green recovery planning and local context variables and adaptive conditions analysis. It sets the context for the reader's review of the case studies and outlines the key learning questions readers should consider.

Keywords Global South Flood risk assessment · Africa climate risk management · South East Asia disaster risk management · Community-driven resilience building · Digitilization of climate risk management · Adaption co-benefit models · Nature-based hazard reduction measures · Comprehensive urban planning

1 Overview to Case Studies

The centerpiece of this book are the thirteen case study papers and research presentations that focus on flood risk assessment and green hazard reduction in eleven countries across the Global South. This work was undertaken by professionals and practitioners as part of student coursework for "Climate Disaster and Restoration," seminars taught by Professor Kwi-Gon Kim, co-editor of this book, in 2020–2022 as part of KOICA-Hallym master's degree program in Response to Climate Change. The program at Hallym University in Korea was created to nurture field-oriented climate change policy experts from the Global South and is funded by the Korean International Cooperation Agency (KOICA) (Table 1).

C. Atkin (✉)
Stanford CodeX Climate Data Policy Initiative, Stanford, USA
e-mail: catherine.atkin@codex.stanford.edu

K.-G. Kim
Seoul National University, Seoul, South Korea
e-mail: kwigon@snu.ac.kr

K. Kim and C. Atkin (eds.), *Cross-Cultural Perspectives on Climate Change Adaptation*, Cities and Nature, https://doi.org/10.1007/978-3-031-50365-8_7

Table 1 Case studies and research presentations grouped by geography

Flood Risk Assessment and Green Recovery Planning Case Studies—Africa			
Ch. 8	Ethiopia	Flood risk Analysis: Causes and Landscape-based Mitigation Strategies in Dire-Dawa city, Ethiopia (Tewelde Alema Fukur)	Environment, Forest and Climate Change Commission, Ethiopia
Ch. 9	Ghana	A Systemic Analysis of Causes, Impacts, and Coping Strategies of the People Related to Flooding in Ghana's Capital City, Accra (Ezekial Atanga Azuure)	Ministry of Energy, Ghana
Ch. 10	Nigeria	A Comparative Assessment of Flood Hazard Susceptibility Modeling in Niger State, North-Central Nigeria (Saadatu Gambo Madaki)	Department of Climate Change, Federal Ministry of Environment, Nigeria
Ch. 11	Uganda	Proposed Study Design for Map-based Approach to Flood Risk Assessment in Bwaiise III in Kampala City, Uganda: An Indicator Approach (Yawe Samuel)	Rakai District Local Government, Kampala, Uganda
Flood Risk Assessment and Green Recovery Planning Case Studies—South and Southeast Asia			
Ch. 12	Bangladesh	Flood Mapping and Damage Assessment using Synthetic Aperture Radar (SAR) data in GIS and Google Earth Engine—A Case Study of 2020 Riverine Flooding in Gaibandha, Kurigram, and Jamalpur Districts of Bangladesh (Siddique Md Abu Bakar)	Ministry of Local Government, Rural Development and Cooperatives, Bangladesh
Ch. 13	Cambodia	Community-based Adaptation to Flood Risk Along Mekong River of Cambodia: A Case Study of Kratie Province (Navin Chea)	Royal University of Agriculture (RUA), Cambodia
Ch. 14	Cambodia	Floods Risk Management in Phnom Penh, Cambodia (Keo Hak)	Provincial Department of Rural Development, Svay Rieng (PDRD/SVG), Ministry of Rural Development, Cambodia
Ch. 15	Jakarta	Flood Cause Analysis Using Remote Sensing and Upper-air Observation Data: Case Study of Jakarta (Azhari Putri Cempaka)	Indonesia Agency for Meteorology, Climatology and Geophysics, Indonesia
Ch. 16	Philippines	Water-Related Disaster Security on Coastal Community (Jordan, Guimara, Philippines) Using Climate and Disaster Risk Assessment (CDRA) (Neil Paul Fara-on)	Department of the Interior and Local Government, Philippines
Ch. 17	Philippines	Flood Risk Assessment in the Philippines: Iloilo City Flood Risk Management Plan (Chris Velmida Perfas Gianzon)	Department of the Interior and Local Government Region VI, Philippines
Flood Risk Assessment and Green Recovery Planning Case Studies—Middle East and South America			

(continued)

Table 1 (continued)

Ch. 18	Afghanistan	Flood Control and Its Management in the Badakhshan Province of Afghanistan (Kambiz Karimi)	National Environmental Protection Bureau, Afghanistan
Ch. 19	Afghanistan	Assessment of Early Warning System in Afghanistan: Avalanches and Flash Floods in Shighnan District of Badakhshan Province, Afghanistan (Mohammad Nazir Safi)	National Environmental Agency, Afghanistan
Ch. 20	Peru	Proposed Study Design for Spatio-temporal Drought Assessment Using the Standardized Precipitation Evapotranspiration Index (SPEI) and Adaptation Over the Mantaro Valley of Peru (Del Piero Rafael Arana Ruedas)	Universidad Continental, Peru

2 Case Study Elements—Flood Risk Assessment and Green Recovery Planning

Case study authors drew upon a variety of new and existing climate risk management frameworks and took an indicator approach to flood risk assessment. This analysis included an assessment of local context variables that impact vulnerability and adaptation capacity as well as the identification of high-impact flooding hazard reduction strategies. The geographic boundaries of the case studies vary with a focus on neighborhoods, cities, and regions in urban areas. A number of students used the Nine-step Conceptual Framework for Climate Disaster Risk Assessment and Green Recovery developed by Professor Kwi-Gon Kim outlined in Chap. "Nine-Step Conceptual Framework for Climate Disaster Risk Assessment and Green Recovery Planning" to conduct their analysis.

In the Chaps. "Flood Risk Analysis: Causes and Landscape-Based Mitigation Strategies in Dire-Dawa City, Ethiopia"–"Proposed Study Design for Spatio-Temporal Drought Assessment Using Standardized Precipitation Evapotranspiration Index (SPEI) and Adaptation Over Mantaro Valley, Peru", readers will be exposed to a rich tapestry of case study content that explores three critical elements of comprehensive climate risk management and green recovery planning including:

1. **Flood Risk Assessment**: Case study authors look at the history of flooding within targe geographies and outline how people, ecosystems, and places have been impacted. They use a range of international and localized data sources to assess risk as a combination of hazards, exposure, and community vulnerability levels to understand the potential for and geographic distribution of adverse consequences from flooding hazards.
2. **Green Recovery Planning**: Based on the flooding risk analysis, many contributors engage in the critical next step of identifying what are the flood risk reduction strategies that can help communities reduce, prevent, and prepare for increased rainfall and other climate-related flooding hazards. A variety of structural and

non-structural flood hazard reduction strategies, including nature-based solutions, are considered and authors identify those they consider having the highest potential for impact.

3. **Local Context Variables and Adaptive Conditions**: Authors identify local context variables and conditions—governmental and policy, economic, social, and technological—that determine a community's vulnerability and adaptive capacity. These variables are explored in both the case study risk assessments and the identification and evaluation of possible flooding hazard reduction measures.

3 Case Study Parameters and Learning Dimensions

Readers will notice that the case studies vary in their focus, structure, and analytical approach. As seminar papers in Dr. Kim's post-graduate course, authors were encouraged to be innovative in their approach to the material, to identify topics and deep dives that interested them, and to bring their unique perspective to their analysis of target communities. In many cases, student authors had to contend with the limited availability of localized data sources. Some of the authors had little prior knowledge of the their case study focus area while others have worked professionally as planners and government staff in and around the communities they studied.

Readers have the unique opportunity to learn from in-country professionals who live and work in these countries as they take a problem-solving approach to the flooding hazards facing Global South communities. They offer important insights into not only what is happening on the ground—including the theory and practice of flood disaster risk management—but also what is needed to strengthen the available tools and increase access to capacity building. Key learning dimensions that readers are encouraged to consider include:

1. What are the key findings from the case studies regarding flood risk assessment—including hazard, exposure, and vulnerability analysis—and green recovery planning?
2. How are the currently available climate risk management planning tools and frameworks used and what does it tell us about opportunities and limitations of their current application and deployment? (Photos 1 and 2).

Photo 1 Dr. Kwi-Gon Kim's 2021 "Climate Disaster and Restoration" Seminar Class - KOICA-Hallym University Response to Climate Change master's degree program

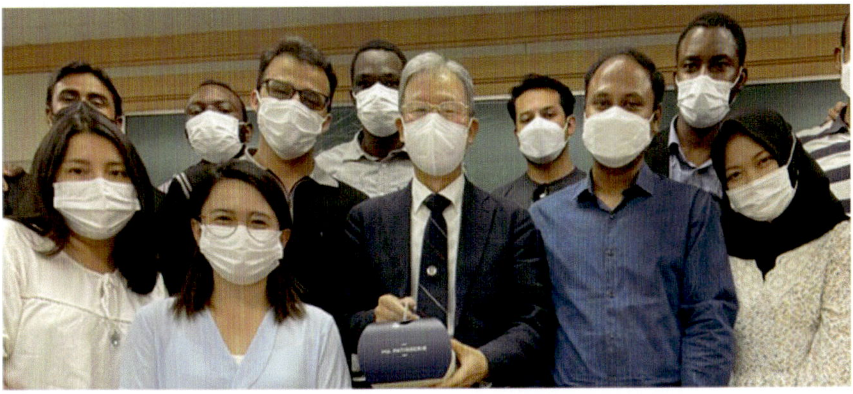

Photo 2 Dr. Kwi-Gon Kim's 2022 "Climate Disaster and Restoration" Seminar Class - KOICA-Hallym University Response to Climate Change master's degree program

Flood Risk Assessment and Green Recovery Planning Case Studies—Africa

Flood Risk Analysis: Causes and Landscape-Based Mitigation Strategies in Dire-Dawa City, Ethiopia

Tewelde Alema Fukur

Abstract In flood prone areas, understanding flood causing factors, assessing the flood induced risks and adopting nature-based mitigation strategies can increase the knowledge andawareness and help individual communities to protect themselves and their properties using appropriate flood management measures before and during flood events. Dire Dawa city is in the foothills of southern mountains from where rivers crossing the city are originated. The multidimensional causes of flood hazard and limited landscape-based mitigation strategies in the study area have exacerbated the impacts of flooding. This study was conducted with the aim of assessing flood causing factors and to propose landscape-based flood mitigation strategies. Case study data was collected from the Ethiopian Meteorology Agency, land use maps of Dire Dawa city administration and government reports. Findings of the study indicated that flood risk in the study area has resulted from multiple factors such as intense rainfall, topography, encroachment to the riverbanks, institutional problems and aggravating factors resulted from power interruption during heavy rain and regime changes. More importantly, flood risk of the study area was found to be sensitive to hourly variation of rainfall distributions and varies based on the location of the sub-watersheds. Based on the prioritization of sub-watersheds' susceptibility to flood risk through morphometric analysis, select sub-watersheds are identified as the most flood risk susceptible watersheds demanding urgent landscape-based conservation measures and sustainable water conservation structures are also identified. The overall result of the study shows that managing the root causes of flooding at the upper catchments and adopting recommended proposed water conservation structures at proposed site helps to sustainably curb flood induced risks of Dire Dawa city.

Keywords Flood causes · Morphometry · Mitigation strategies · Climate risk assessment · Climate risk management · Disaster risk management · Adaptation · Resilience · Vulnerability · Adaptive capacity · Indicator analysis · Hazard reduction measures · Nature-based solutions · Comprehensive urban planning · Early warning systems

T. Alema Fukur (✉)
Environment, Forest and Climate Change Commission, Addis Ababa, Ethiopia
e-mail: teweldealema@gmail.com

K. Kim and C. Atkin (eds.), *Cross-Cultural Perspectives on Climate Change Adaptation*, Cities and Nature, https://doi.org/10.1007/978-3-031-50365-8_8

1 Introduction

Floods are known as the most frequent and devastating natural disasters in both developed and developing countries (Erena et al. 2018). It is one of the leading natural hazards worldwide both in terms of the frequency of occurrence and the resulting damages to human lives, the environment and economic assets (Erena and Worku 2018).

The occurrence of a flood depends on meteorology, topography, land use, soil type and antecedent moisture conditions (Alemayehu 2007). The multi-dimensional causes of the flood make it less predictable and aggravate its impacts worldwide. As flood risk and its impacts increase over time, there has been more emphasis by flood research scholars on understanding the root causes of flood risk and identifying mitigation strategies. Many of the previous studies conducted on causes of flood risk and mitigation strategies confirmed that flood triggering factors are mainly derived from meteorological, hydrological and anthropogenic effects (Tegegne et al. 2018).

Background of the Study

Dire Dawa city is in the foothills of the mountains located in the eastern part of Ethiopia. The flooding in Dire Dawa occurs because of its geographical location, topography and rainfall pattern of the city watershed. Flood hazards have been part of the history of Dire Dawa. Even though records on urban floods of Dire Dawa city are irregular and incomplete, historical information reveals that the first damaging flood risk was recorded in 1945. Since then, flood has been affecting the city at different times. Different authors have investigated the causes of floods in Dire Dawa city concerning geomorphology, hydrology and land-use changes within the city boundary (Alemayehu 2007) (Fig. 1).

Objectives of the Study

The objectives of this case study include:

- To explore the causes of flood risk in Dire Dawa city.
- To identify flood vulnerable sub-watersheds in the study area.
- To propose landscape-based suitable and sustainable water retention structures.

2 Site Location and Characteristics

The study area, Dire Dawa City, is in the great East African rift valley between 9°25′ N and 9°45′ N latitude and 41°40′ E and 42°10′ E longitude. The city is located along the foothills of surrounding mountains drained by tributaries of five major rivers, Dechatu, Goro, Laga Hare, Butuji and Melka Jabdu. Dechatu, the largest river, flows across the city.

The Dire Dawa watershed covers an area of 698.5 km^2. The elevation of Dire Dawa ranges from 1000 to 2400 m above sea level. Dangago is the highest mountain

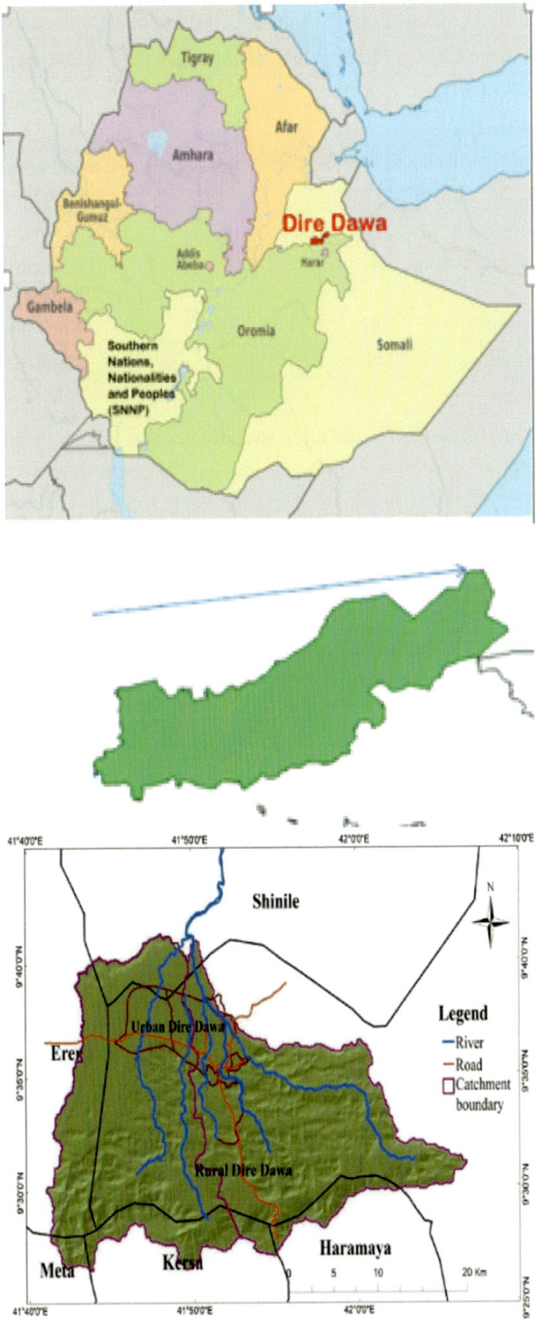

Fig. 1 Map of Dire Dawa city watershed. *Source* Erena and Worku (2018)

Table 1 Summary of damage associated with 2206 flood that includes in Dire Dawa city

Village/city	Human loss	Livestock loss	Property loss	Remarks
Metokomia	2	Not stated	Not stated	9027 people were displaced
Biftu Geda	5	11	10 ha crops	
Dire Dawa	339	Not stated	200 houses, roads and bridges	
Ija Anani	2	21	178 ha crops	
Harala Balina	4	27	80 ha crops and 40 houses	
Lazaret	22	900	Not stated	
Shinile	Not stated	200	3 houses	
Total	374	1100	268 crops, 243 houses	

Source Compiled from Dire Dawa city flood assessment unpublished report (2006–2016)

located at the upper part of the watershed. Based on the 2019 estimates, the Dire Dawa city population is more than 520,000 with an annual growth rate of 2.7%. The recall of elders and flood related reports of the city indicate that major floods have occurred in 1945, 1968, 1977, 1983, 1985,1995, 2003, 2005, 2006, 2010 and 2016 due to flash floods coming from surrounding mountains. Among these, the 2006 flood caused the most catastrophic damage including the death of 339 human people in Dire Dawa and displacement of 90,127 throughout the region (Table 1).

3 Methods

This study considered both primary and secondary data to achieve the intended objectives of the study. In addition to primary and secondary data sources, satellite imagery analysis was used. Considering the relationship between regime change and land use policy of the country, four land use maps from satellite remote sensing at about ten years intervals (1985, 1995, 2006 and 2016) were selected for land use land cover change analysis.

In addition to this, the precipitation-based cause of flood hazard analysis was also employed for Dire Dawa (1953–2014), Dangago (1986–2014) and Haramaya (1960–2013) meteorological stations to detect climate variability. The frequency distribution of hourly rainfall was assessed by identifying the number of days having maximum hourly rainfall exceeding 30 mm. The daily rainfall distribution of each year and standardized hourly rainfall index was computed and examined for Dire Dawa, Haramaya and Dangago stations.

$$Z = (x_i - \overline{x}) \Big/ \Sigma d$$

If $Z < -1$, it would be below normal
If $-1 \leq Z \leq 1$, it would be normal
$Z > 1$, it would be above normal
Where, Z = Standardized minimum rainfall anomaly
Σd = standard deviation
$\overline{x}$ = Mean hourly rainfall records
x_i = Observed hourly rainfall.

When dealing with extremes, it is important to employ the probability of return period. If 'P' is the probability that the event will be equaled or exceeded in a particular year, then the return period 'T' may be expressed.

$$T = N + 1/m$$

where

m is order or rank of the event
N is total number of event in the data
T is return period.

In different methods, the probability of occurrence of a particular extreme rainfall is important, and such information is obtained by the frequency analysis of rainfall data.

$$P = 1/\mathrm{T}$$

where

T is return period
P is probability.

To meet the objective of developing a landscape-based flood mitigation strategy, the analysis of morphometric parameters was employed. The watershed prioritization has been done based on the linear and shape aspects of morphometric analysis. The linear parameters such as drainage density, stream frequency, bifurcation ratio and texture ratio have a direct relationship with flooding. On the contrary, shape parameters such as elongation ratio, compactness constant, circulatory ratio and form factor have an inverse relationship with flooding. Prioritization rating of all the sub watersheds was carried out by calculating the compound parameter values (Table 2).

Following the identification of flood susceptible sub-watersheds, FAO and Integrated Mission for Sustainable Development (IMSD) guidelines were employed to identify suitable sites where to adopt water conservation technologies (Table 3).

Table 2 Formula for computation of morphometric parameters

No.	Morphometric parameter	Formula/definitions/methods
Linear aspect		
1	Stream order	Hierarchical order
2	Stream length (Lu)	Length of the stream (km)
3	Mean stream length (Lsm)	Lsm = Lu/Nu, where, Lu = Total stream length of order 'U' Nu = Stream length of next higher stream order
4	Stream length ratio (Rl)	Rl = Lu/Lu − 1, where, Lu = Total number of stream segment of order 'U' Lu − 1 = Stream length of next lower order
5	Bifurcation ratio (Rb)	Rb = Nu/Nu + 1, where, Nu = Total number of stream segments of order 'U' Nu + 1 = Number of segments of the next higher order
Areal aspect		
6	Drainage density (Dd)	Dd = L/A, where, L = Total length of stream, A = Area of the watershed
7	Stream frequency (Fs)	Fs = N/A. Where, N = Total number of stream, A = Area of the watershed
8	Texture ratio (T)	T = N1/P, where N1 = Total number of first order stream, p = Perimeter of watershed
9	Form factor (Rf)	Rf = $A/(Lb)^2$, where, A = Area of the watershed, Lb = Maximum basin length
10	Circularity ratio (Rc)	Rc = $4ðA/P^2$, where, A = Area of the watershed P = Perimeter of the basin, Đ = 3.14
11	Elongation ratio (Re)	Re = $2\sqrt{(A/ð)}/Lb$, where A = Area of the watershed Lb = Maximum basin length, Đ = 3.14

Table 3 Water conservation site identification criterion

Structure type	Soil	Permeability	LULC	Slope	Drainage
Terracing	Sandy clay, clay loam, sandy loam	Medium, high	Bush land and shrub land	5–30	1st, 2nd order
Percolation Tank	Silty loam/Clay loam	High	Bare/shrub land	< 10	2nd/3rd/4th order
Pond	Sandy loam silty loam	Low	Shrub	< 5%	1st order
Check dam	Sandy clay loam	Low	Bare/shrub	< 15%	5th and 6th order
Soil and stone bunds	Silty loam	Low	Bare/shrub	< 10	3rd and 4th order

4 Result and Discussion

4.1 Causes of Flood Hazards

Intense Rainfall

There are three meteorological stations (Dire Dawa, Haramaya and Dengago) that fall inside the watershed. The meteorological analysis of this study shows that flood-inducing rainfall is coming from the surrounding highlands located outside of Dire Dawa city. The records of Dire Dawa meteorological station show that the station has experienced marked decline of rainfall trend analysis (Y = − 0.001x + 60.22) and positive trend for Dangago (Y = 0.040x − 24.69) and Haramaya (Y = 0.0257x − 462.1) stations which are in the upper. This shows that the effect of rainfall in Dire Dawa station is relatively modest. Hence, flood causing rainfall is coming from upper catchments out of the political boundary of Dire Dawa city (Figs. 2, 3 and 4).

Flooding of the study area is the result of hourly rainfall variability. More importantly, variations arising from hourly distribution of rainfall play a significant role. The standardized daily rainfall anomaly shows that sometimes years with high hourly rainfall can still have a decline in daily rainfall pattern that falls below mean hourly rainfall records. This confirms that the hourly variation of rainfall distribution that determines the occurrence of flood risk is working independent of hourly, monthly and yearly rainfall distributions. The detailed monthly rainfall analysis of Dire Dawa meteorological station showed that severe and heavy rainfall was recorded in 1956, 1961, 1963, 1964, 1983, 1996 and 2010 at Dire Dawa meteorological stations. The

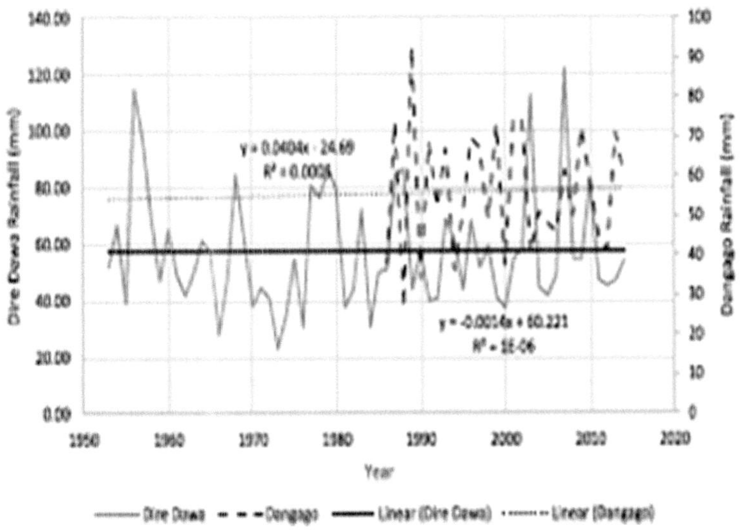

Fig. 2 Daily rainfall distribution of Dire Dawa, Haramaya Station (1960–2013) (1953–2014)

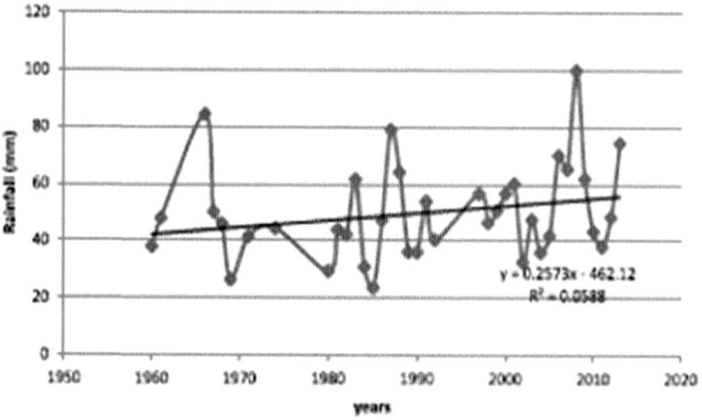

Fig. 3 Daily rainfall distribution of Dangago (1986–2014) station

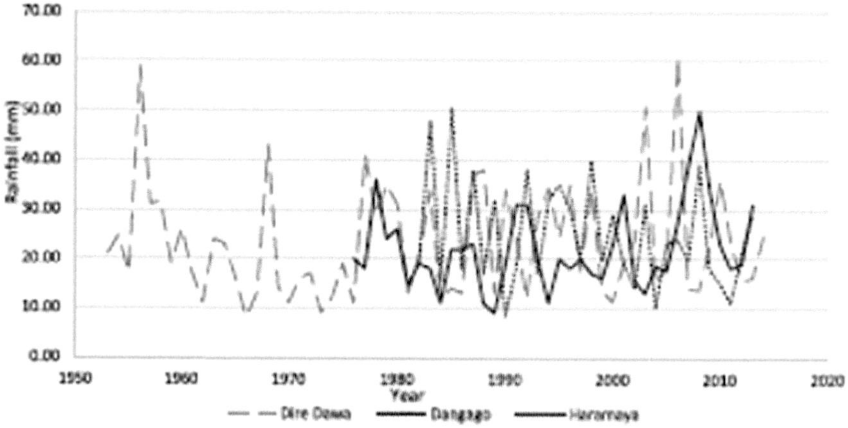

Fig. 4 Hourly rainfall distribution of Dire rainfall station

comparison of these higher raining years with historically identified flood event years shows that most of these events have no causative relationship. However, the hourly distribution of extreme rainfall shows that the result is high for the periods identified as flood event years. For example, the 61 mm/h. the amount of rainfall recorded in 2006 confirms that the figure represents the highest ever record of hourly rainfall during the study period at all stations (Fig. 5). This implies that the hourly variation of rainfall condition of this station is a potential driver of flood hazards of the city.

Similarly, the result of hourly variation of rainfall in other stations complements the scenario of Dire Dawa station. At both Dengogo and Haromaya stations, the high daily rainfall records have no strong relation with identified flood event years. On the contrary, the maximum hourly rainfall distribution of both stations shows that the hourly distribution result has a relationship with identified flood event years. The

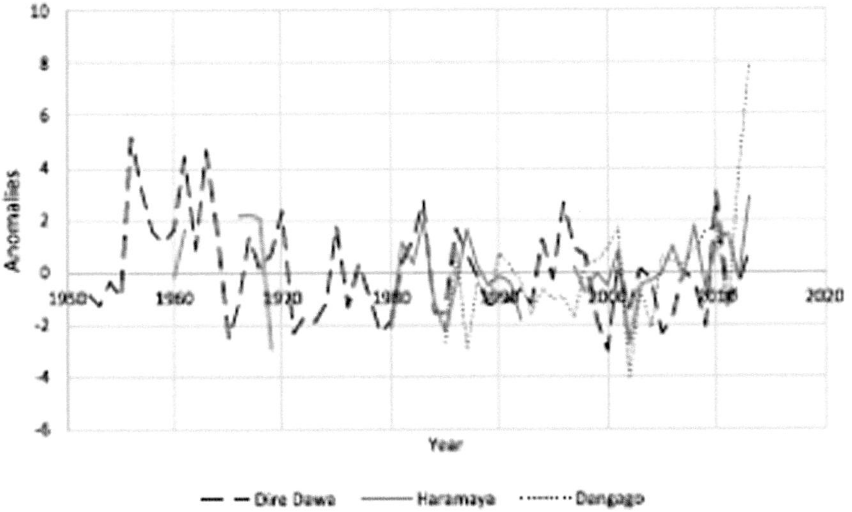

Fig. 5 Standardized daily extreme Dawa, Haramaya and Dangago anomalies of Dire Dawa, Haramaya and Dangago stations

trend analysis result of hourly rainfall data distribution shows that the likelihood of getting more than 30 mm of hourly maximum rainfall happened every 5, 4 and 3 years cycle for Haramaya, Dire Dawa and Dangago stations, respectively. The cumulative return period of the three stations shows that rainfall affects the city at every four-year interval.

Topography

Dire Dawa's watershed comprises different elevation points that range from 1000 to 2400 m above sea level. Dire Dawa city, where different rivers meet, is the lowest point with an altitude of 1000 m above sea level. The location of Dire Dawa at the foot of mountain chains has exacerbated the flooding of the city (Fig. 6).

Land Use Change and Urban Expansion

To explore the land use change and growth direction of Dire Dawa city, the analysis of four decades of satellite images was conducted (Table 4). The overall land use land cover change of the study area shows that the sizes of forest, bush and grass lands are continuously declining over the periods. On the contrary, settlements and bare land are significantly increasing (Fig. 7).

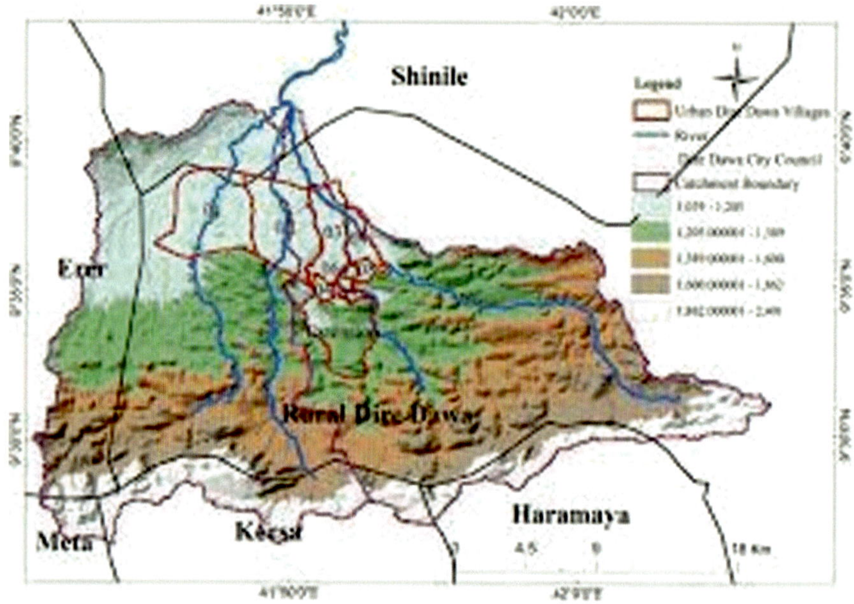

Fig. 6 Topographic map of Dire Dawa watershed

Table 4 Land use land cover changes of the watershed (1985, 1995, 2006 and 2016)

Land use (km^2)	1985	1995	2006	2016
Forest	63.73	50.01	48.71	37.09
Settlements	17.27	27.60	30.06	44.66
Bush and grass	345.90	316.91	282.27	236.24
Agriculture	139.34	102.73	63.02	73.64
Bare land	115.61	171.37	261.29	308.45
River side	20.76	34.37	17.59	2.83

4.2 Institutional Factor

Due to the geographical location and topography, Dire Dawa city is serving as an outlet for floods coming from different districts of Oromia Regional State. The sources of floods are outside of Dire Dawa administration. The trans-regional nature of floods in Dire Dawa city demands the integration and collaboration of institutions from different regions across political boundaries. The institutional gap in flood management is a major challenge. In particular, there is no cross institutional connections across political boundaries. Secondly, there are no intuitions that specifically deals with flood issues neither at city nor at federal levels.

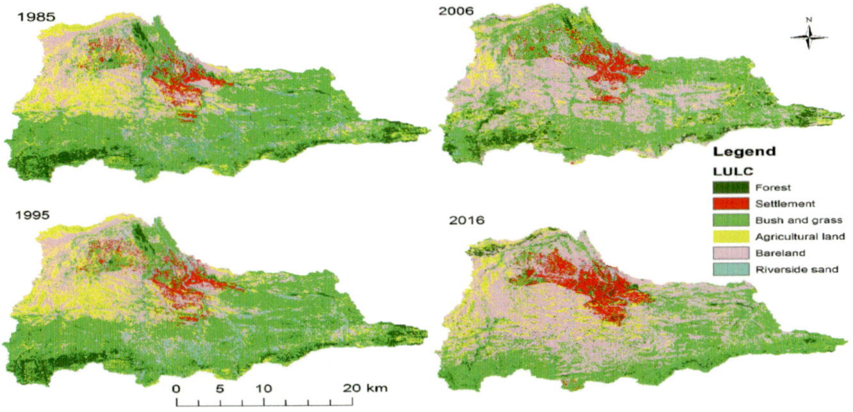

Fig. 7 Impact of land use change and urban expansion over time in Dire Dawa city

4.3 *Aggravating Factors*

Regime change and related land use changes: Land use changes such as inappropriate, misuse and informal occupation of protected areas were affected by political instability. Political transition and regime change allow irresponsible peoples to colonize forests and other buffer zones left for rehabilitation purposes. The vegetation cover of the highlands of Dangago during the rule of the Emperor (1954–1974) was moderately dense. Following the collapse of Imperial rule, massive forests were colonized and destroyed by communities. In the 1980s, the military government launched a large-scale reforestation campaign on the upper catchments of Dire Dawa city including Dangago. However, upon the fall of the military government in 1991, large-scale deforestation and human settlement took place within the rehabilitated area especially during the transition period (1991–1995).

Power interruption during heavy rain: Interruption of power during the rainy season has exacerbated the degree of risks and damages caused by flooding. The condition is worse when the rain takes place during night time. The lack of light during night time makes the people unable to identify the volume and intensity of the flood. This made the flood to be less understood so that the mobility of the people to rescue their property at expense of their life will occur.

4.4 Morphometric Analysis of the Watershed

Stream Orders

The 4th order which possibly can generate high discharge and velocity is located at sub-watersheds where Dechatu River is located in the center of Dire Dawa city (Fig. 8) (Table 5).

Stream Length (Sl)

The analysis of cumulative stream lengths helps to know the drainage characteristics and movements of the water at different basins (Altaf et al. 2013). Furthermore, the analysis of stream length of individual stream order helps to estimate the amount of discharges traveling downstream. Stream orders 3rd and 4th are passing through the middle of the city carrying large volumes, making the risk of flooding very high and challenging to handle (Table 6).

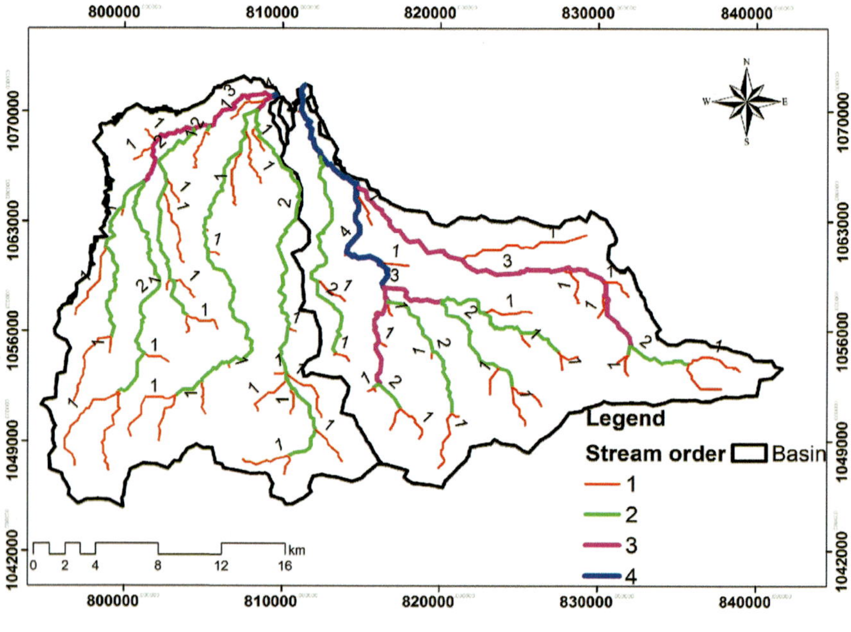

Fig. 8 Stream order distribution of study area watershed

Table 5 Stream orders of each sub-watershed

Watershed	No. of streams					Stream length (km)				Cumulative stream length (km)			
	1st	2nd	3rd	4th	Nu	1st	2nd	4th	3rd	1st	2nd	3rd	4th
1	3	1			4	1.8	3.91	0.00	0.00	1.8	5.7	5.75	5.75
2	14	3	1		18	11.5	6.49	0.81	0.00	11.5	18.0	18.79	18.79
3	6	1			7	2.7	0.77	0.00	0.00	2.7	3.4	3.43	3.43
4	36	7	1		44	5.5	22.49	24.78	0.00	35.5	58.0	82.74	82.74
5	53	13	2	1	69	53.9	24.55	8.05	13.97	53.9	78.4	86.50	100.47
6	6	1			7	3.0	1.35	0.00	0.00	3.0	4.3	4.32	4.32
7	19	6	1		26	27.1	13.59	11.05	0.00	27.1	40.7	51.74	51.74
8	6	2	1		9	7.6	1.76	1.43	0.00	7.6	9.3	10.75	10.75
9	16	3	1		20	12.8	3.10	14.09	0.00	12.8	15.9	30.00	30.00
10	21	4	1		26	19.1	4.29	15.07	0.00	19.1	23.4	38.43	38.43
11	2	1			3	2.6	1.78	0.00	0.00	2.6	4.4	4.41	4.41
12	3	1			4	3.6	4.94	0.00	0.00	3.6	8.5	8.54	8.54
13	10	3	1		14	9.6	9.11	4.55	0.00	9.6	18.7	23.29	23.29
14	8	1			9	10.0	8.25	0.00	0.00	10.0	18.2	18.24	18.24
15	11	4	1		16	5.4	4.33	7.41	0.00	5.4	9.8	17.18	17.18
16	7	1			8	4.4	5.58	0.00	0.00	4.4	10.0	9.99	9.99
17	15	4	1		20	8.5	4.09	10.33	0.00	8.5	12.5	22.87	22.87
18	36	8	1		45	32.9	17.69	18.82	0.00	32.9	50.6	69.38	69.38
19	22	3	1		26	8.0	3.93	1.42	0.00	8.0	11.9	13.37	13.37
20	15	5	1		21	13.8	9.35	11.23	00.0	13.8	23.2	34.41	34.41
21	17	5	2	1	25	19.8	6.21	2.48	6.49	19.8	26.0	28.46	34.96
22	21	4	1		26	19.9	8.59	9.45	0.00	19.9	28.5	37.9	37.9
23	15	3	1		19	11.8	8.30	8.07	0.00	11.8	20.1	28.2	28.2
24	4	1			5	6.9	5.54	0.00	0.00	6.9	12.4	12.40	12.40
25	8	1			9	6.0	5.94	0.00	0.00	6.0	11.9	11.93	11.93
26	2	1			3	1.3	2.43	0.00	0.00	1.3	3.7	3.74	3.74
Total	376	87	18	2	483								

5 Texture Ratio

Texture ratio (T) represents the total number of stream segments of all orders per unit perimeter of the basin. Based on texture ratio, basins are classified into different classes; coarse (2–4), fine (5–8) and very fine (> 8) texture ratios. The texture ratio of the study area is less than 2 over all the sub-watersheds which show that the whole study area watershed is classified under very coarse texture that imply high infiltration.

Table 6 Watershed parameter analysis

Watershed	Area (km^2)	Perimeter (km)	Sl (km)	Dd (km/km^2)	Sf
1	10.7	22.6	5.04	0.54	0.37
2	17.9	38.8	6.75	1.05	1.01
3	3.6	16.4	2.72	0.95	1.94
4	77.8	92.5	15.56	1.06	0.57
5	95.7	97.3	17.50	1.05	0.72
6	7.7	18.8	4.18	0.56	0.91
7	44.6	53.3	11.35	1.16	0.58
8	10.5	22.6	5.00	1.02	0.85
9	29.6	56.5	8.98	1.01	0.68
10	39.3	51.8	10.60	0.98	0.66
11	3.7	17.9	2.76	1.19	0.81
12	6.6	24.0	3.84	1.29	0.60
13	21.1	32.8	7.42	1.10	0.66
14	17.8	30.9	6.72	1.03	0.51
15	20.4	31.9	7.28	0.84	0.78
16	10.4	21.4	4.97	0.96	0.77
17	25.7	39.5	8.30	0.89	0.78
18	78.1	69.9	15.59	0.89	0.58
19	5.5	15.1	3.45	2.44	4.74
20	39.2	42.1	10.55	0.88	0.54
21	33.8	39.5	9.69	1.03	0.74
22	38.3	47.5	10.41	0.99	0.68
23	30.4	39.1	9.12	0.93	0.63
24	13.3	27.0	8.54	0.93	0.38
25	14.3	24.6	5.94	0.84	0.63
26	2.4	13.8	2.14	1.58	1.27

Bifurcation Ratio

Bifurcation ratio (Rb) is the ratio between the numbers of streams of any given order to the number of streams of next higher order. Bifurcation ratio is an important parameter to estimate where flood can take place in the watershed. The mean bifurcation ratio of study area sub-watersheds ranges from 2 to 8. The value of Rb tells about the geologic nature and shape of the watershed. When the value of Rb is more than 5, the basin is elongated and geological structure controls the basin.

Stream Frequency (Sf)

Stream frequency is an expression of the total number of streams of all orders to total unit area. The analysis of stream frequency reflects permeability, infiltration capacity and relief of watersheds and runoff process. Basins with high stream frequency are characterized by low infiltration capacities. The result of this study shows that the highest stream frequency occurs at the final outlet of the watersheds located at the extreme downstream, where flood problems are more severe.

Drainage Density (Dd)

Dd is the total length of all streams per unit area. It expresses the closeness of spacing of streams and drainage efficiency of the watershed. The overall Dd of study area ranges between 0.54 and 2.44 indicating that the study area watershed is characterized by low drainage density.

Elongation Ratio (Re)

The elongation ratio is the radius of the minimum bounding circle with the radius of the maximal inscribed circle, i.e., L = r/R. Re is used to assess the shape of the basin which can tell about the flooding condition of the basin. Basins with higher elongation ratio are tending to have high infiltration capacity and low runoff. Under varieties of climate condition and geologic type, the value of elongation ratio lies between 0.6 and 1 (Altaf et al. 2013).

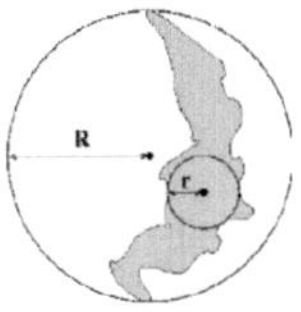

In the study area, the elongation ratio ranges from 0.6 to 0.810 except for sub-watershed 24 with Re of 0.48 (Table 7). This indicates that the whole watershed is elongated in shape implying low infiltration capacity that leads to high soil erosion which induces flash flooding.

Circulatory Ratio (Rc)

Rc is expressed as the area of the basin to the area of a circle having the same circumference as the perimeter of the basin. Rc expresses basin shape, rate of infiltration and the time needed to reach the basin outlet. Higher Rc shows that the basin could have delayed time to peak flow while the lower value shows shorter time to peak. Circulatory ratio of the study area (Table 7) varies from 0.1 to 0.3. The result of Rc shows that all sub-watersheds are elongated in shape indicating that the basins are characterized by elongated shape with low infiltration that can cause flood risks.

Form Factor (Rf)

Rf is the ratio between watershed area and the square of watershed length. Form factor is used to forecast the flow intensity of a given basin. The value of the form factor parameter is reversely proportional to flooding. The computed value of the form factor of the study area lies between 0.3 and 0.52. This articulates that all sub-watersheds are elongated in shape and are susceptible to flooding.

Table 7 Morphometric parameter analysis

Watershed	Rc	Re	Rf	T	Cc	Bifurcation ratio (Rb)			
						1 to 2	2 to 3	3 to 4	Mean Rb
1	0.26	0.73	0.42	0.18	1.94	3.0			3.0
2	0.15	0.71	0.39	0.46	2.59	4.7	3		3.8
3	0.17	0.79	0.49	0.43	2.43	6.0			6.0
4	0.11	0.64	0.32	0.48	2.96	5.1	7		6.1
5	0.13	0.63	0.31	0.71	2.81	4.1	6.5	2	4.2
6	0.27	0.75	0.44	0.37	1.92	6.0			6.0
7	0.20	0.66	0.35	0.49	2.25	3.2	6		4.6
8	0.26	0.73	0.42	0.40	1.96	3.0	2		2.5
9	0.12	0.68	0.37	0.35	2.93	5.3	3		4.2
10	0.18	0.67	0.35	0.50	2.33	5.3	4		4.6
11	0.15	0.79	0.49	0.17	2.62	2.0			2.0
12	0.14	0.76	0.45	0.17	2.63	3.0			3.0
13	0.25	0.70	0.38	0.43	2.01	3.3	3		3.2
14	0.23	0.71	0.39	0.29	2.07	8.0			8.0
15	0.25	0.70	0.39	0.50	2.8	1.99	4		
16	0.29	0.73	0.42	0.37	1.87	7.0			7.0
17	0.21	0.69	0.37	0.51	2.20	3.8	4		
18	0.20	0.64	0.32	0.64	2.23	4.5	8		6.3
19	0.30	0.77	0.46	1.72	1.82	7.3	3		5.2
20	0.28	0.67	0.35	0.50	1.90	3.0	5		4.0
21	0.27	0.68	0.36	0.63	1.92	3.4	2.5	2	
22	0.21	0.67	0.35	0.55	2.17	5.3	4		4.6
23	0.25	0.68	0.37	0.49	2.00	5.0	3		4.0
24	0.23	0.48	0.18	0.18	2.09	4.0			4.0
25	0.30	0.72	0.40	0.37	1.84	8.0			8.0
26	0.16	0.81	0.52	0.22	2.53	2.0			2.0

Compactness Constant/Coefficient (cc)

Cc is the ratio of perimeter of watershed to circumference of circular area, which equals the area of the watershed. A perfectly circular shape watershed could have a high compactness constant that implies that the basin experiences high infiltration. The computed value of Cc of the study area shows that sub-watershed 4 has a maximum value 2.96 while sub-watershed 19 has a minimum value 1.82. This shows that most of the sub-watersheds are not circular in shape and hence susceptible to flooding.

Form Factor (Rf)

Rf is defined as the ratio between watershed area and the square of watershed length. Form factor is used to forecast the flow intensity of a given basin. The value of the form factor parameter is reversely proportional to flooding. The computed value of the form factor of the study area lies between 0.3 and 0.52 (Table 7). This articulates that all sub-watersheds are elongated in shape and are susceptible to flooding.

Compactness Constant/Coefficient (cc)

Compactness coefficient is the ratio of perimeter of watershed to circumference of circular area, which equals the area of the watershed. A perfectly circular shape watershed could have a high compactness constant that implies that the basin experiences high infiltration. The computed value of Cc of the study area shows that sub-watershed 4 has a maximum value 2.96 while sub-watershed 19 has a minimum value 1.82. This shows that most of the sub-watersheds are not circular in shape and hence susceptible to flooding.

5.1 *Prioritization of Sub-watersheds for Conservation Activities*

Taking different factors into consideration, it is not possible to take the conservation of the whole area at a time and all sub-watersheds are not equally exposed to flooding. The prioritization of the sub-watershed was determined based on linear and shape parameters of morphometric analysis. The morphometric result-based rank of sub-watershed was described in Table 8 and Fig. 9.

Based on compound parameter result of morphometric analysis, the study area sub-watersheds are classified in to five categories: extremely high (8.0–9.9), very high (10.0–11.9), high (12.0–13.9), moderate (14.0–15.9) and low priority (16.0–17.9). Accordingly, SW-5 and SWS-18 fall in the highest priority and sub-watershed 1, 11, 12 and 26 fall in the low priority category.

5.2 *Water Conservation Site Identification*

Landscape-based strategies are important to curb the problem of flooding in the study area. Accordingly, 6257 ha were proposed for terracing activities. In the same way, 36 sites were identified for construction of ponds, 19 for percolation tanks, 9 for storage tanks, 8 for Nala bunds and 5 sites for check dams. These sites were identified using soil, permeability, slope, land use, land cover and stream orders by combining with flood sensitive sub-watershed generated from morphometric parameters. As shown in Fig. 10, most of the locations proposed for water and soil conservation activities are located at upper catchments out of the city boundary. For example,

Table 8 Prioritization of watershed-based morphometric analysis rank results

Watershed	Rbm	Dd	Fs	T	Rf	Rc	Cc	Re	CP	Rank
1	21	18	26	24	1	20	7	18	16.875	23
2	18	14	4	13	24	6	21	15	14.375	18
3	6	25	2	14	25	8	19	25	15.5	20
4	5	3	22	12	10	1	26	4	10.375	3
5	12	1	12	2	22	3	24	2	9.75	2
6	6	21	5	18	20	22	5	21	14.75	19
7	11	4	20	10	7	10	17	5	10.5	4
8	24	19	6	16	16	19	8	19	15.875	22
9	13	10	14	20	13	2	25	11	13.5	14
10	9	5	16	8	14	9	18	6	10.625	6
11	25	24	7	25	6	5	22	24	17.25	24
12	21	22	19	26	3	4	23	22	17.5	25
13	20	12	15	15	9	16	11	13	13.875	17
14	1	15	24	21	2	15	12	16	13.25	13
15	19	13	8	7	17	18	9	14	13.125	12
16	3	20	10	17	12	24	3	20	13.625	15
17	17	11	9	6	21	12	15	12	12.875	11
18	4	2	21	3	11	11	16	3	8.875	1
19	8	23	1	1	26	26	1	23	13.625	15
20	14	6	23	9	4	23	4	7	11.25	7
21	23	8	11	4	19	21	6	9	12.625	9
22	9	7	13	5	15	13	14	8	10.5	4
23	14	9	18	11	8	17	10	10	12.125	8
24	23	17	25	23	18	14	13	1	15.625	21
25	9	16	17	19	5	25	2	17	12.75	10
26	14	26	3	22	23	7	20	26	19	26

many percolation tanks are proposed at one of the most flood vulnerable sites (SW-18) of upper catchments, while least is proposed for the SW-5 which is located within the urban boundary.

Table 8 shows that most of the sources are coming from upper catchments. As major areas of SW-5 are in urban boundaries, less percolation tanks were proposed at down streams of this sub-watershed. In general, the location of conservation measures confirms that effective utilization of proposed structures at proposed site can significantly minimize the amount of flood water entering the city boundaries

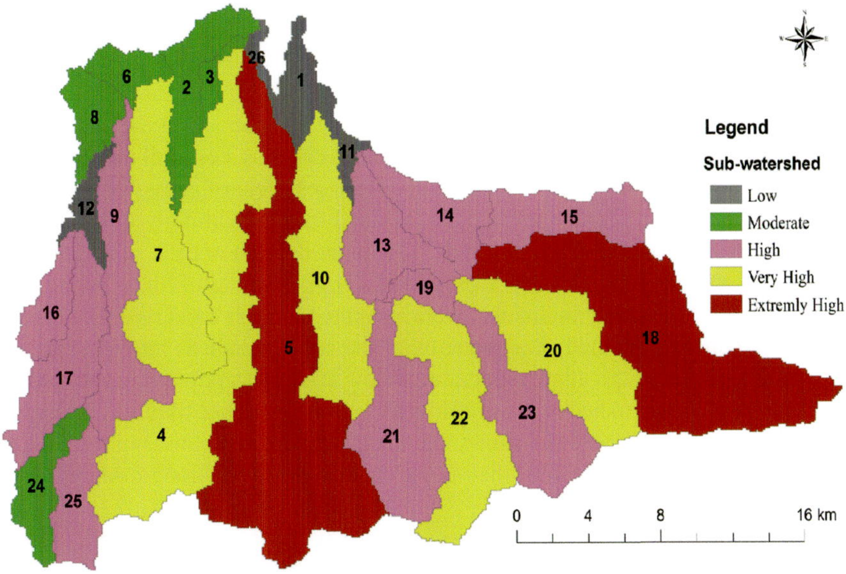

Fig. 9 Map of prioritized sub-watersheds sites

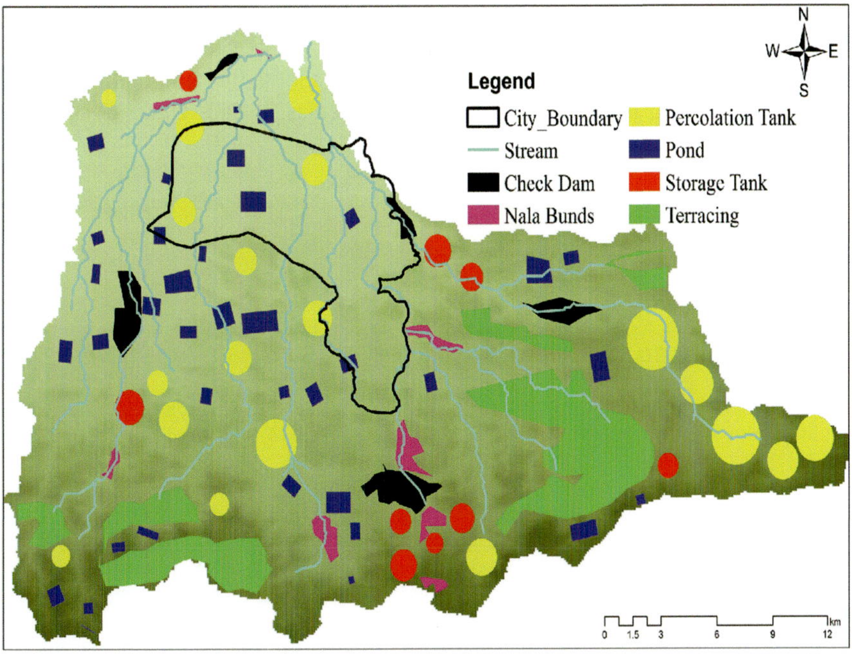

Fig. 10 Proposed map of water conservation

6 Conclusion

The findings of the study show that flood risk in Dire Dawa city is caused by multiple factors. These causes are fueled by aggravating factors such as regime change and power interruption. Most importantly, the study area flooding was attributed to hourly rainfall variability. Following the comprehensive assessment of causes of flood risk, sub-watersheds' susceptibility to flood risk was analyzed. Accordingly, the watershed was divided into 26 mini sub-watersheds. The quantification of each sub-watershed was carried out using morphometric parameters. Higher values of linear parameters and lower values of shape parameters were attributed to high flood risk. The results of these parameters enabled categorization of the sub-watersheds into five classes, extremely high, very high, high, moderate and low flood susceptible zones. Based on these classes, sub-watersheds were prioritized for conservation activities. The overall result of morphometric analysis shows that managing the root causes of flooding at upper catchments can minimize the volume and intensity of flood-causing water before it enters the city boundary.

Finally, site selection for water conservation was carried out using soil, permeability, slope, land use, land cover and stream orders and by combining these overlays with classes generated from morphometric parameters. Accordingly, different sites were proposed for landscape-based flood mitigation measures such as construction of terracing, check dam, percolation tank, storage tank, ponds and soil and stone bunds. To fully curb the problem of flooding, it is highly recommended that the proposed water conservation structures should be adopted as one of the components of sustainable flood risk management strategies in Dire Dawa city and its surroundings.

References

Alemayehu D (2007) Assessment of flood risk in Dire Dawa Town, Eastern Ethiopia, using GIS. Unpublished, 10. http://etd.aau.edu.et/dspace/bitstream/123456789/1659/1/DANIELALEMAYEHU.pdf

Altaf F, Meraj G, Romshoo SA (2013) Morphometric analysis to infer hydrological behaviour of Lidder Watershed, Western Himalaya, India. Geogr J 2013:1–14. https://doi.org/10.1155/2013/178021

Erena SH, Worku H (2018) Flood risk analysis: causes and landscape based mitigation strategies in Dire Dawa city, Ethiopia. Geoenviron Disasters 5(1). https://doi.org/10.1186/s40677-018-0110-8

Erena SH, Worku H, De Paola F (2018) Flood hazard mapping using FLO-2D and local management strategies of Dire Dawa city, Ethiopia. J Hydrol Reg Stud 19(June):224–239. https://doi.org/10.1016/j.ejrh.2018.09.005

Tegegne D, Esa E, Nebere A (2018) Flood hazard assessment using multi-criteria evaluation approach in Dembiya Woreda, Amhara Region, Ethiopia. Ethiopian Renaissance J Soc Sci Hum 5(2):18–31

A Systemic Analysis of Causes, Impacts, and Coping Strategies of the People Related to Flooding in Ghana Capital City, Accra

Ezekiel Atanga Azuure

Abstract Because of its social, economic, and environmental consequences, urban flooding has become a serious issue in many regions of the world. Every year, floods destroy livelihoods, property, infrastructure, and lives in Ghana, making many people homeless. The purpose of this work is to comprehend the present status of flood research in Ghana, with an emphasis on how the academic community has tackled the causes, effects/impacts, and coping mechanisms used by individuals in urban areas. The author examines academic databases such as SCOPUS, Web of Science, Springer, Taylor & Francis, Science Direct, and Google Scholar for current studies based on a complete literature assessment and individual co-author in-depth expertise in research and practice in Ghana. The author's findings, based on 15 papers, show that poor urban planning and development of poor and insufficient drainage systems, a lack of environmental awareness, and excessive rainfall are the leading causes of urban flooding in Ghana. The most often cited impacts/effects are physical costs, economic infrastructure devastation, and health issues. The most often reported coping techniques are property removal and protection and drain construction. The evaluation also identifies major research gaps in Ghana and proposed a new topic for future research direction and practice.

Keywords Ghana and climate · Accra and climate · Climate risk management · Disaster risk management · Adaptation · Resilience · Vulnerability · Adaptive capacity · Indicator analysis · Hazard reduction measures · Nature-based solutions · Comprehensive urban planning · Early warning systems

E. A. Azuure (✉)
Ministry of Energy, Accra, Ghana
e-mail: eazuure@yahoo.com

K. Kim and C. Atkin (eds.), *Cross-Cultural Perspectives on Climate Change Adaptation*, Cities and Nature, https://doi.org/10.1007/978-3-031-50365-8_9

1 Introduction

Background

According to NASA, the average earth's surface temperature has risen by 0.93 0.07 °C since the 1880s (Kim and Kang 2021). With the increasing trend of global warming, the frequency and magnitude of maximum rainfall rise, resulting in floods (Hirabayashi et al. 2013; Avand et al. 2021). According to the IPCC study, the water cycle is becoming more vigorous as the climate warms, which means more powerful storms and flooding. In July 2021, the entire planet watched as heavy rains turned into floods that wiped away centuries-old houses in Europe, created landslides in Asia, and submerged subways in China. More than 900 individuals were killed in the devastation. Western states in North America were fighting flames amid an extreme drought that was impacting the water and electricity supply. Water-related risks may be extremely damaging, and the influence of climate change on catastrophic water-related occurrences like these is becoming increasingly clear.

The growing urbanization of African towns has also increased the demand for land. People with little means who cannot afford to live in flood-free regions choose to dwell in flood-prone areas, which are usually inexpensive. Persons living in poverty are more vulnerable to floods because they have few choices and they normally accept and adapt to flooding (Winter and Karvonen 2022). Flooding damages in Sub-Saharan Africa exceed US $300 billion, and given the poor and restricted infrastructure, the low capability of local governments (e.g., human and financial resources), and weak coordination of essential parties in flood control, urban flooding has become one of the primary problems to deal with in Africa (Mensah and Ahadzie 2020).

The massive expansion of many Ghanaian cities in the absence of rigorous adherence to development laws has made them vulnerable to a variety of natural calamities. Floods have become a recurring problem in Ghana's major towns and communities located in low-lying areas. Low-lying locations are naturally vulnerable to flooding (Attipoe 2014). Rivers, basins, valleys, lakes, and low-elevated terrain are examples of low-lying areas. As a result, cities and communities in these low-lying areas can be classed as potentially flood-prone.

In addition, development has encroached on natural spaces. When buildings are placed on streams, paved surface areas hinder precipitation from being absorbed. Areas intended for runoff retention (swamps and valleys) have been utilized for infrastructure construction.

The Objective of the Study

- This study aims to assess the present status of flood research, with an emphasis on causes, effects/impacts, and coping techniques, as well as identify gaps in the existing literature to influence future research and practice.

Research Questions

- What are the impacts and effects of the flood on the economy and lives of the affected people?
- What strategies are people using to adapt to the impacts of flood and climate change?

2 Literature Review

In general, previous studies have confirmed that the Accra Metropolis has modest slopes up to roughly 22%. Settlements within or near main rivers in Accra, as well as those within their tributaries, are prone to flooding in the event of heavy rains (Mensah and Ahadzie 2020). Earlier studies have also found that the majority of Accra's drains are either inadequate, late, silted, or incorrectly connected. It is also true that Accra's higher population density has increased the number of impervious surface areas through pavement and buildings. Flooding caused by siltation is a vital component, which many inhabitants in the capital city refuse to practice. Dumping garbage into drains diminishes their ability to transport runoff into the sea. Accra's significant water bodies include the Odaw and Onyasia rivers, as well as the Korle lagoon (Asumadu-Sarkodie et al. 2015). These bodies of water collect and convey surface runoff from their catchments into the sea. However, many have converted them into garbage dumps.

A flood combined with an explosion at the Ghana Oil filling station on June 3, 2015, took over 152 lives (Asumadu-Sarkodie et al. 2015), bringing the total number of deaths between 1968 and 2015 to 567. Flooding has plagued Accra since 1968 and continues to this day.

The city has a population of roughly 3.86 million people, and floods killed 415 people between 1968 and 2014 (Asumadu-Sarkodie et al. 2015).

The blocking of the Odaw river's mouth, which prevented water from flowing into the sea, was a major cause of the June 3, 2015, flood. Many lives and possessions were also lost as a result of the repeated floods that occurred between 2016 and 2019.

In flood vulnerability assessments, drainage density is an important factor. An analysis of drainage density is performed to evaluate how successfully a catchment or basin is drained by river networks or streams.

It is calculated by dividing the entire length of rivers and/or stream networks in the watershed or catchment by the total area of the drainage basin. A drainage density map is an important component in flood and groundwater recharge studies. It shows the locations where runoff will accumulate. Runoff often flows from highlands to lowlands through valleys, basins, rivers, lakes, and lagoons.

It is consequently critical to evaluate the drainage density and elevation patterns of the Accra Metropolis to assess the city's vulnerability to floods. Applied Water Science (2021). This information will be critical for municipal planners and interventions targeted at alleviating Accra's recurring flooding.

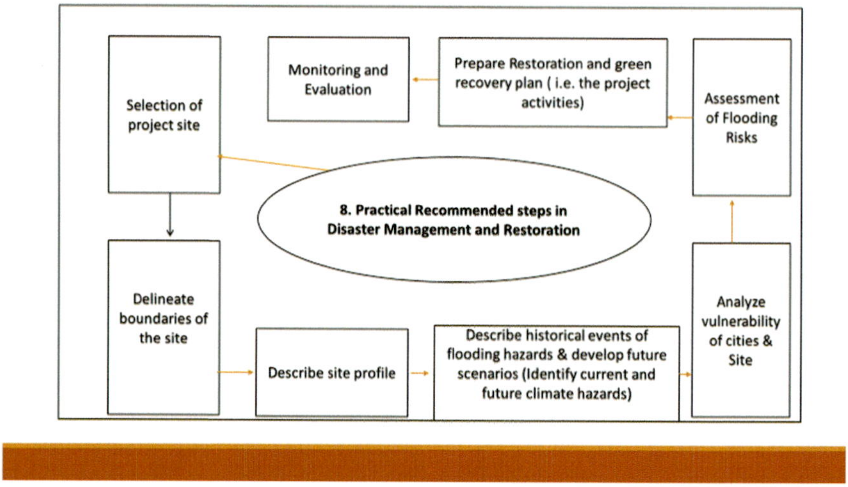

Fig. 1 Dr. Kim's disaster management and restoration framework

Using existing literature, the study also gives a detailed evaluation of the number of lives and properties lost due to flood-related incidents. The AMA's population is over 2.5 million people, according to the 2010 population and housing census, with a 4.3% annual population growth rate between 2010 and 2020 (Fig. 10).

3 Method and Materials

A Case Study in the Greater Accra Metropolis

According to the Ghana Statistical Service, the population of Accra rose from 190,000 when the country got independence in 1957 to 5,055,883 as of 2020. The city is low-lying and located between latitude 0006′ W and longitude 0535′ N, with a maximum elevation of 200 ft. above sea level and an area of 181 km^2. The landscape in the region ranges from level to moderately rising with elevations of up to 246 ft. in the foothills, as well as occasional isolated hills and rock outcrops. The region has two wet seasons and is in the dry coastal equatorial climatic zone. The first season lasts from March to July, while the second season lasts from September to November. The annual rainfall average is 810 mm. Short storms typically produce a large amount of rain, resulting in annual local floods, particularly in locations where drainage systems are deficient or clogged.

Accra Metropolis Is Known to Experience Flash Flooding

This study aims to assess the present status of flood research, with an emphasis on causes, effects, and coping techniques, as well as identify gaps in the existing literature to influence future research and practice. Twenty-one peer-reviewed papers from 2009 to 2019 were analyzed for this paper and colleagues with extensive expertise in urban floods in Ghana, both in research and practice, critically analyzed and improved chosen papers as part of the review to increase the validity of the findings. The term "secondary data" is used in this study to refer to existing research data that is reviewed to obtain answers to research questions that are unrelated to the original research purpose.

The researcher followed the instructions for a standard literature review provided by Cronin et al., which included doing a literature search, gathering, reading, and analyzing the literature.

The author employed phrases and keywords that were related to the study when doing an electronic search. Each search engine included items published in the previous ten years (from 2009 to 2019). The year chosen was crucial in ensuring that the research covered contemporary literature and debate about the topic area. The author used academic databases such as Google Scholar, Scopus, Web of Science, Springer, and Taylor & Francis to conduct his research (Mensah and Ahadzie 2020). For example, Scopus search terms included "coping techniques" AND "Ghana" AND "flood" AND "adaptation" AND "causes" AND "mitigation" AND "impacts" OR "effects" to find relevant material.

Data Acquisition and Analysis

A qualitative data analysis program (NVivo 10) was utilized to store textual information in the content analyses of the selected articles. The short-listed papers were examined for themes and issues that emerged from the articles. The frequency of publishing in journal outlets, the year of publication, research techniques, and the location where the study was done were all examined further. These were then loaded into an Excel spreadsheet for simple descriptive analysis (Fig. 2).

Delineation of Flood-Prone Areas

The method employed here is comparison of historical data from previous research on urban flooding in cities of Ghana and risk assessment and prediction modeling (Dekongmen et al. 2021) (Fig. 3).

Accra is Ghana's capital city and the seat of government for the Greater Accra Region. This city is the Government of Ghana's administrative, political, and commercial capital. With a population of about 5 million people and an annual growth rate of 3.1%, and land area of 3245 km^2, Accra is Ghana's biggest and fastest-expanding metropolis (Fig. 4).

Drainage Density

The Digital Elevation Model (DEM) of the study was used to analyze the direction of flows and flow accumulation in the area. The drainage density map of the area was generated from the DEM using the ArcMap environment through the Line Density tool. The Line Density file generated was further reclassified into four classes from low to very high (from low converging to very high converging areas) using the equation below: $DD = \Sigma Le/A$ where Le is the total distance of all the streams and rivers, and A is the total area of the drainage basin or the watershed (Fig. 5).

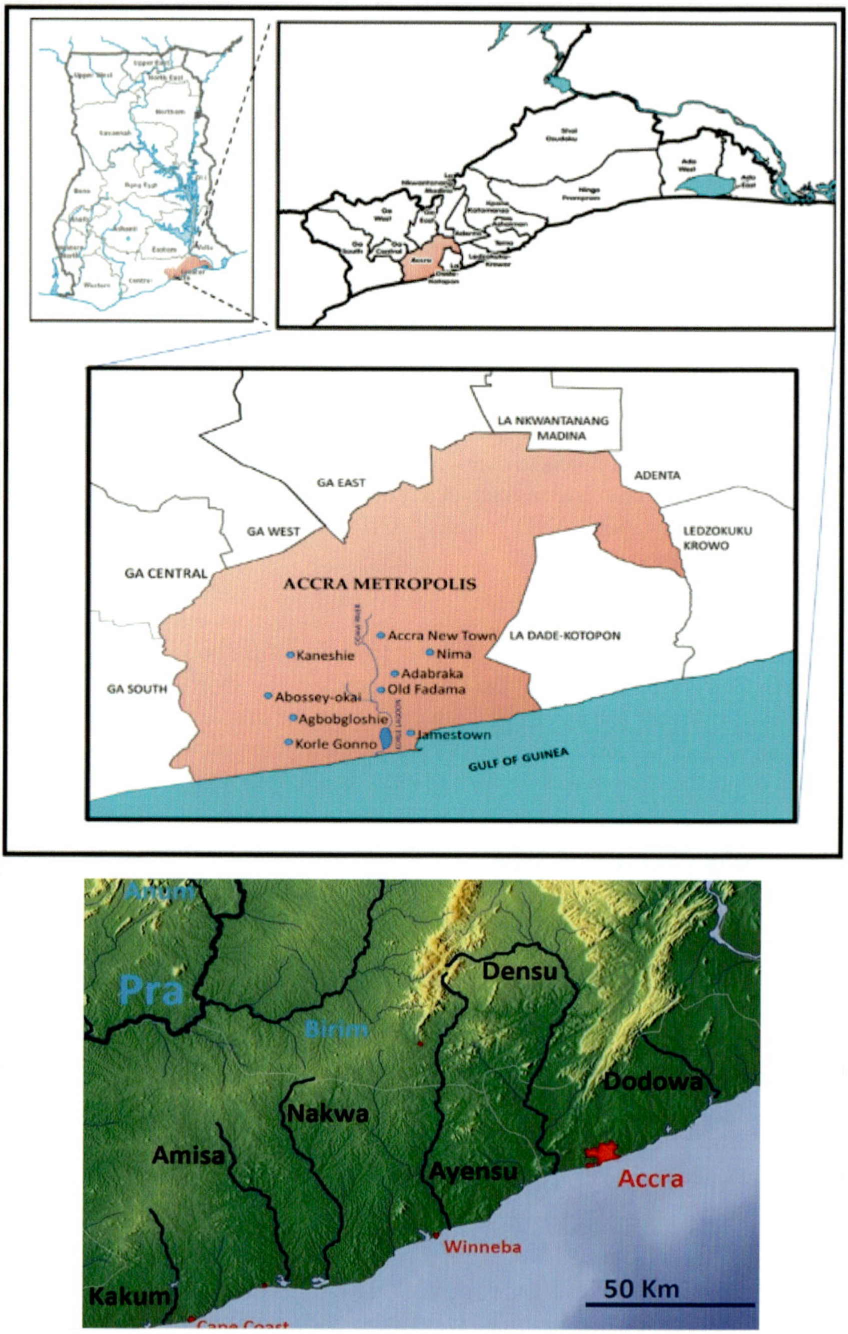

Fig. 2 Maps of Accra Metropolis showing the study region

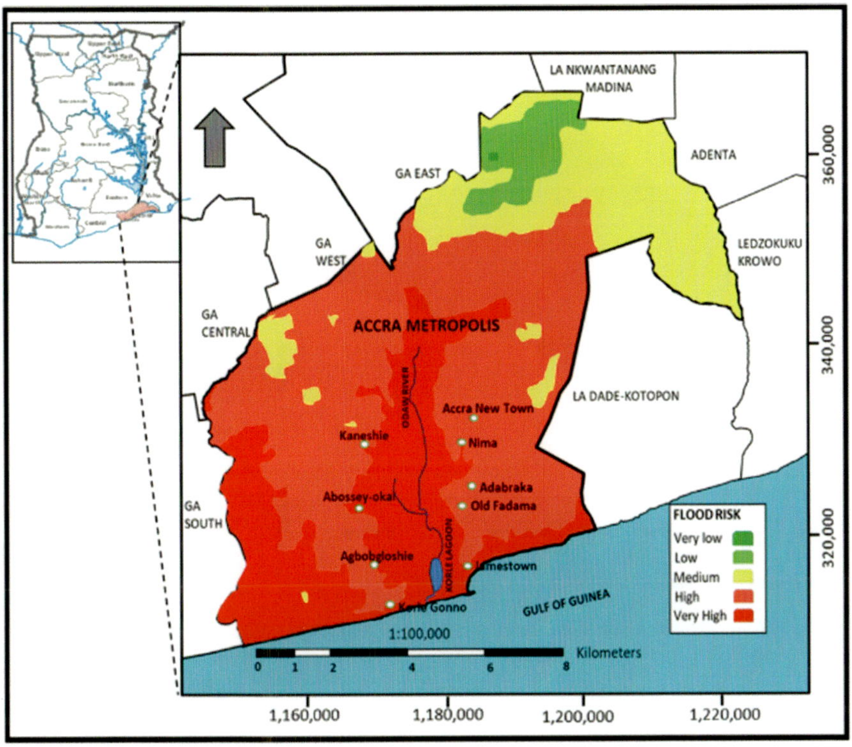

Fig. 3 Map of Greater Accra showing the study areas and vulnerable to flooding

Contour Characteristics in Accra

The contour map of the study was generated using the Digital Elevation Model in the ArcMap environment, using the contour command tool. The output polyline assigned was 10 m, which generated five classes of the contour lines. The five classes were further reclassified into three classes (Dekongmen et al. 2021) (Fig. 6).

The unregulated informal urbanization and occupancy in flood plains have been blamed for Accra's chronic flood threats (Karley 2009).

For example, in Karley's (2009) study on flooding and urban planning in Accra, he concluded: "There is no indication that extraordinary rainfall has recently occurred that may explain the increased incidences of flooding being observed." (Fig. 7).

Flood-prone areas and types of floods in the Greater Accra Plains. Kagblor a spatio-temporal study of urbanization and flooding in the Greater Accra Metropolitan Area (GAMA) of Ghana.

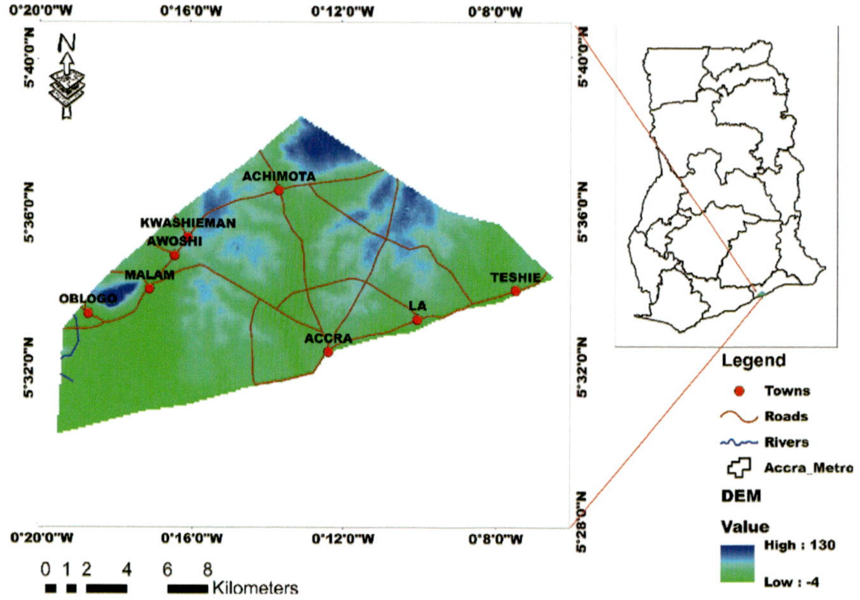

Fig. 4 Drainage density map of the study area (Dekongmen et al. 2021)

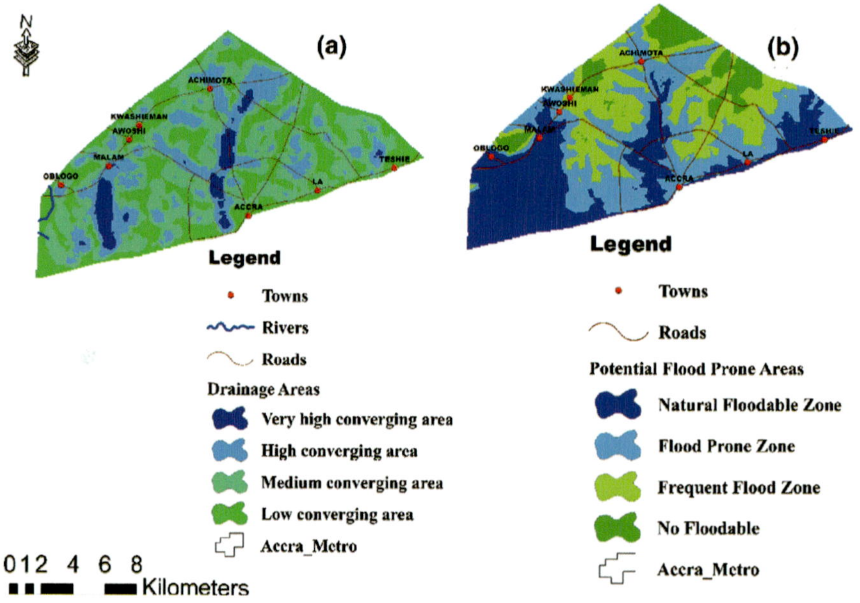

Fig. 5 Maps of drainage and flood-prone zones

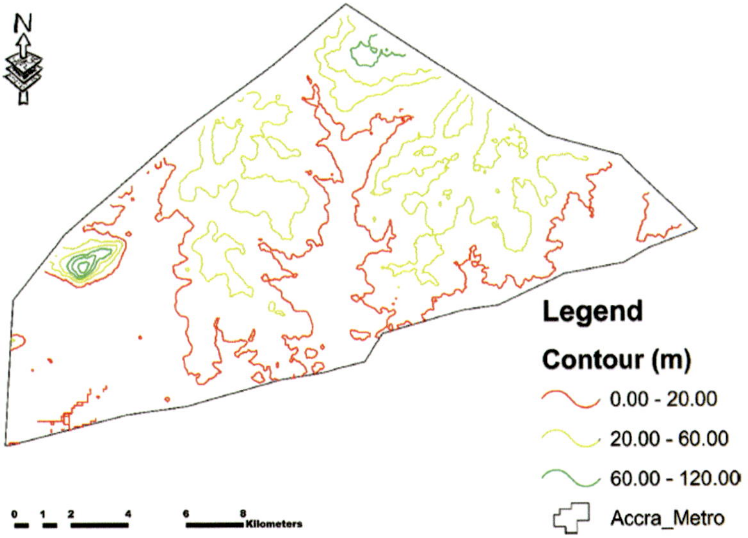

Fig. 6 Contour map of the study region. Accra's uncontrolled urbanization and slum growth in flood-prone regions

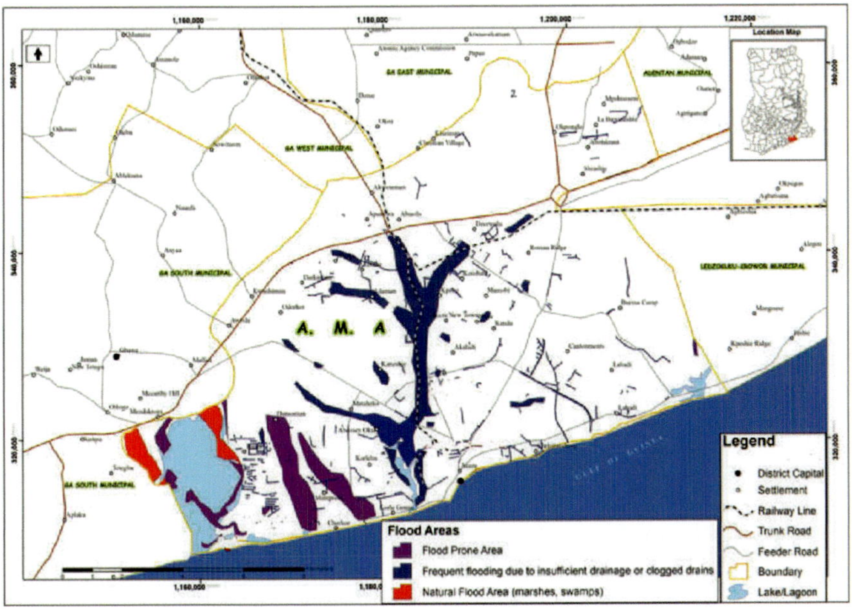

Fig. 7 Flood-prone areas and types of floods in the Greater Accra Plains

Fig. 8 Sediment load and waste materials within drainage channels inhibiting the flow of water

4 Results and Discussion

4.1 Causes of Flooding in Accra

The inadequacy of drainage systems to collect rainwater for safe disposal is the core of the problem. Flooding in Accra has also been linked to poor spatial planning and drainage system flaws (Karley 2009); major city growth, which prevents permeation by asphalt; unofficial property development practices; and poor physical development control and waste management practices in the city (Karley 2009) (Fig. 8; Table 1).

4.2 Annual Population Growth Estimates

In 2010, and 2015, respectively, 10 and 152 people died in floods, owing to the growing population. The Accra Metropolitan Assembly's population has been rapidly expanding. Because Accra serves as the country's capital, around 1 million people commute daily to the city for various services, bringing the metropolis' population to 3 million in 2010. Overall, due to the construction of various infrastructures on streams and the depletion of wetlands and surface runoff retention zones, the AMA hydrological pattern and drainage system have been damaged by population growth (Figs. 9 and 10).

5 Future Scenarios

Floods in Accra are expected to increase due to population increase and climate change impacts. Flooding is likely to increase due to climate change which causes irregular rain patterns and constant population growth which increases both residential areas and waste generations.

A wing of government has been established to tackle any disasters in the future whether they be from anthropogenic or natural causes. The Government of Ghana has

Table 1 Causes impacts and adapting strategies to flooding in Accra City

Causes	Natural anthropogenic	Uncontrolled conversion of vegetated land, poor coordination among state agencies in flood management; neglect of city authorities and official insensitivity to communities' problems; unregulated physical expansion; uncontrolled urbanization and slum clearance sand mining for development; urban growth; indiscriminate waste dumping; dwellings built in wetlands; loss of surrounding vegetation; closeness of structures to floodplain and rivers. Poor and insufficient drainage infrastructure: poor drainage; a poor or non-existent storm drainage system; drainage problems; drainage conditions; drainage system failure indiscriminate waste disposal; poor waste disposal; incorrect waste and refuse disposal; residential waste in streams and drains; waste into the drainage system; indiscriminate waste disposal; poor waste disposal; improper waste and refuse disposal Heavy rain; heavy and unexpected rain; river and lagoon overflow; heavy rainfall; high rainfall occurrences; unusually heavy rainfall events; extended and intense character of rainfall; excess rainfall; torrential rains; heavy rain; intense rainfall; landslides
Impacts	Negative effects	Loss of household things; building collapse; disruption and loss of services and companies; loss of human and animal life; destruction of livelihoods; lost sources of livelihood, injuries; destruction of household possessions, damage and destruction of crops and farmlands; asset losses; missed work days and medical bills; housing repairs; labor income losses, unable to go to work; high cost of upgrading houses; loss of business; damage of storm drains, community roads, bridges, schools, and health institutions cholera pandemic; disease outbreak; drowning; personal injury; watery disease outbreak; dirty and stinky surroundings; cholera or dysentery epidemic; dirt; foul odor; harmed persons a malaria epidemic; rubble from floods psychological aftermath, social stigmatization of residents; anxiety

(continued)

Table 1 (continued)

Coping mechanisms	Reactive tactics	Relocation and property protection include: relocating to stay with friends; evacuating victims and property; early transfer of objects to safe areas; temporarily finding sanctuary; and transferring assets to neighbors. Considering moving during floods; packing items on shelves and high levels; locating valuable items on shelves; stocking assets to a safe place; attempting to build platforms in rooms to keep desirable characteristics; raising foundations; sandbags to block water flows; mounting beds on concrete blocks; elevating land before building, walls, and entry points of the house; rebuilding destroyed houses with cement; building flood steps; putting stones at the river's edge; filling their compounds with coarse gravel Drain construction includes the building of temporal drains, temporary drains, storm drains, the removal of structures in rivers, the redirection of flood water out of the neighborhood, and the creation of temporary channels Self-built channels: building additional channels surrounding the home; establishing stormwater channels; clearing the barrier
Preventative strategies	Recovery methods/flood mitigations	Trash disposal; gutter clearing; community clean-up activities; cleaning of clogged local drains; arranging spraying exercises; waste management; desilting of clogged drains Support from the government, friends, and family: relying on city officials and other stakeholders for relief products; relying on the media; friends, relatives, governmental and non-governmental organizations; extension information; personal funds to repair damages

enacted a policy for national general desiltation of drainage channels and cleaning of the cities on the first Saturday of every month to adapt and mitigate floods disasters. Though the government has made progress in preservation of wetlands as a powerful nature-based solution along the rivers in Accra because of the encroachment of poor people settling in the area, the flooding vulnerability is increasing (Pabi et al. 2021).

6 Conclusions and Recommendations

- The study discovered that locations with very high drainage density and high drainage density were inside the most floodable zones.
- This discovery is compatible with the idea that all runoff flows from highland locations to lowland areas, causing floods.

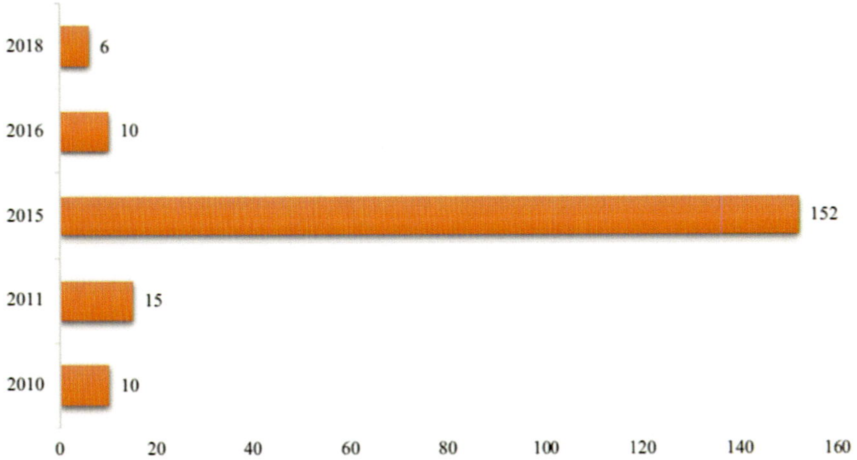

Fig. 9 Major impacts of floods from 2010 to 2018 flood year (Songsore 2017)

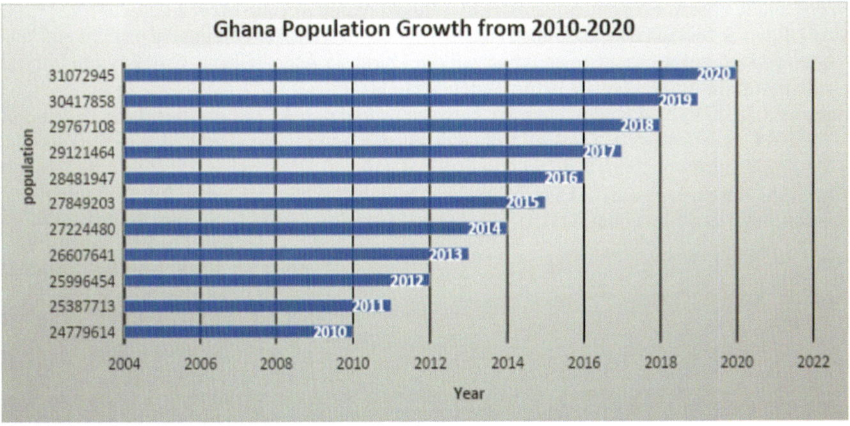

Fig. 10 Ghana population growth and influence on flooding (World Bank Group Open Data)

- The mild contour slopes were also discovered to correlate with the city's natural floodable and flood-prone zones.
- It is recommended that the government should expedite policy regulations to prevent the illegal siting of structures in the city.
- Reconstructions of drainage systems should also be undertaken to ensure the large volumes of water can find their way to the sea.

References

Asumadu-Sarkodie S, Phebe AO, Jayaweera Herath MPC (2015) Flood risk management in Ghana: a case study in Accra. Adv Appl Sci Res 6(4):196–201

Attipoe SK (2014) An assessment of flood mitigation measures in Accra, Ghana

Avand M, Moradi HR, Lasboyee MR (2021) Spatial prediction of future flood risk: an approach to the effects of climate change. Geosciences 11(25):1–20. https://doi.org/10.3390/geosciences11010025

Dekongmen, BW, Kabo-bah AT, Domfeh MK, Sunkari ED, Dile YT, Antwi EO, Gyimah RAA (2021) Flood vulnerability assessment in the Accra Metropolis, southeastern Ghana. Appl Water Sci 11(7). https://doi.org/10.1007/s13201-021-01463-9

Hirabayashi Y, Mahendran R, Koirala S, Konoshima L, Yamazaki D, Watanabe S, Kim H, Kanae S (2013) Global flood risk under climate change. Nat Clim Change 3(9):816–821. https://doi.org/10.1038/nclimate1911

Karley NK (2009) Flooding and physical planning in Urban Areas in West Africa: situational analysis of Accra, Ghana. Theor Empir Res Urban Manag 4(13):25–41

Kim J, Kang J (2021) Analysis of flood damage in the Seoul Metropolitan government using climate change scenarios and mitigation technologies. Sustainability 13(105):1–28. https://doi.org/10.3390/su13010105

Mensah H, Ahadzie DK (2020) Causes, impacts and coping strategies of floods in Ghana: a systematic review. SN Appl Sci 2(5):1–13. https://doi.org/10.1007/s42452-020-2548-z

Pabi O, Egyir S, Morgan E (2021) Flood hazard response to scenarios of rainfall dynamics and land use and land cover change in an urbanized river basin in Accra, Ghana. City Environ Interact 12:100075. https://doi.org/10.1016/j.cacint.2021.100075

Songsore J (2017) The complex interplay between everyday risks and disaster risks: the case of the 2014 cholera pandemic and 2015 flood disaster in Accra, Ghana. Int J Disaster Risk Reduct 26:43–50. https://doi.org/10.1016/j.ijdrr.2017.09.043

Winter AK, Karvonen A (2022) Climate governance at the fringes: peri-urban flooding drivers and responses. Land Use Policy 117:106124. https://doi.org/10.1016/j.landusepol.2022.106124

A Comparative Assessment of Flood Hazard Susceptibility Modeling in Niger State, North-Central Nigeria

Saadatu Gambo Madaki

Abstract Flood in Nigeria is becoming an annual event; it arises as coastal flood, flash floods, river flood, and urban flood. Years back states and cities in Nigeria experienced flood calamities that undermined the government's ability to protect their residents. Heavy rains caused the country's most disastrous flood catastrophe in 2012, which was classified as the worst in 40 years and devastated 33 states including Niger State of the 36 states, causing the loss of life and property. The findings of this study revealed that many topographic and hydroclimatic elements connected with the region's geomorphological complexity are responsible for flood occurrence in Niger State. Anthropogenic variables, such as unplanned development and agriculture, exacerbate the impact of natural causes, according to field observations conducted as part of this strategy. To determine the extent of the phenomenon in the watershed using a combination of qualitative and quantitative methods, detailed field observation, climate analysis, and literature review were used to select and weigh the parameters related to the triggering factors, and the AHP model was used to calculate the flood hazard vulnerability index by integrating and combining these parameters. The generated flood hazard vulnerability map developed in the case study gives a detailed insight into the watershed's highly and extremely vulnerable sections. Thus, the vulnerability of the location was analyzed, and flood risk was equally assessed based on the "*Standard Methodology for Climate-Related Disaster and Green Recovery*" adopted for the study.

Keywords Niger and climate change · Nigeria and climate change · Flood risk · Climate risk management · Disaster risk management · Adaptation · Resilience · Vulnerability · Adaptive capacity · Indicator analysis · Hazard reduction measures · Nature-based solutions · Comprehensive urban planning

S. Gambo Madaki (✉)
Department of Research, Innovation and Outreach, National Council on Climate Change, Abuja, Nigeria
e-mail: saadatu.gambo@natccc.gov.ng

K. Kim and C. Atkin (eds.), *Cross-Cultural Perspectives on Climate Change Adaptation*, Cities and Nature, https://doi.org/10.1007/978-3-031-50365-8_10

1 Introduction

Floods, particularly flash floods, are one of the most serious geohazards facing road networks and their users around the world. Floods can arise unexpectedly after brief periods of heavy rain, resulting in huge human and economic losses (Sampson et al. 2015). Flooding is one of the most common and destructive geohazards in Nigeria over the last several decades. These events have harmed the country's economic and social growth in numerous places (Adeloye and Rustum 2011; Cirella and Iyalomhe 2018). Flood hazard assessment is a frequent study that determines the quantitative or qualitative classification, location, and extent of present or potential flood risks in a specific area or on transportation infrastructure like roads, railroads, or subways (Abdelkarim et al. 2020). Flood susceptibility maps are crucial for designing flood mitigation measures, assisting decision-makers and authorities in the installation of protective structures, and reducing indirect transportation infrastructure damage through emergency evacuation plans (Neal et al. 2012). Although the phenomena cannot be halted, the accompanying damages can be significantly mitigated if flood-prone locations are precisely identified (Pedrozo-Acuña et al. 2017).

Hence, to manage future hazards, mitigate their impact, and ensure the infrastructure's safety, a suitable approach to understanding and predicting flash floods in the watershed and the vital road project is critical. The use of a climate risk assessment methodology has never been used in this region, emphasizing the importance of this research, particularly because it concerns a major road project with important socioeconomic implications. Despite the absence of data in the research area, this assessment can be regarded as a meaningful primary infrastructure susceptibility assessment. Authorities can utilize the findings to design protection policies to safeguard the safety of the infrastructure and its users.

2 Significance of the Study

Most of the existing research has focused on examining flooding on a regional or watershed scale, with few studies taking into account the vulnerability of transportation infrastructure to flooding (Kalantari et al. 2014). To manage future hazards, mitigate their impact, and ensure the infrastructure's safety, a suitable approach to understanding and predicting flash floods in the watershed and on vital roads is critical. Despite the absence of data in the research area, this assessment can be regarded as a meaningful primary structural vulnerability assessment. Authorities can utilize the findings to design protection policies to safeguard the safety of the environment and communities.

Research Objectives

Overall Objective:

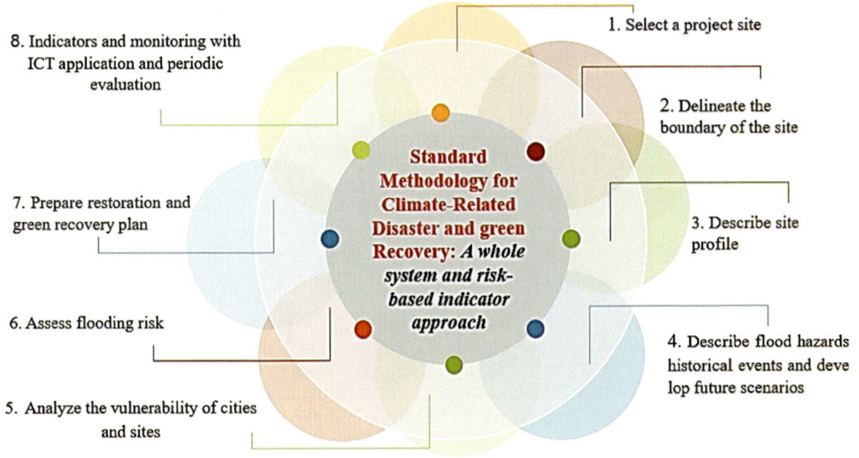

Fig. 1 Climate-related disaster risk assessment and green recovery framework (Dr. Kwi-Gon Kim)

- To do a comparative assessment of flood hazard vulnerability modeling in Niger State, Central Nigeria.

Specific Objectives:

- Analyze flood hazards in the watershed of Niger State;
- Identify triggering variables, their spatial distribution, and map flood-prone areas;
- Identify a suitable approach to understanding and predicting flash floods in the watershed, while considering environmental vulnerability.

3 Methodological Framework

Toward achieving the objectives of the study and answering the research questions, this study will trail the eight (8) steps approach, new standard methodology for climate-related disaster and green recovery approach (Fig. 1) by Dr. Kwi-Gon Kim, to guide the research direction:

4 Selection and Description of Case Study Area: Niger State, Nigeria

The Niger State study area is in Nigeria's North-Central region. Mayomi et al. (2014) asserted that Niger State's vulnerability to floods was based on two factors: the location of the great part of the state's inhabitants in lower topographies and the location of the country's three Hydroelectric Power (HEP) stations in the state: Jebba and Kainji

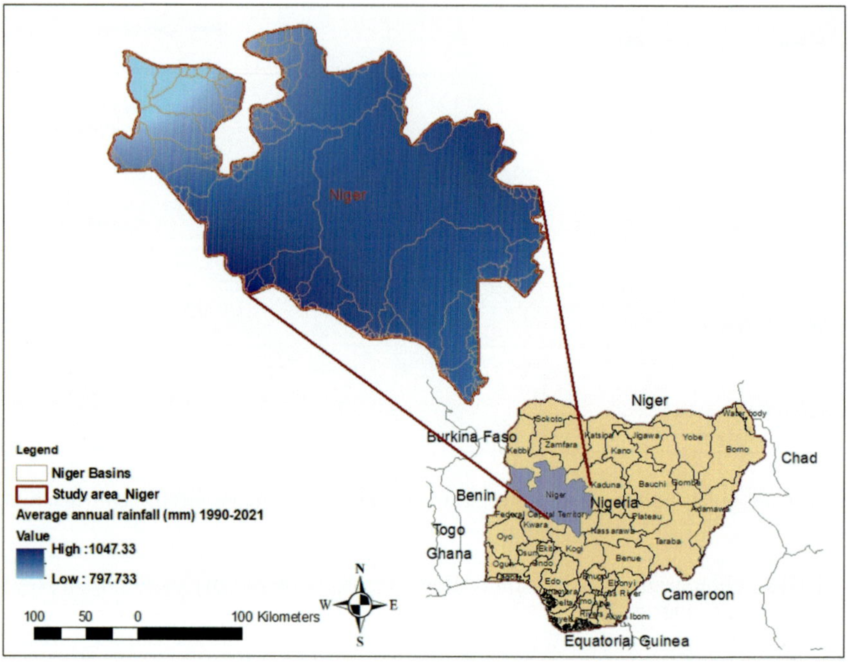

Fig. 2 Geographical location of the study area

dams and Shiroro dam on the Kaduna River. The State stretches from 10° 0′ 0″ N to 6° 0′ 0″ E, spans 76,363 km^2, and is divided into 25 local government areas (Fig. 2). Niger is Nigeria's second-largest state with a landmass (72,200.14 km^2), following Borno State (74.83 km^2). The State is characterized by two distinct seasons: wet (annual average precipitation: 982.88 mm) and dry (annual average temperature: 32.6 °C max and − 22.4 °C min) seasons.

It has a watershed area of 76,363 km^2, with 355 sub-basins and a population of about 5,556,247 (2022). The Niger River is the state's name. Niger State is home to the famed Gurara Falls, and the Gurara Local Government Area is named after the Gurara River, which runs through the area. There is also Kainji National Park, Nigeria's largest national park, which features Kainji Lake and the Borgu Game Reserve.

5 Historical Hazards Events and Future Scenario

Flood in Nigeria is becoming an annual event; it arises as coastal flood, flash floods, river flood, and urban flood. Years back states and cities in Nigeria experienced flood calamities that undermined the government's ability to protect their residents (Komolafe et al. 2015). Heavy rains caused the country's most disastrous flood

Natural Hazard 1900–2020	Subtype	Events Count	Total Deaths	Total Affected	Total Damage ('000 USD)
Drought	Drought	1	0	3,000,000	71,103
Epidemic	Bacterial Disease	27	17,278	163,378	0
	Viral Disease	24	8,233	182,474	0
Extreme Temperature	Cold Wave	1	18	Data not available	0
	Heat Wave	1	60	Data not available	0
Flood	Flash Flood	6	330	109,165	7,805
	Riverine Flood	28	1,110	10,275,064	636,717
Insect Infestation	Grasshopper	1	Data not available	Data not available	0
	Locust	1	Data not available	Data not available	0
Storm	Convective Storm	3	54	16,012	1,000
Landslide	Landslide	3	32	1,800	0

Source: Climate Risk Profile: Nigeria (2021): The World Bank Group.

Fig. 3 Natural disasters in Nigeria, 1900–2020. *Source* Climate Risk Profile: Nigeria (2021, p.12): The World Bank Group

catastrophe in 2012, which was classified as the worst in 40 years and devastated 33 states including Niger State of the 36 states, causing the loss of life and property (UNCHA 2012). Mayomi et al. (2014) opined that as a primary cause of floods, the yearly floods brought on by spill-water from dams across the Rivers Niger and Kaduna along the Niger-Benue trough have threatened the survival of the locals for some time (Mayomi et al. 2014). Figure 3 illustrated data from the Emergency Events Database (EM-Dat database) indicating the series of natural disasters the country has encountered including, floods, droughts, storms, epidemics, and landslides.

Figure 4 highlights a study by the World Bank, depicting Nigeria's annual average precipitation which is predicted to remain comparable to historical records, although, in the highest emissions scenario, RCP8.5, rainfall is expected to decrease drastically in the north and increase in the south along the coast.

6 Analyzing the Vulnerability and Assessing the Flood Risk of the Location

Prior to analyzing the vulnerability and assessing the flood risk as part of steps 5 and 6 of Dr. Kim's Standardized Methodology for Climate-Related Disaster, the following steps are implemented to analyze the vulnerability and assess the flood risk of the location:

(1) The key flood triggering factors in the watershed will be identified by comprehensive field observations and climatic analysis; (2) parameters relevant to the indicated components were chosen and mapped utilizing remote sensing, geological, and climatic datasets in a geographic information system (GIS) environment; (3) to establish the weight values for each class, the mapped parameters will be ranked and then placed into the AHP model; and (4) a flood hazard susceptibility index will be

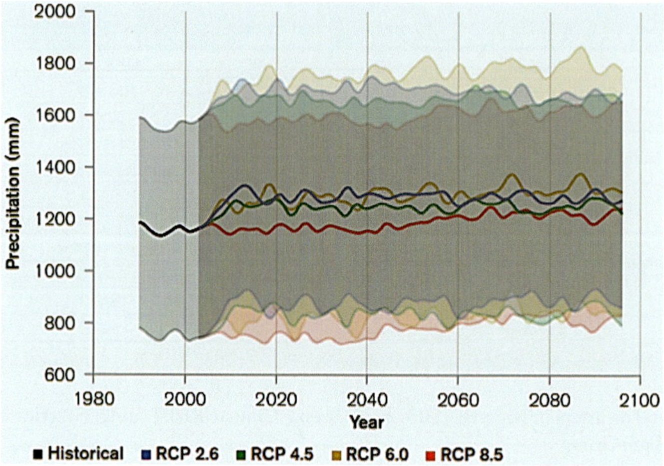

Fig. 4 Projected average precipitation in Nigeria (reference period, 1986–2005). *Source* Climate Risk Profile: Nigeria (2021, p.10): The World Bank Group

derived from the data and utilized to create a flood hazard susceptibility map for the research area. The approach is depicted in the following picture (Fig. 5).

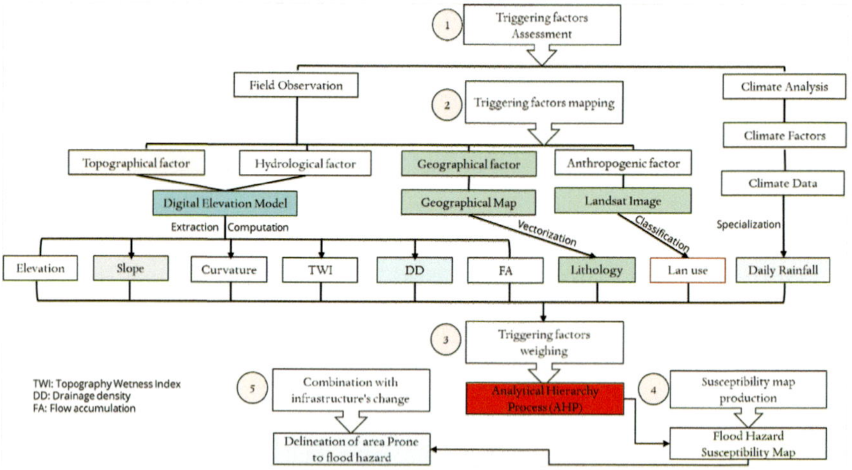

Fig. 5 Methodological flowchart showing the approach for vulnerability and assessment of the study area. *Source* Author's Elaboration

6.1 Climatic Analysis

The Nigerian Meteorological Agency (NIMET) provided rainfall and temperature data from three meteorological stations for this study (Table 1). The chosen stations are in or near the watershed to better understand the climatic parameter fluctuation in the research area. The dataset covers 31 years (1990–2021) and includes daily rainfall (mm), total monthly rainfall (mm), and temperature series. This is to analyze and assess the risk in the location.

The variation in average monthly rainfall (AMR) and average monthly temperatures (AMT) was determined and shown in Figs. 6 and 7. The findings point to two distinct climatic periods in the research area: one cold and rainy and the other hot and dry (Fig. 8).

The rainfall amount related to flash food occurrences (January 1995, January 2008, February 2013, and April 2019) exceeds 70 mm, according to statistical analysis of the rainfall data. When these results were compared to the GUMBEL equation's frequencies and return periods, it was discovered that decadal rainfall (P10) is the

Table 1 Characteristics of the selected meteorological stations used in this work

Station	Latitude	Longitude	Elevation (m)
Abuja	11.379	3.563	430
Niger	8.167	7.674	343
Kogi	11.36	7.698	224
Kwara	8.15	3.56	281

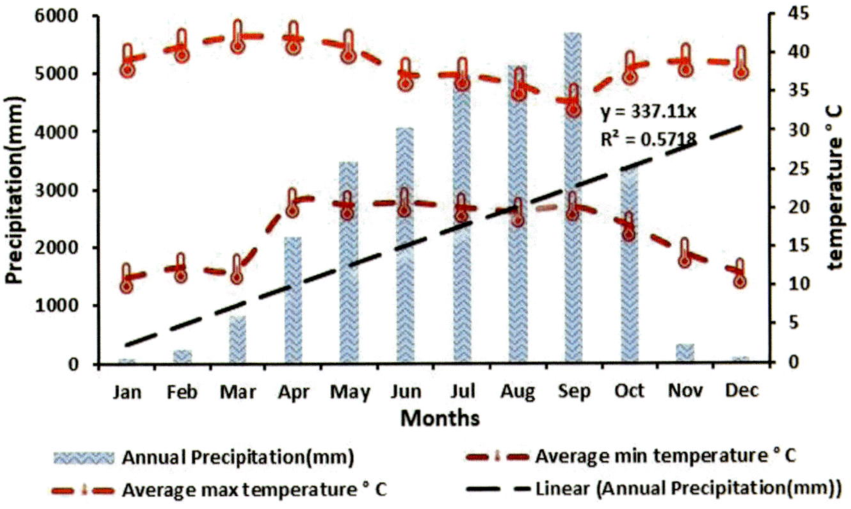

Fig. 6 Long-term (1990–2021) monthly average precipitation and the maximum and minimum temperature at the meteorological station in Nigeria. *Source* Nigerian Meteorological Agency (NIMET)

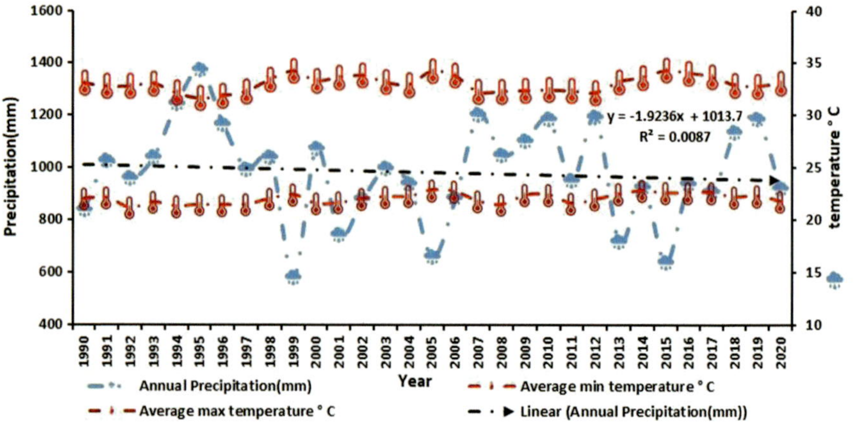

Fig. 7 Long-term (1990–2021) yearly average precipitation and the maximum and minimum temperature at the meteorological station in Nigeria. *Source* Nigerian Meteorological Agency (NIMET)

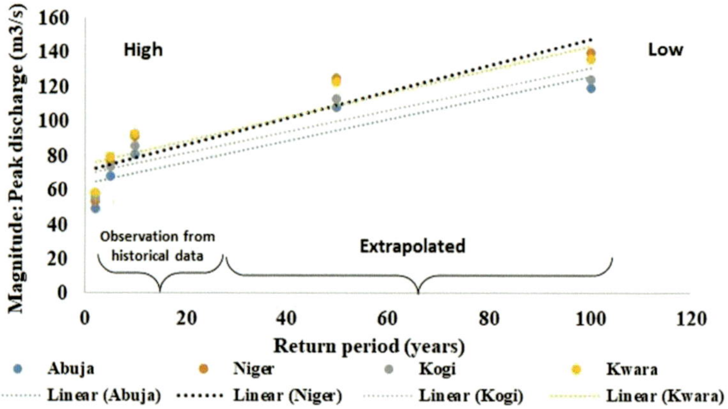

Fig. 8 Frequency and return period of maximum daily rainfall (mm). *Source* Author's Elaboration

primary driver of flash flood occurrence in the watershed (Table 2). These findings point to (P10) as the climatic factor that causes the flash flood phenomenon in the research area.

6.2 *Vulnerability Map Production*

The raster data (Fig. 9) were reclassified according to AHP categories using the ArcMap reclassify tool in ArcGIS 10.3 software after the weights and ranks for each

Table 2 Assessment's selected triggering factors parameters

S. No.	Factor's nature	Parameters
1	Topographic	Elevation
		Slope
		Curvature
2	Hydrologic	Topographic wetness index (TWI)
		Drainage density
		Flow accumulation
3	Geologic	Lithology
4	Climatic	Daily maximum rainfall
5	Anthropogenic	Land use

parameter and subclass were determined.

$$\begin{aligned}\text{FHS index} = {} & (\text{Elevation} \times 0.27 + \text{Slope} \times 0.18 + \text{Drainage density} \times 0.19 \\ & + \text{Daily maximum rainfall} \times 0.12 + \text{Flow accumulation} \times 0.09 \\ & + \text{Land use} \times 0.06 + \text{Lithology} \times 0.04 + \text{TWI} \times 0.03 \\ & + \text{Curvature} \times 0.02).\end{aligned}$$

The flood hazard vulnerability map in Niger State will be created using the flood hazard susceptibility index values, which will be divided into five groups. The vulnerability classes will be very low, low, moderate, high, and very high (Fig. 9).

6.3 *Flood Hazard Vulnerability Map*

Mayomi et al. (2014) study revealed that plain terrain occupies a 33% greater area in Niger State than the other terrain classifications. Despite this, more than half (58.43%) of the State's land area is in the Niger valley or plains, both of which are classed as "very vulnerable" and "vulnerable" to flood disasters on the flood hazard map (Fig. 10). This aligns with the findings of Mayomi et al. (2014), which stated that the topography of Niger State has made it susceptible to floods and their related hazards. Hence, Fig. 10, the flood hazard map, is the coverage of the flood hazard susceptibility index values and classes, which are divided into classes as shown in Fig. 9.

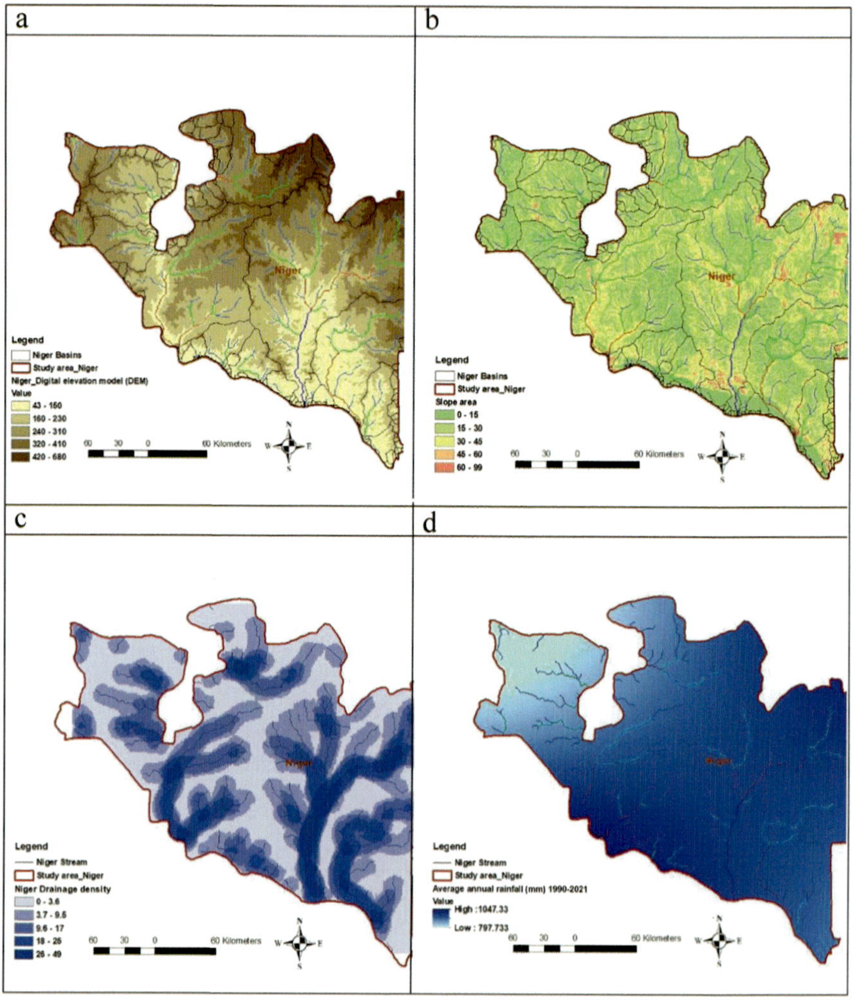

Fig. 9 Triggering factors vulnerability maps for the study area. *Source* Author's Elaboration

7 Recommendations and Conclusion

Recommendations

- The flood hazard vulnerability map and flood zoning described in this study are critical for protecting and preserving the area. They will also assist planners and decision-makers in determining the alignment of the remaining land along the route.
- Strengthening National Disaster Preparedness by developing national and state contingency plans, updating national and state databases on past disasters, and improving disaster risk awareness and alerts.

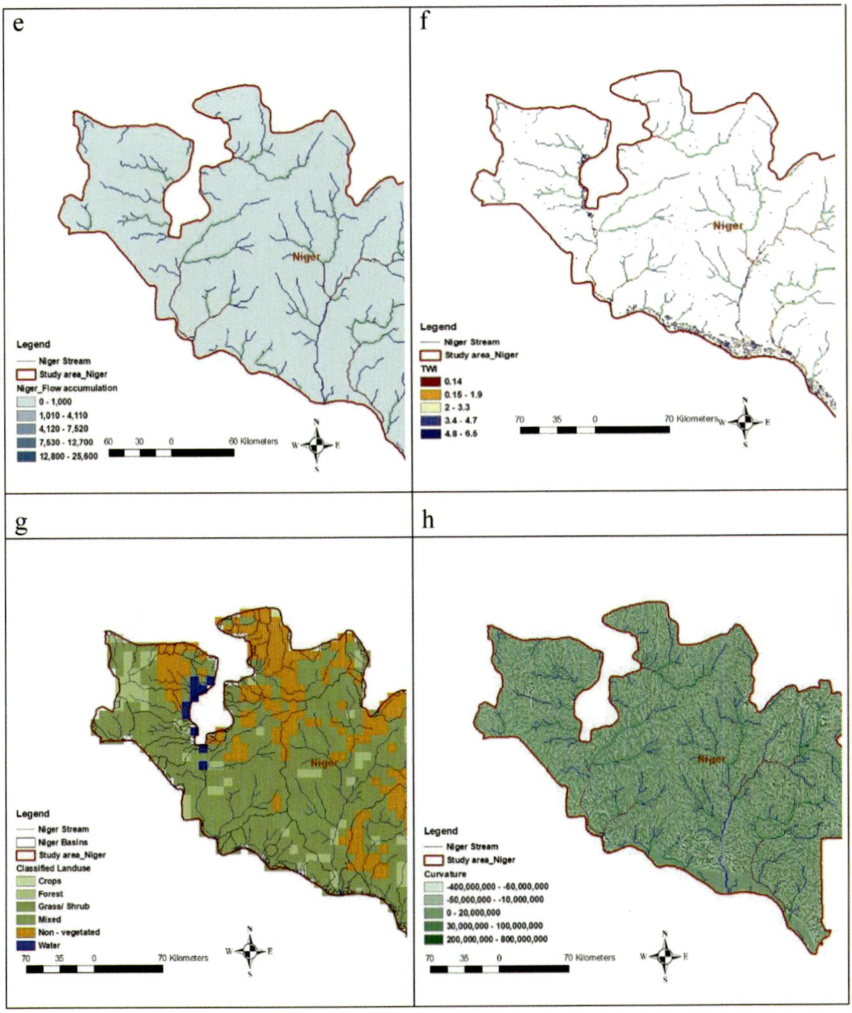

Fig. 9 (continued)

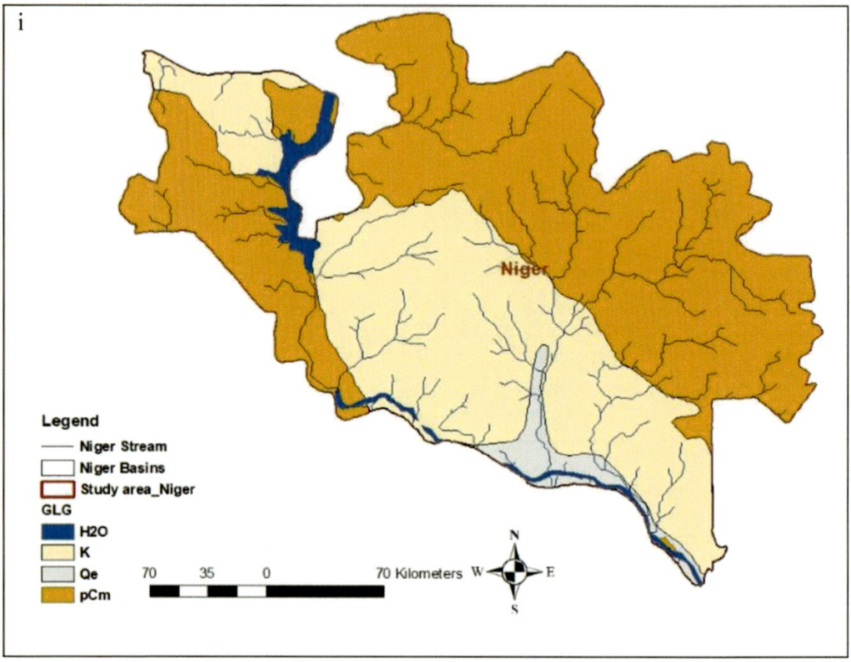

Fig. 9 (continued)

- Developing adequate flood control infrastructures including identifying aging or weak dams and building adequate dam monitoring, buffer zones, and maintenance.
- Capacity building to improve the lack of flood risk awareness.

Conclusion

The findings of this study revealed that many topographic and hydroclimatic elements connected with the region's geomorphological complexity are responsible for flood occurrence in Niger State. Plain, Atlas, and Plateau are three distinct morphostructural units. Anthropogenic variables, such as unplanned development and agriculture, exacerbate the impact of natural causes, according to field observations conducted as part of this strategy. The decadal maximum daily precipitation (P10) was discovered as the climatic element that caused floods in this region by the climatic study. To determine the extent of the phenomenon in the watershed using a combination of qualitative and quantitative methods, detailed field observation, climate analysis, and literature review were used to select and weigh the parameters related to the triggering factors, and the AHP model was used to calculate the flood hazard vulnerability index by integrating and combining these parameters. The generated flood hazard vulnerability map gives a detailed insight into the watershed's highly and extremely vulnerable sections. Thus, the vulnerability of the location was

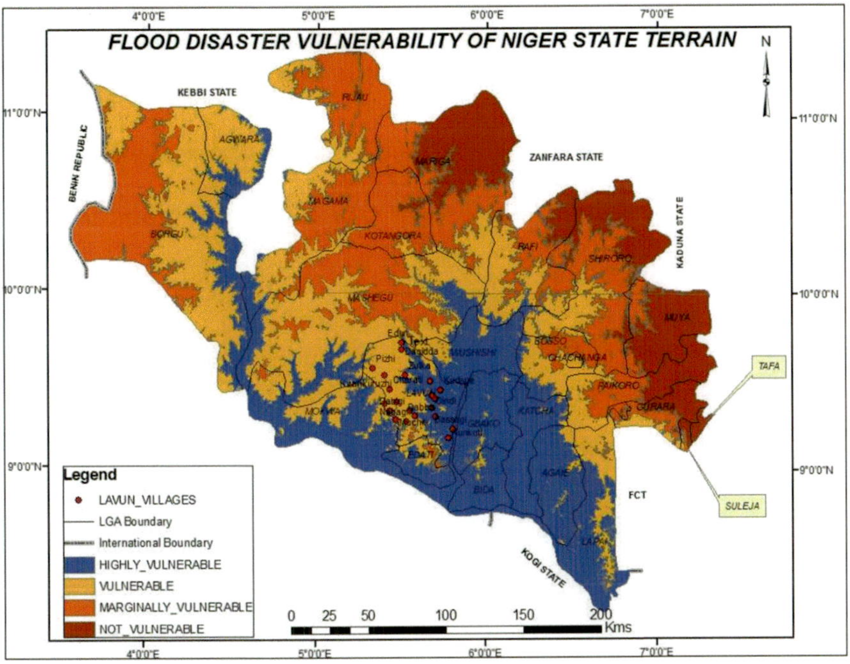

Fig. 10 Flood hazard (disaster) vulnerability map of the case study area (Niger State). *Source* Mayomi et al. (2014, p. 127)

analyzed, and flood risk was equally assessed based on the "*Standard Methodology for Climate-Related Disaster and Green Recovery*" adopted for the study.

References

Abdelkarim A, Al-Alola SS, Alogayell HM, Mohamed SA, Alkadi II, Ismail IY (2020) Integration of GIS-based multicriteria decision analysis and analytic hierarchy process to assess food hazard on the Al-Shamal train pathway in Al-Qurayyat region, Kingdom of Saudi Arabia. Water 12(6):1702

Adeloye AJ, Rustum R (2011) Lagos (Nigeria) flooding and influence of urban planning. Proc Inst Civ Eng Urban Des Plann 164(3):175–187

Cirella GT, Iyalomhe FO (2018) Flooding conceptual review: sustainability-focalized best practices in Nigeria. Appl Sci 8(9):1558

EM-DAT: The emergency events database—Universite catholique de Louvain (UCL)—CRED, D. Guha-Sapir, Brussels, Belgium. http://emdat.be/emdat_db/

Kalantari Z, Nickman A, Lyon SW, Olofsson B, Folkeson L (2014) A method for mapping flood hazard along roads. J Environ Manage 133:69–77

Komolafe AA, Adegboyega SAA, Akinluyi FO (2015) A review of flood risk analysis in Nigeria. Am J Environ Sci 11(3):157

Mayomi I, Kolawole MS, Martins AK (2014) Terrain analysis for flood disaster vulnerability assessment: a case study of Niger State, Nigeria. Am J Geogr Inf Syst 3(3):122–134. https://doi.org/10.5923/j.ajgis.20140303.02

Neal J, Villanueva I, Wright N, Willis T, Fewtrell T, Bates P (2012) How much physical complexity is needed to model flood inundation? Hydro Process 26(15):2264–2282
Pedrozo-Acuña A, Moreno G, Mejía-Estrada P, Paredes-Victoria P, Breña-Naranjo JA, Meza C (2017) Integrated approach to determine highway flooding and critical points of drainage. Transp Res D Transp Environ 50:182–191
Sampson CC, Smith AM, Bates PD, Neal JC, Alfieri L, Freer JE (2015) A high-resolution global flood hazard model. Water Resour Res 51(9):7358–7381
The World Bank Group (2021) Climate risk profile: the world bank group. World Bank Publications, The World Bank Group, USA. 15918-WB_Nigeria Country Profile-WEB.pdf
UNCHA (2012) Floods situation report, UN Office for the Coordination of Humanitarian Affairs, v 2

Proposed Study Design for Map-Based Approach to Flood Risk Assessment in Bwaise III in Kampala City, Uganda: An Indicator Approach

Yawe Samuel

Abstract Bwaise III is a parish found in Kawempe division which is one of the five divisions that make up Kampala city. Kampala is Uganda's biggest city as well as the capital of the country with an estimated population of 1,610,500. Kampala is a city facing many challenges including high unemployment, urban density and limited financial resources, energy and food insecurity. In addition, climate-related weather changes threaten the social development of the country and the sustainable structure of the city. This paper presents a comprehensive proposed study design to focus on flood disasters at local level in one of Kampala city divisions of Kawempe which will be scaled down to a parish level of Bwaise III. The study includes a proposal to use GIS and other software to produce a number of maps that will aid in decision-making and planning. Flood zone maps, location maps and other inventory maps will be generated and watershed and the different water catchments will also be identified. The study aims to carry out a risk assessment to determine the extent of vulnerabilities and provide solutions for disaster recovery and management in this area under study based on an indicator approach.

Keywords Uganda and climate change · Kampala and flooding · Flood risk management · GIS and vulnerability · Disaster risk management · Adaptation · Resilience · Vulnerability · Adaptive capacity · Indicator analysis · Hazard reduction measures · Nature-based solutions

1 Introduction

This paper presents a comprehensive proposed study design to focus on flood disasters at local level in one of Kampala city divisions of Kawempe which will be scaled down to a parish level of Bwaise III. While the area is experiencing a number of climate disasters it will focus on flooding in the region. The study includes the use

Y. Samuel (✉)
Rakai District Local Government, Kampala, Uganda
e-mail: yawehsamuel@gmail.com

K. Kim and C. Atkin (eds.), *Cross-Cultural Perspectives on Climate Change Adaptation*, Cities and Nature, https://doi.org/10.1007/978-3-031-50365-8_11

of GIS and other software to produce a number of maps that will aid in decision-making and planning. Flood zone maps, location maps and other inventory maps will be generated. Watershed and the different water catchments will also be identified. Using a risk assessment framework, the study will identify project goals, objectives and approaches. There will be an analysis of the existing situation, identifying the hazardous zones, a look into the socio-economic strata with a view to devise means of how to revive and restore affected communities. The study aims to carry out a risk assessment to determine the extent of vulnerabilities and provide solutions for disaster recovery and management in this area under study based on an indicator approach.

Background on the Study Area

Bwaise III is a parish found in Kawempe division which is one of the five divisions that make up Kampala city. Kampala is Uganda's biggest city as well as the capital of the country. It has a total area of approximately 195 km^2, with 33 km of shoreline on Lake Victoria. Kampala has an estimated population of 1,610,500. The population under 30 years of age represents 77.8% of the total population.

Kampala is a city facing many challenges including high unemployment, urban density and limited financial resources, energy and food insecurity. In addition, climate-related weather changes threaten the social development of the country and the sustainable structure of the city. Despite all the available evidence of climate risks, the Kampala government has not focused attention on flooding as a risk. Therefore, this study will highlight why there is a need for future Kampala development plans to consider flood risk hazards amidst the changing climate while creating resilient communities and cities.

Objectives and Structure of the Study

The objectives of the study are:

- To identify and understand the level of flood risk in different areas of Bwaise (Kampala) using indicators.
- To pinpoint the flood-related problems Bwaise/Kampala has and develop suitable solutions aimed at improving flood risk management and resilience.

The study will look at numerous factors such as hazardous zones, eco-systems and all other factors that can be used to assess flood risks in the area of study. Indicators will also be selected to support the factors to give a better understanding of the flood risk affects the different set of components. Therefore, a comprehensive framework from flood risk assessment will be discussed including the development of flood risk recovery plans.

Table 1 Elements of flood risk management model

Procedure	Framework (methods and techniques)
Objective	Literature
Methodology • Steps 1–10	Flood risk management steps
Case study • Vulnerability groups and flood zones	Flood risk management steps
Recommendations • Identification of outputs and outcomes	Structural and non-structural measures

2 Framework/Conceptual Framework

The following conceptual and methodological framework for flood risk assessment will be used to develop estimates of likely future damages and losses. An indicator approach will be taken and multiple data sources will be used including Geographical Information Systems data and a statistical analysis of flood probabilities and populations at risk will be done and a number of flood-related maps and figures will be generated (Table 1).

3 A Case Study

3.1 Site Selection

This section will consist of boundary maps, location and watershed maps among other useful flood risk thematic maps for Bwaise III in Kampala City.

3.2 Site Characteristics

This section will be supported by maps and figures-based information to give the frequency of flood events and scenarios with in the study area. Also, more modelling will be intertwined with flood types while considering areas of low, medium and high flood risk areas.

3.3 SWOT Analysis

This section will identify the issues within the study area. For example, topographic maps which show reservoirs and other features, analyze how to use block chain technology for flood inventory and Kampala wetland mitigation needs, avoidance, mitigation and restoration options (Fig. 1).

Kampala's proximity to Lake Victoria, makes it one of the areas which receive a lot of rainfall throughout the year. This has been fuelled by climate change and its adverse impacts on infrastructures and humans.

River flooding: The fact that Bwaise III is located close to River Lubigi makes the area vulnerable to river flooding mostly during the rainy seasons.

Surface water flooding: Bwaise III suffers from the surface water flooding due to its location which is exacerbated by the poor waste management system.

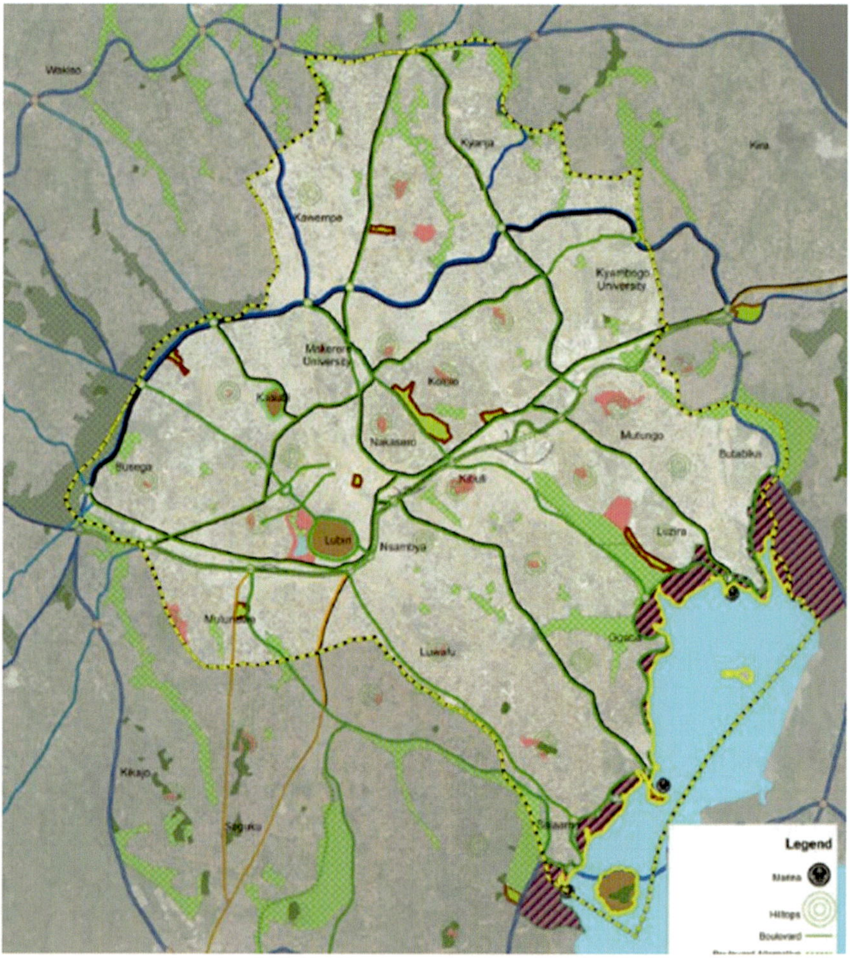

Fig. 1 Shows Kampala city's natural resources and open spaces. *Source* KCCA (2012)

Fluvial flooding: River Lubigi passes in the middle of Bwaise III and this river has a number of other secondary rivers/tributaries that pour into it. Most of these are joined by the drainage channels that carry water from the upstream of residential and other areas, altered from blockage or restriction of flow with in the river and drainage channels.

Wetlands: A wetland map of Lubigi will be prepared and analyzed to help produce a wetland restoration map. This will be linked to city's wetland system and the outcome will be the preparation of a digitalized wetland connectivity map which will help to identify water flow and retention by different wetlands to guide decision-making and actions towards flood risk management.

Therefore, it will establish how wetlands, rivers, people and biodiversity connects with each other (Fig. 2).

3.4 Land-Use Plan

Under this section, the study will try to investigate how different land uses have been assigned over time. The study will also make a comparison on the changes in land use zoning; old versus new to ascertain how different land uses have been exposed to flood risk over time.

It will consider more components such as urbanization/urban sprawl, growth strategies of Kampala/Bwaise and zoning among others (Figs. 3, 4 and 5).

4 Flood Risk Assessment Design

When undertaken the flood risk assessment exercise will include:

Step 1: Flood hazard

This will be composed of different flood hazard maps, flood zone maps and areas plus infrastructures exposed to floods. (It should be noted that due to lack of access to some flood maps, the maps will not be attached to this document but will be mentioned.)

Figure 6 shows that each of the scenarios shows a number of flood hot spots, numbered 1–5. These are areas that experience chronic flooding and where specific planning measures and development control are especially important. Hotspots 1 and 2 result from the confluence of drains where relatively shallow flooding over a large area (0.2–1 m) can readily occur. It is advisable to designate these areas as flood retention zones, to encourage the retention of grassed surfaces and to prevent further building there. The flood level at Hotspot 3 is dependent on the final state of the primary drain and its dimensions. As the valley is relatively narrow here large stretches of the ring road may be submerged with extreme rainfall.

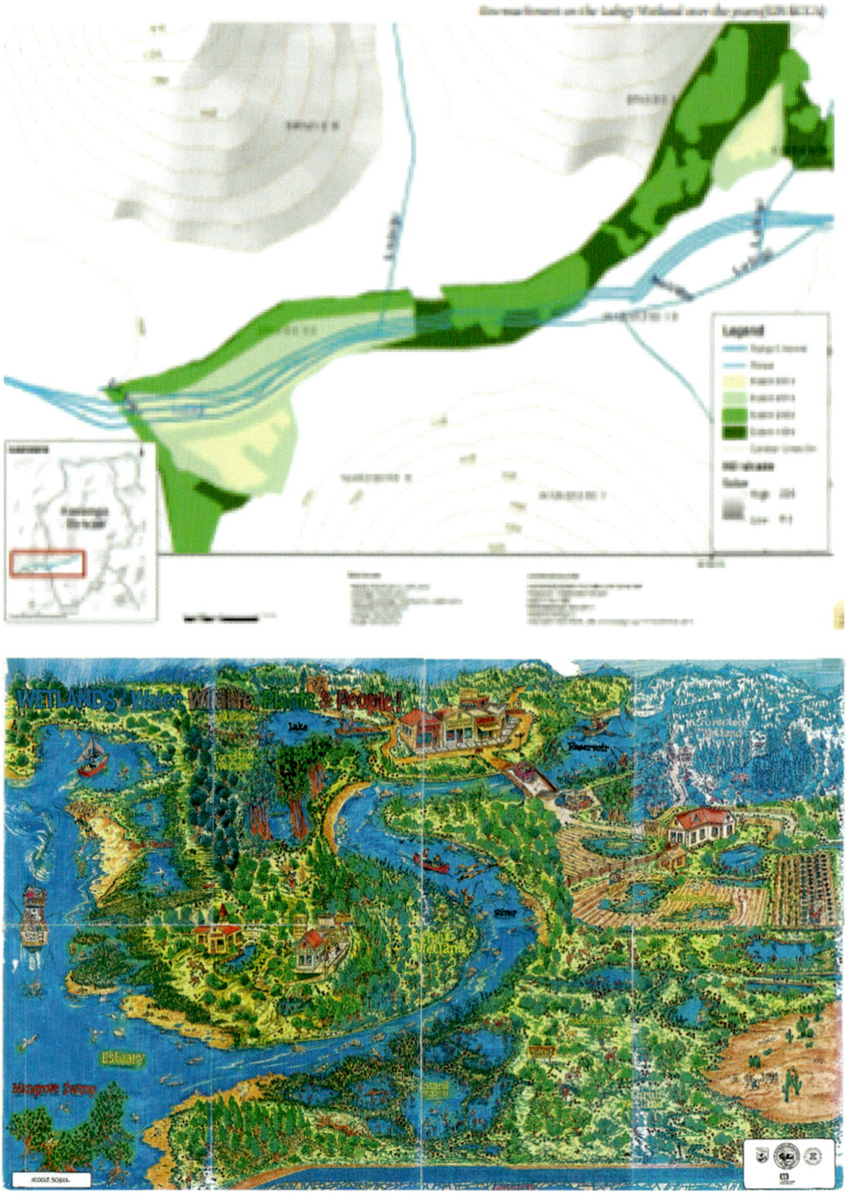

Fig. 2 Wetlands; connectivity of water, wildlife, plants, people (build-up). *Source* Prof. Kim Integrated Solutions to wetland connectivity; an IoT/blockchain-based approach, DMZ, Korea (2018)

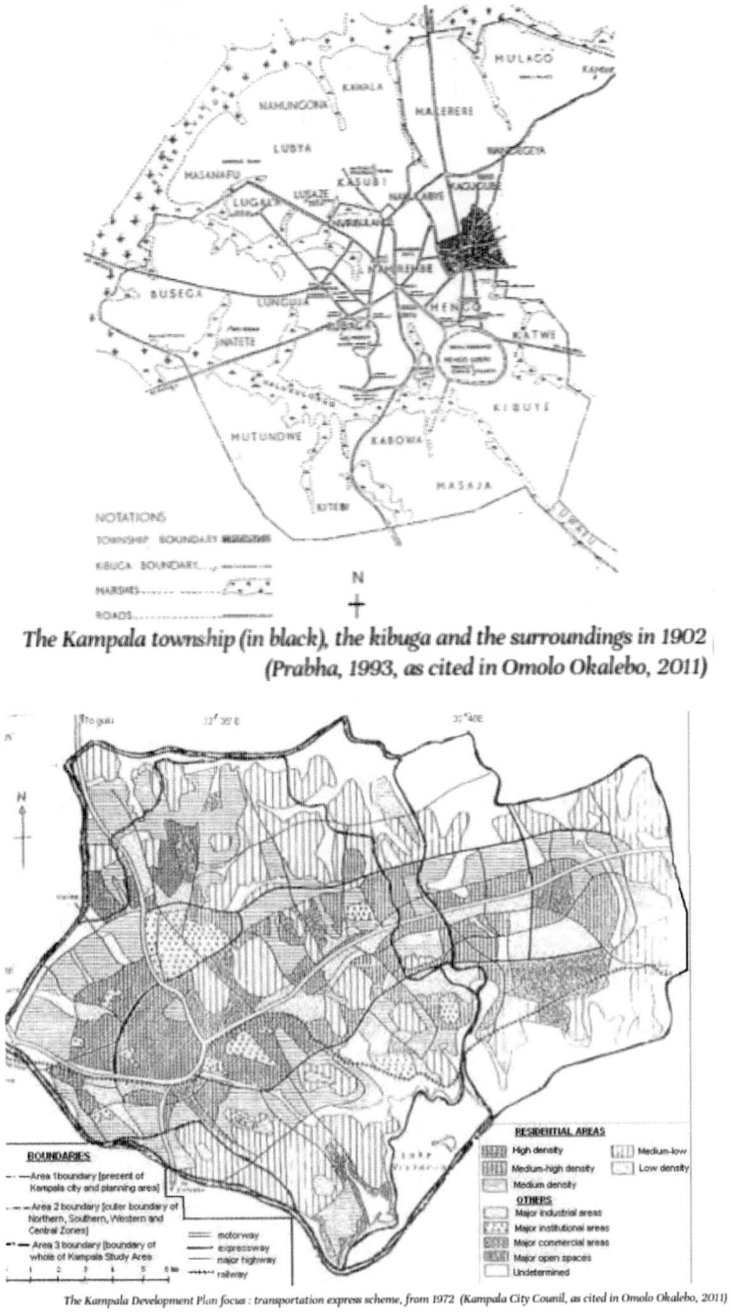

Fig. 3 Land use zoning

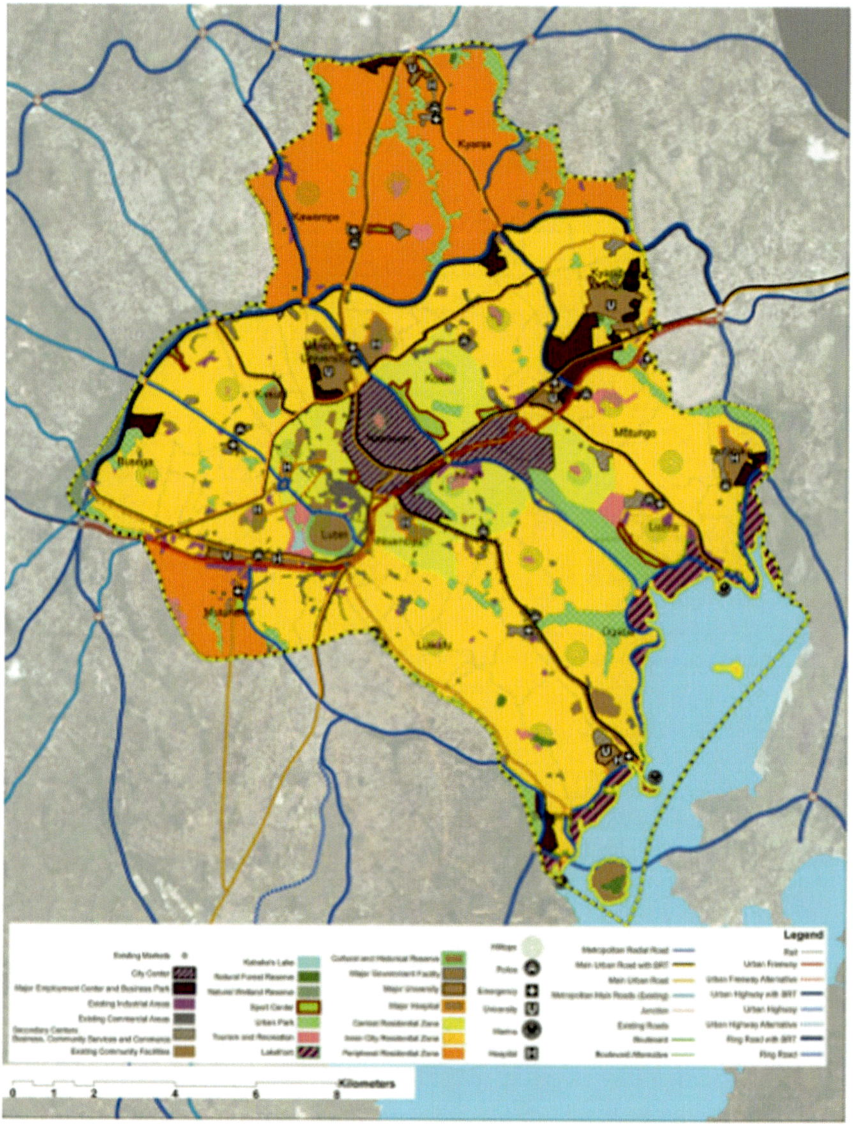

Fig. 4 Urbanization/urban sprawl. *Source* KCCA (2012), Kampala Physical Development Plan

Step 2: Delineation of boundary of landscape district and place unit (location map)

See Fig. 7.

Step 3: Delineation of boundary of catchment area (location map)

See Figs. 8 and 9.

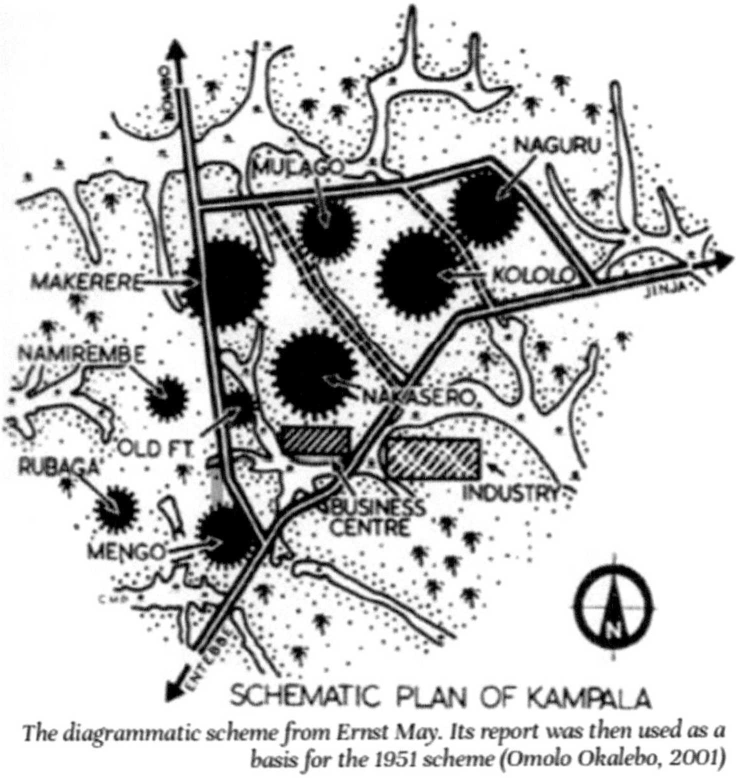

Fig. 5 Schematic plan of Kampala

Step 4: Preparation of an inventory map for each factor showing how it varies over the study area

In this step, data for each of the already identified indicators will be collected at particular spatial levels within the study area, i.e. watershed, sub-catchment and micro-catchment. This will help to produce a wetland connectivity map that will show how different catchments are connected to each other to come up with realistic measures (Figs. 10, 11 and 12).

Step 5: Creation of composite maps by overlay

This involves the overlay of different maps using different indicators such as contour, buildings, health facilities, etc.

Step 6: Analysis of composite maps to make inferences relevant to flooding risk assessment

When considering exposure of built up land to flood, some spatial scenarios do introduce clearer differences (Fig. 13); specifically, when development is assumed

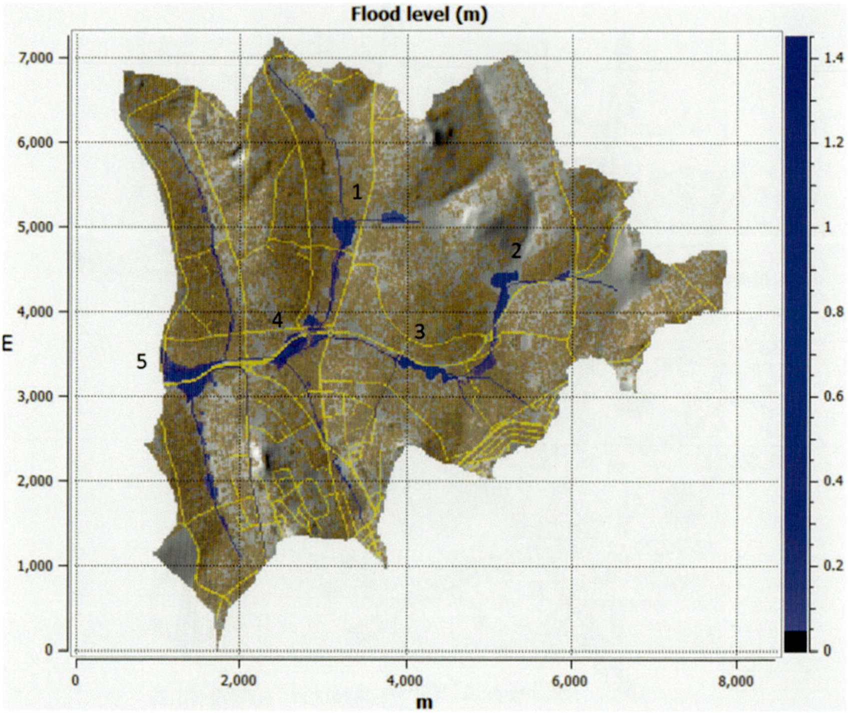

Fig. 6 Flood hotspots in Bwaise; upper Lubigi catchment based on multiple scenarios analyses. *Source* Sliuzas et al. (2013a)

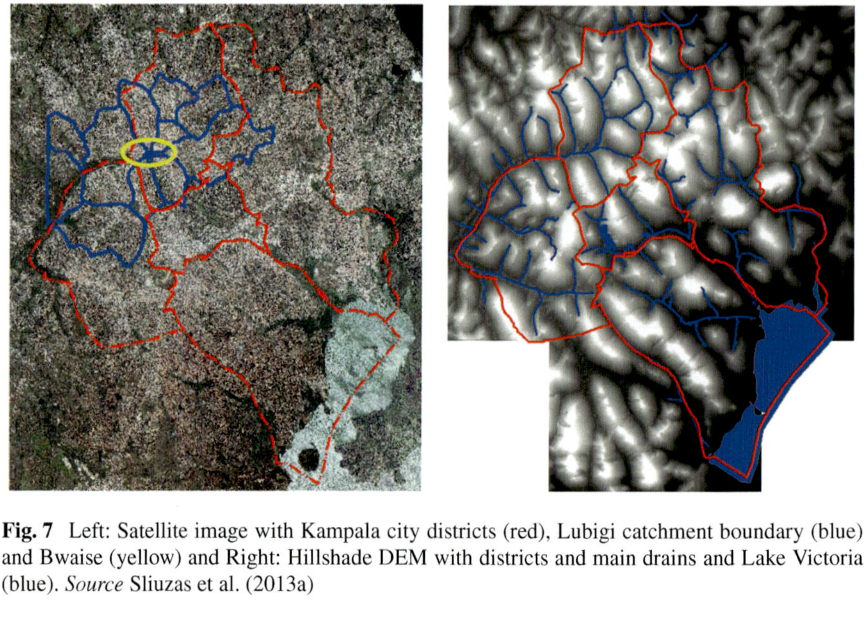

Fig. 7 Left: Satellite image with Kampala city districts (red), Lubigi catchment boundary (blue) and Bwaise (yellow) and Right: Hillshade DEM with districts and main drains and Lake Victoria (blue). *Source* Sliuzas et al. (2013a)

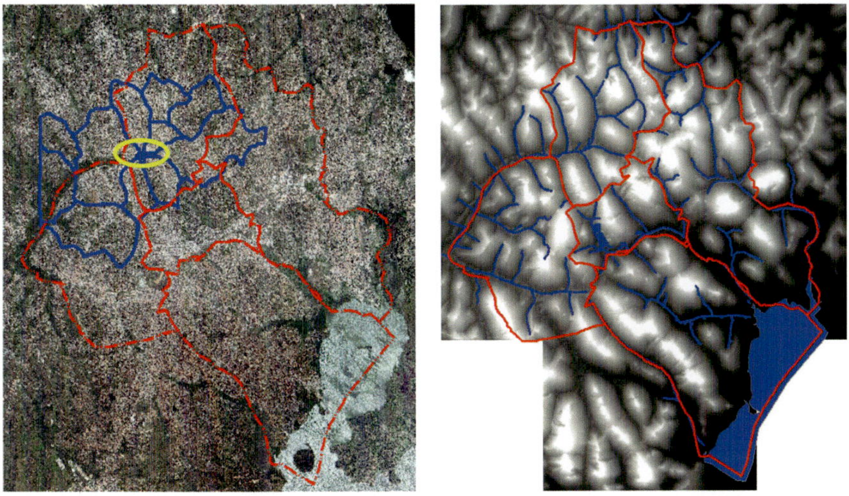

Fig. 8 Left: Satellite image with Kampala city districts (red), Lubigi catchment boundary (blue) and Bwaise (yellow) and Right: Hillshade DEM with districts and main drains and Lake Victoria (blue). *Source* Sliuzas et al. (2013a)

to avoid areas flooded in the base year—a reasonable rule, since most urban actors would like to shun these locations—or when it is assumed to occur from higher to lower elevations and exposure of built up land to flood depths greater than 25 cm decreases substantially (Fig. 14).

For the calibration data (2004–2010), Pérez-Molina (2014) concluded wetland and flood in the base year caused similar effects. However, when 2010 is used as the base year and since saturation is nearly reached by 2020 for trend or high population growth, the consequences of these factors diverge when simulating into the future.

Step 7: Evaluating of risk (probability of damage and losses) and risk mapping

This will involve flood risk probability ranking using the interconnection matrix where different factors will be assigned values through a pairwise comparison technique. (It should be noted that this table is prepared for demonstration purpose to bring out the picture the researcher wanted to portray due to lack of access to enough data/information about the study area).

This kind of grid system modelling scenario can be used for Bwaise to identify areas which fall under categories of high, medium and low areas vulnerable to flooding. Thus, it helps to determine which areas are hazardous (Fig. 15; Table 2).

This can be done for other factors by assigning relative values against each other.

Also, an interconnectivity matrix can be used under this step to determine the flooding levels (Table 3).

Thus, based on the different cell values, we can rank depending on highest (most risk area), medium to least (lowest risk) as indicated by the arrow in the figure above.

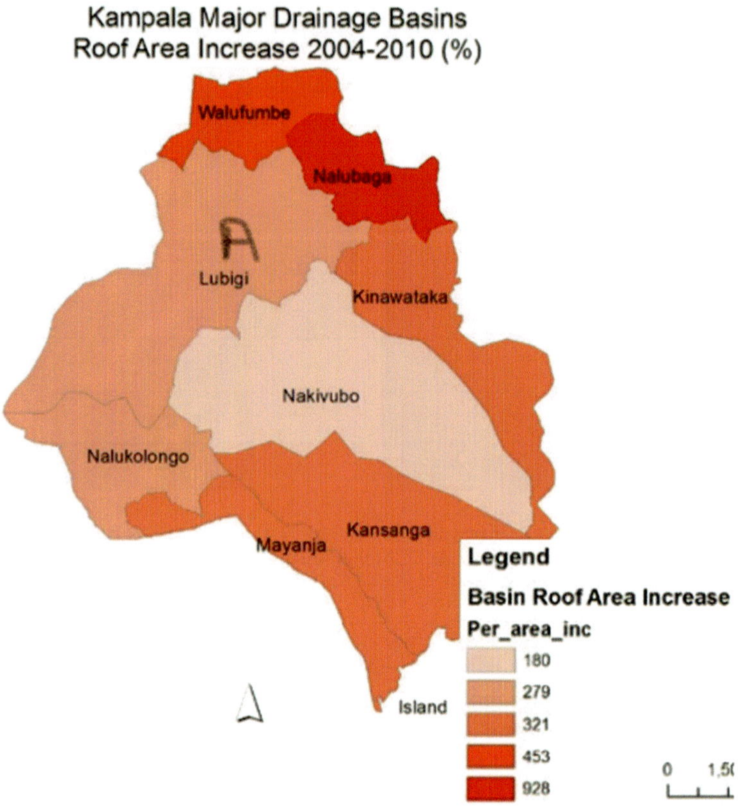

Fig. 9 Increase in roof area per catchment, 2004–2020, and "A" represents the study area. *Source* UN (2013)

Step 8: Planning for appropriate remediation and strategies and hazard reduction measures

Under this step, flood control measures will be chosen based on spatial levels and location based on the tasks handled among others as shown in Table 4.

Different locations and watersheds/catchments call for different flood control and management approaches. But in most cases, it is a combination of hard and soft structural methods and non-structural measures. In weighing the relative efficacy of possible measures some of the factors include:

- Nature of the risk
- Development locations that may be affected
- Information available
- Broad flood risk management options
- The likely impact of climate change
- Strategic recommendations.

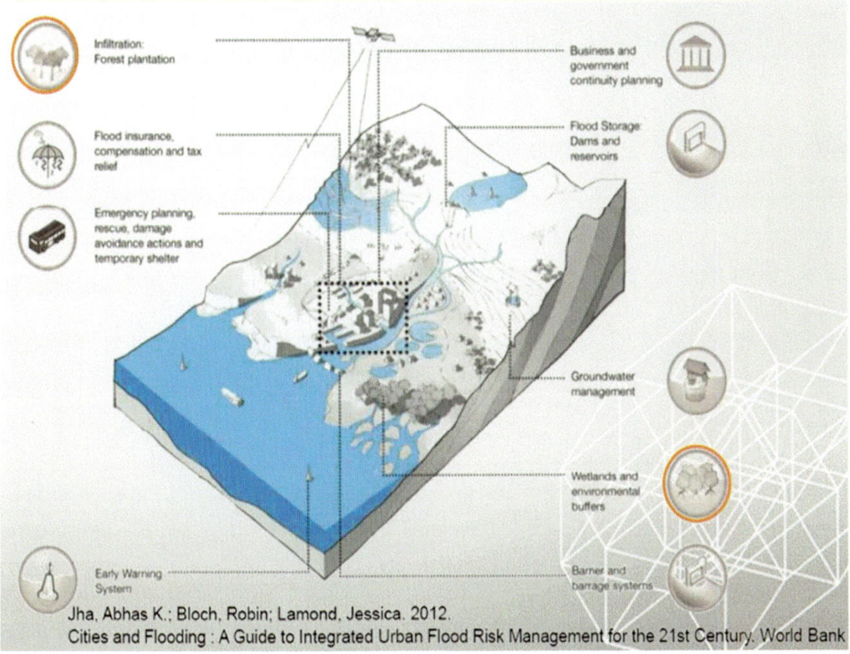

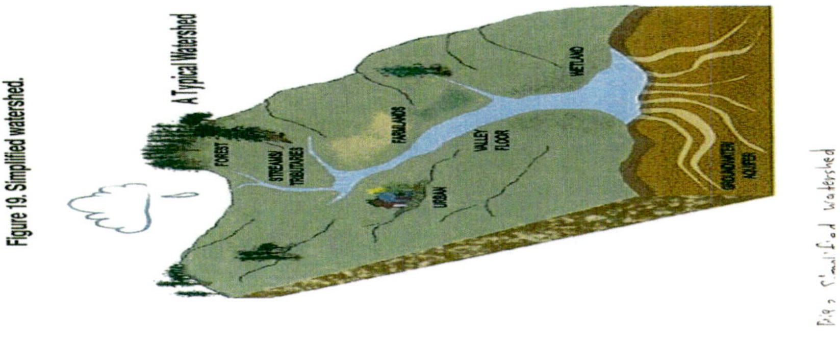

Fig. 10 Simplified watershed

Step 9 Monitoring with ICT applications and periodic evaluations

This involves the use of prediction of future hazards using big data, IoT-Internet of things, AI and algorithms (Fig. 16).

Figure 17 shows how ICT applications can be used to monitor flood and other climate hazards to aid in carrying out periodic evaluation and prediction of future events and shocks. This involves the use of sensors, CCTV cameras among other equipments.

Fig. 11 Watershed area for Gangsung. *Source* Kwi-Gon, Kim, Integrated Water Resource Managements and Smart Water Grid (2015)

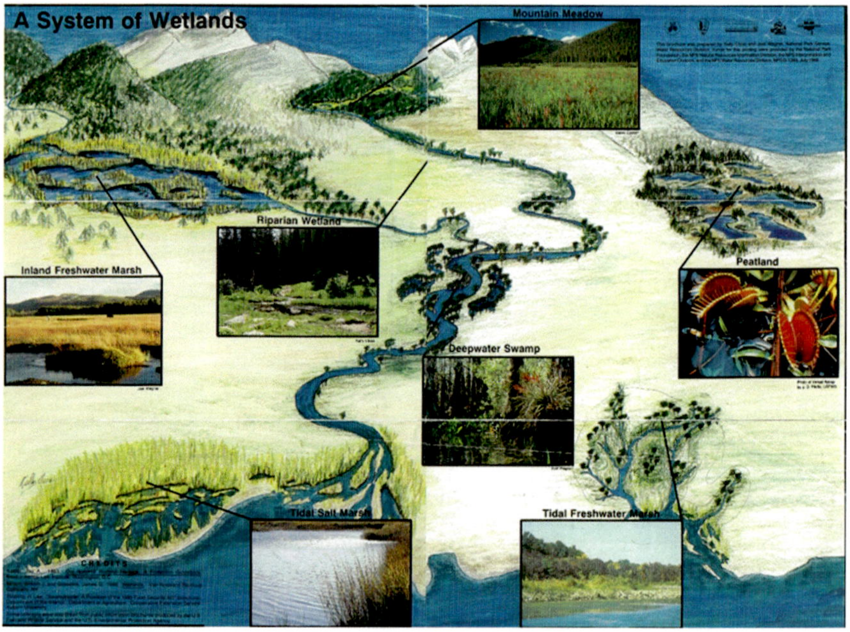

Fig. 12 System of wetlands and their connectivity

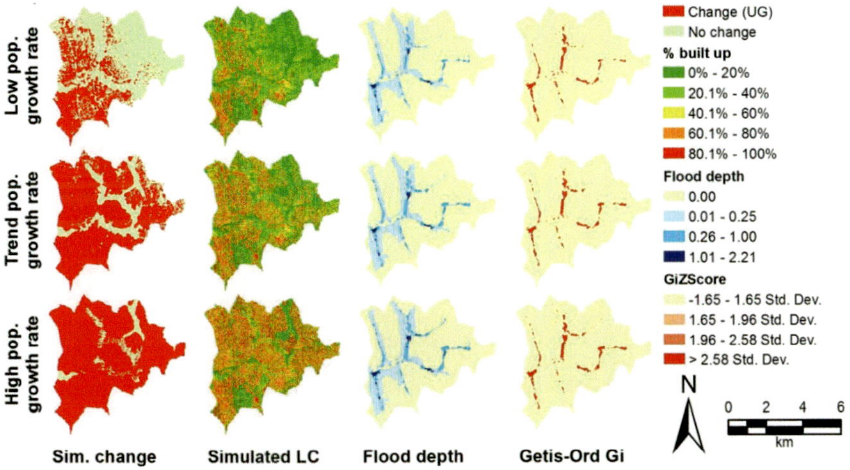

Fig. 13 Shows the spatial trend (wetland) land cover scenarios and flood maps for low, trend and high population growth. *Source* Sliuzas, Simulating Flood Impacts in Kampala, Uganda: When Do Land Patterns Matter (2015)

5 Expected Findings for Kampala City

Hypothesis: Land use plan for Kampala does not consider flood risk and so the development of new plans should be a high priority.

Growth/Urbanization Control

Future growth patterns will be compared with flood risk. This will involve projections and climate predictions so as to tell/identify the areas which will be prone/threatened by flood risks in the future (Fig. 18).

Also under this section, the climate-related flood risk assessment should include a study on the historical patterns of rainfall intensity overtime and also flood damage data among others as indicated in Tables 5 and 6.

Hazard Reduction Measures: This section will be comprised of the already existing measures as well as those that will be proposed by this study. They will be arranged in a format shown in Table 7.

Details of Hazard Reduction Measures

The proposed measures will be categorized as shown in Table 8 where structural methods are those that seek to modify the damage susceptibility of the developed areas and locations.

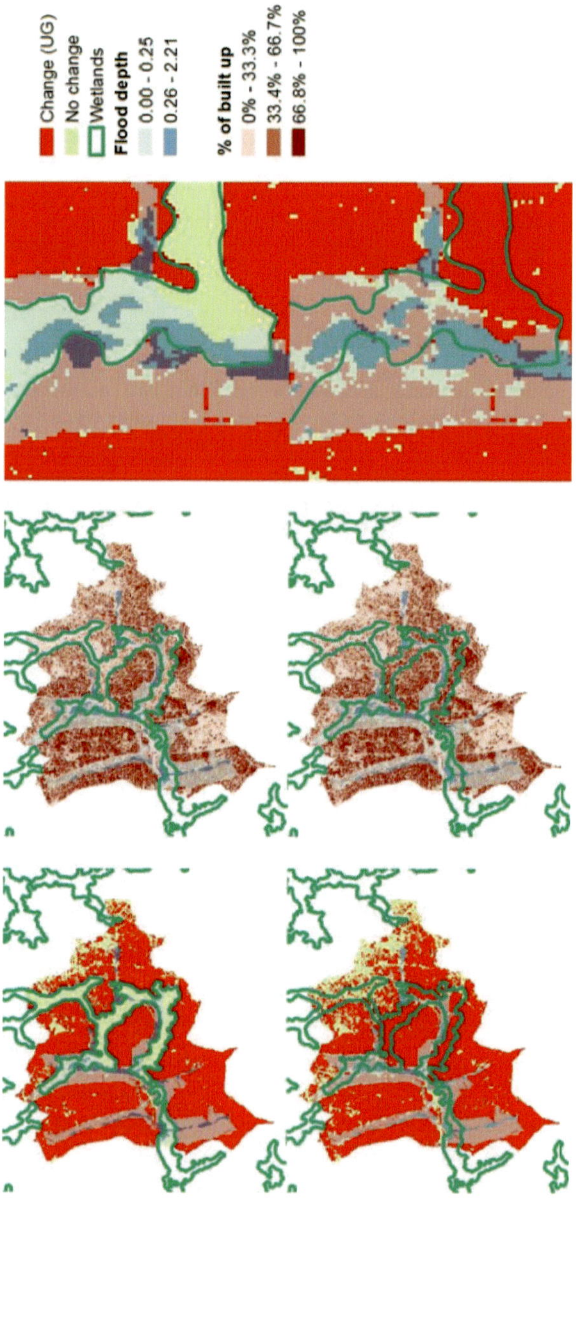

Fig. 14 Shows the trend (wetland) versus trend (flood) scenarios for trend growth population

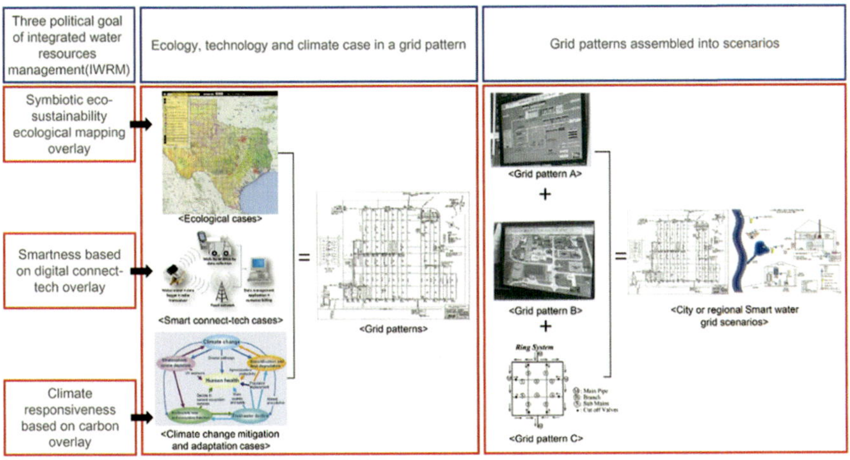

Fig. 15 Alternate smart water grid development scenario model. *Source* Kim (2017)

Table 2 Pairwise comparison technique using the ranking system

Factor	Assigned value			Sum	FIC
Hazard area	1	0	–	1	0.1
Soil erosion potential	0	–	0.5	0.5	0.05
Steep slope	–	1	0.5	1.5	0.15
				3	0.3

1 = very important, 0.5 = important and 0 = less important

Table 3 Inter-connection matrix

Cell No.	Numerical value	Value	Weighting value	
			FIC	Cell value
Cell 1	Hazardous—low	5	0.1	0.5
	Soil erosion—high	1	0.05	0.05
	Steep slope—medium	3	0.45	0.45
	Total			1.00
Cell 2	Low			

Table 4 Planning for appropriate remediation and strategies and actions/measures for flood control

Task	Activities	Model	Outcomes	Outputs
Flood hazards and rainfall predication (magnitude and intensity)		Hydrological modelling	To improve weather prediction capacities	Need for super computers
Vulnerability assessment (identification of vulnerable groups)		Vulnerability and model assessment		
Risk probability (susceptibility) assessment		Sustainability model		
Mitigation and adaptation	Policy measures		Planning and management	
	Structural measures			
	Non-structural measures			

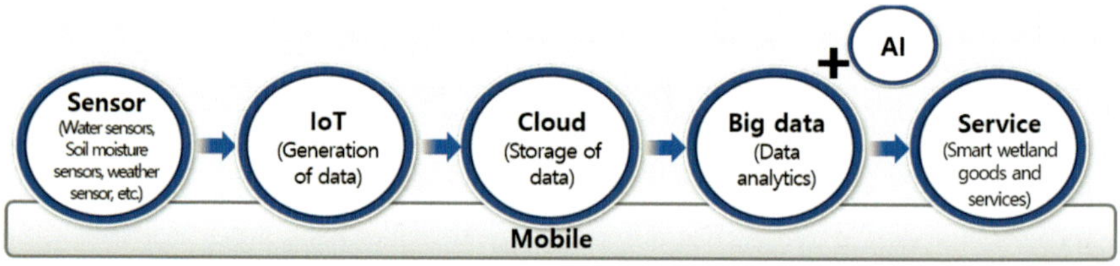

Fig. 16 Shows how ICT, big data and AI as well as IoT can be arranged for monitoring for flood risk management. *Source* Schwab (2016)

6 Structure of Conclusions and Recommendations Section

The suggestions and conclusions will be based on the study findings to make resource-wise recommendations that will help in solving the flood risk problem with an aim of creating resilient communities while taking into consideration a number of solution approaches such as planning-based solutions, digital technology-based solutions and nature-based solutions among others. Therefore, these recommendations and suggestions will be used to solve the flood problem in Kampala city.

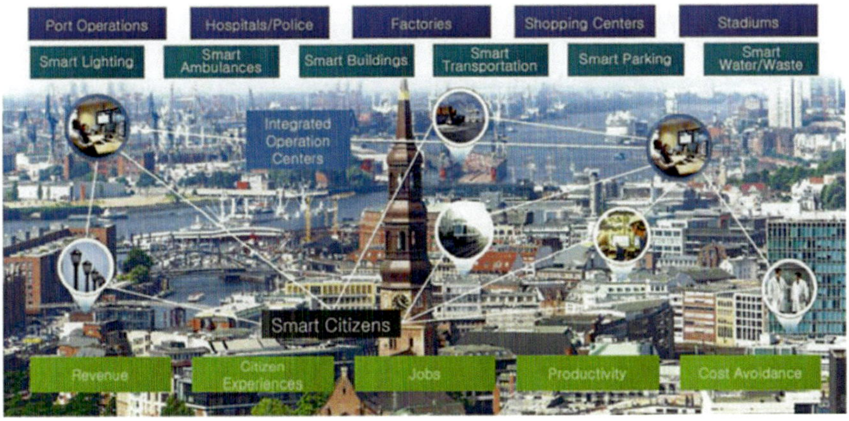

Fig. 17 Image of a smart/connected city. *Source* Fellor (2016)

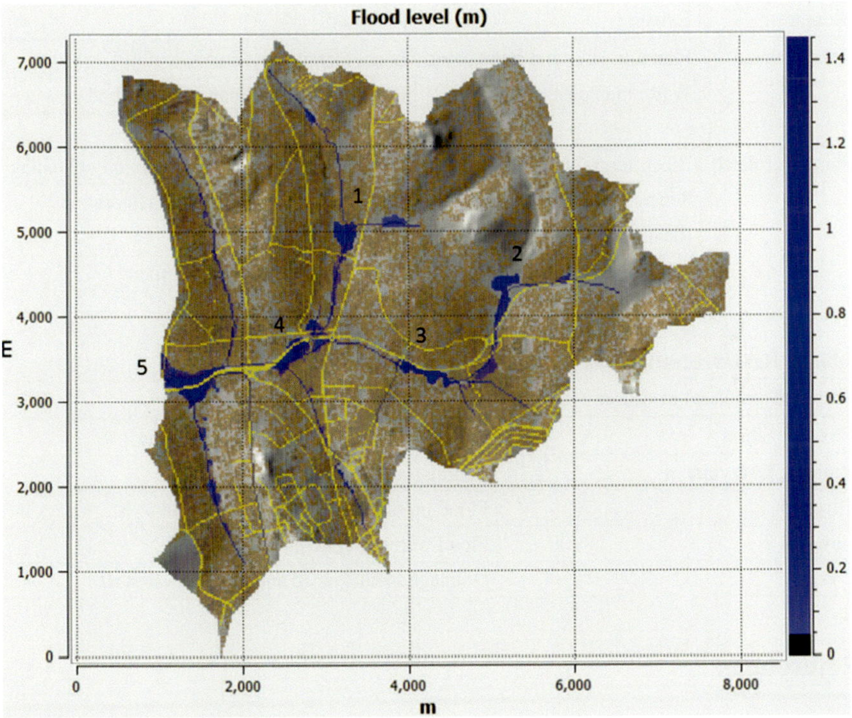

Fig. 18 Flood level of Bwaise III District, Kampala city, Uganda. *Source* Sliuzas et al. (2013b)

Table 5 Rainfall intensity and temperature

1900	1910	2010	2020
Rainfall			
Frequency			
Temperature			

Table 6 Flood damage data

Affected items	Years	Number of cases
People		
Crops		
Buildings		
Roads		

Table 7 Design of measures table

Measures	Existing	Proposed
Policy related	Planning laws and regulations	Local identity creation
	Waste management	Flood control and management fund
		Climate smart traffic system
Planning related	Biostructural technique	Cultural and ecological tourism city planning
	Sewage treatment plant	Ecological conservation and restoration
	Reuse of rainwater	Habitat-sensitive design
		Climate smart urban basic plan

Table 8 Hazard reduction measures table

Proposed structural measures	Proposed non-structural measures
Channelization	Flood plain land acquisition/relocations
Building reservoirs	Zoning (land use planning)
Diversions	Flood proofing
Levees	Flood warning systems
Dykes	Drainage maintenance and public awareness

References

Fellor G (2016) Internet of everything: revitalizing cities, citizens, and urban systems, June 6, 2016, unpaged

KCCA (2012) Kampala physical development plan: updating kampala structure plan and updating the kampala GIS unit, Draft final report

Kim K-G (2015) Integrated water resource manaements and smart water grid

Kim K-G (2017) Development of an integrated smart water grid model as a portfolio of climate smart cities

Kim K-G (2018) Low carbon smart cities: tools for climate resilience planning
London flood risk appraisal
New York flood assessment
Pérez-Molina E (2014) Modeling urban growth and flooding interactions with cellular automata in Kampala, Uganda. Unpublished master's thesis. Faculty of Geo-Information Science and Earth Observation (ITC), University of Twente, Enschede, The Netherlands
Schwab K (2016) The fourth industrial revolution. Penguin Random House, UK
Seoul flood management case study
Sliuzas E (2015) Simulating flood impacts in Kampala, Uganda: when land patterns matter
Sliuzas RV, Flacke J, Jetten, VG (2013a) Modelling urbanization and flooding in Kampala, Uganda. In: Proceedings of the 14th N-AERUS/GISDECO conference, enschede, The Netherlands
Sliuzas RV, Lwasa S, Jetten VG, Petersen G, Flacke J, Wasige J (2013b) Searching for flood risk management strategies in Kampala. In: Planning for resilient cities and regions: proceedings of AESOP-ACSP joint congress, 15–19 July 2013, Dublin, Ireland. 10 p
UN-Habitat (2013) Flood risk assessment strategies and actions for improving flood risk management in Kampala, Final report of integrated flood management project Kampala

Flood Risk Assessment and Green Recovery Planning Case Studies—South Asia and Southeast Asia

Flood Mapping and Damage Assessment Using Synthetic Aperture Radar (SAR) Data in GIS and Google Earth Engine—A Case Study of 2020 Riverine Flood in Gaibandha, Kurigram, and Jamalpur Districts of Bangladesh

Siddique Md Abu Bakar

Abstract Floods are one of the greatest hazards in Bangladesh, however, in most cases, household-level risk-reduction strategies are inadequate for ensuring a livelihood resilient to floods. To formulate effective risk-reduction policies and programmes for riverine areas, it is crucial to measure flood risk at the local level. The monsoon floods coupled with prolonged inundation and the ravages of the COVID-19 pandemic have exacerbated the negative impacts on the flood-affected people. The creation of inundation maps are a very important tool to prevent or reduce flood hazards and risks by providing useful and reliable information to the population and remote sensing (RS)-based data provide multi-resolution satellite data for flood inundation mapping and flood risk zones identification, which are the first steps for the formulation of any flood management. RS technologies have made it easier to map large areas under flooding and provide early warnings of the disaster that helps in providing support in crisis situations and minimizing damage. Space-based sensors are comparatively more effective than ground-based techniques in providing near real-time data for monitoring the extent of floods and satellite platforms and other advances in RS have provided a wide range of applications of satellite data. The case study author conducts a flood extent mapping and damage assessment using Synthetic Aperture Radar (SAR) data captured by the Sentinel-1 satellite. The case study author recommends that Google Earth Engine, a free, cloud-based tool, is very useful for developing countries like Bangladesh to assess damages in emergency situation in a cost-effective way and that Synthetic Aperture Radar (SAR) data can also be utilized for estimating the extent of flooding in affected areas.

Keywords Bangladesh and flooding · Bangladesh and synthetic aperture radar (SAR) data · Google Earth Engine and climate risk assessment · Climate risk

---

S. M. A. Bakar (✉)
Ministry of Local Government, Rural Development and Cooperatives, Dhaka, Bangladesh
e-mail: absiddiquetp@gmail.com

K. Kim and C. Atkin (eds.), *Cross-Cultural Perspectives on Climate Change Adaptation*, Cities and Nature, https://doi.org/10.1007/978-3-031-50365-8_12

management · Disaster risk management · Adaptation · Resilience · Vulnerability · Adaptive capacity · Hazard reduction measures

1 Introduction

Floods are one of the greatest hazards in Bangladesh. It is assumed that people who reside in a riverine area have adapted to flood pulses. However, in most cases, household-level risk-reduction strategies are inadequate for ensuring a livelihood resilient to floods. This is because riverine people are exposed to recurrent floods, which increases their vulnerability to floods. To formulate effective risk-reduction policies and programmes for riverine areas, it is crucial to measure flood risk at the local level.

The flooding in Bangladesh during monsoon season is very common and frequently happens. Consequently, people have been experiencing tremendous damage to properties and infrastructures, and human casualties are high (Talukdar et al. 2020). While floods are one of the most devastating disasters from nature, for developing nations like Bangladesh, the impacts of flooding are particularly devastating. Due to the dynamic and complex nature of the flooding, the prediction of flooding sites is usually very difficult for flood management. The monsoon floods of the year 2020 had a major negative impact on the Northern, North-Eastern and South-Eastern regions of Bangladesh. The flood impacted 30 districts of Bangladesh with moderate to severe impact on 15 districts (CARE 2020). The highest peak of the flooding was predicted to be at the Bahadurabad point, with 71% probability of high flooding. The peak of the flood was anticipated to hit on the 18th July 2020. The monsoon floods coupled with the ravages of the COVID-19 pandemic had an exacerbating effect on the flood-affected people (CARE 2020).

The inundation maps help in preventing or reducing flood hazards and risks by providing useful and reliable information to the population (Demir and Kisi 2016). In this context, remote sensing (RS)-based data provides multi-resolution satellite data for flood inundation mapping and flood risk zones identification, which are the first steps for the formulation of any flood management strategy (Sahoo and Sreeja 2017). Space technology has had a positive impact on the preparedness, prevention, and relief phases of flood disaster management (Jeyaseelan 2003). RS technologies have made it easier to map large areas under flooding and provide early warnings of the disaster that helps in providing support in crisis situations and minimizing damage. Space-based sensors are comparatively more effective than ground-based techniques in providing near real-time data for monitoring the extent of floods (Tiwari et al. 2020). RS analysis can be routinely used to map the flood extent either in near real-time or retrospectively (Cohen et al. 2019). For emergency response, it is highly valuable as it offers continued observation about the extent over large geographical areas. For the effective production of flood inundation maps, the selection of suitable sensors is a major challenge (Islam et al. 2010). The satellite platforms and other advances in RS have provided a wide range of applications of satellite data. Several studies have utilized optical satellite data and related flood products (e.g. MODIS-based NRT) to monitor and assess flood impacts across several river basins (Amarnath 2013). It is true that optical sensors data is problematic in flood studies. Therefore, several researchers emphasized the utility of Synthetic Aperture Radar (SAR) data for flood studies. SAR sensors can easily penetrate clouds and severe rains conditions during flood and post-flood events. SAR data are therefore preferred in real-time flood damage assessment as they discriminate between land and water precisely (Matgen et al. 2011).

Therefore, the study aims to conduct a flood extent mapping and damage assessment using Synthetic Aperture Radar (SAR) data captured by the Sentinel-1 satellite.

2 Background of the Study

As a riverine floodplain country, Bangladesh is considered to be one of the most flood-prone countries in the world. Between 1971 and 2018, floods accounted for 29% of the country's natural disasters (EM-DAT 2019). Floods also accounted for the largest share of Bangladesh's economic losses caused by any disaster in the last 48 years. During this period, the number of people that were affected and became homeless due to floods was significantly higher than for any other type of disaster. The floods in 1974 killed 28,700 people and affected another 38 million people, with estimated damages of $579.2 million (EM-DAT 2019). The 1998 floods resulted in a record amount of damage, totaling around $4.3 billion (EM-DAT 2019). The floods of August 2017 in the northern districts affected around 6.9 million people, with more than 114 reported casualties (Philip et al. 2019). These figures indicate the level of flood risk for riverine communities in Bangladesh. Although the number of flood-related fatalities decreased significantly from 29,375 during the 1970s to 2404 during the 2000s (EM-DAT 2019), the socio-economic impact of floods has risen as a result of demographic changes, increasing the share of crop and property losses (Brammer 2016). Therefore, flood risk assessment is necessary to understand exposure, people's socio-economic vulnerability, and their capacity to respond to floods.

Objectives of the Study

- To analyze and generate flood inundation map using Synthetic Aperture Radar (SAR) dataset acquired by Sentinel-1 mission satellite.
- To assess flood damage using flood inundation maps at the population level.

Research Questions

(1) Which areas in study districts were inundated by flood water?
(2) What are the permanent and semi-permanent water bodies?
(3) Is the applied method successful in extracting data from SAR images?
(4) How many people from different age-sex group were affected?

Structure of the Study

This study was organized following the risk and damage assessment framework developed by Professor Dr. Kwi-Gon Kim and includes elements like affected area selection, remotely sensed data selection, criteria for remote sensing data considering the flood hazard, geographical factors selection, SAR data analysis, flood inundation map generation, damage assessment using population data, and an identification of future implication of these methods.

Both indicators and assessment method are selected considering the vulnerability and institutional capacity of developing countries, more specifically Bangladesh. Therefore, a comprehensive framework for flood inundation mapping and damage assessment in a time and cost-effective way is discussed to improve the post-flood management efforts in a remote and resource constraint environment.

Conceptual Framework

In order to estimate damage and identify the inundated area, the following methodological and conceptual framework was used. Both remotely sensed satellite data and population data was used in this study. For technical analysis, following the UN-Spider method, the author has used the powerful Google Earth Engine tool for Synthetic Aperture Radar data to identify inundation extent and has used ESRI ArcGIS software for analytical map generation. Population in the inundated area were projected using population census data from Bangladesh Bureau of Statistics (BBS) (Fig. 1).

3 A Case Study

For this study, the author has selected Bangladesh with a focus on the devastating riverine flood that occurred in 2020. Severe floods that struck Bangladesh during the last week of June 2020, driven by prolonged and intensified heavy monsoon and upstream water, affected 5.4 million people in the northern, central, and north-eastern part of the country. Around 37% of the country's total areas were flooded affecting 33 districts and it was considered as the longest flooding period in the last 22 years. Until the beginning of October 2020, due to monsoon raining and heavy rainfall in upstream, people in many districts suffered from multiple spells of floods. There was widespread damage to housings and reduced access to clean and safe water, hygiene and sanitation facilities as well as access to livelihoods in most of the affected districts (Fig. 2).

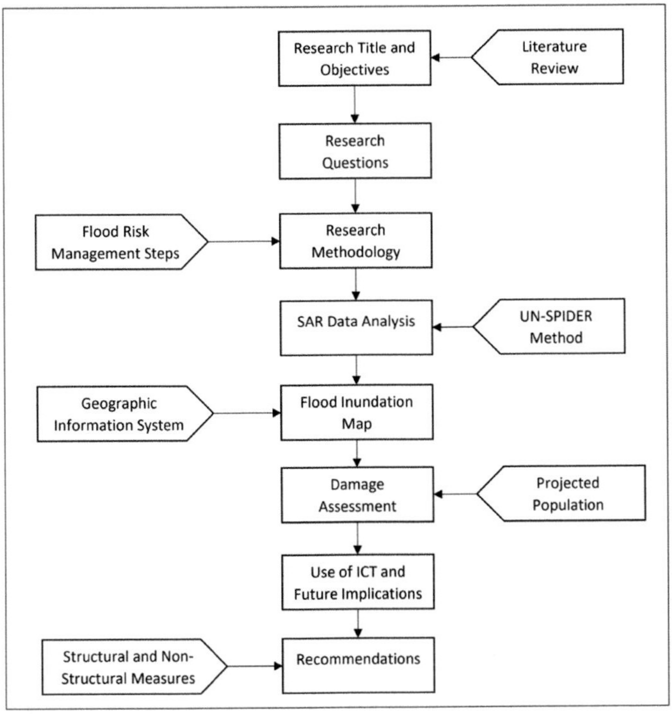

Fig. 1 Conceptual framework of the study

4 Methodology

Methodological steps for this study are described below. For this study, the author attempted to follow the 9-steps for flood risk and damage assessment developed by Professor Dr. Kwi-Gon Kim.

Step-1: Site Selection

According to a situation update report by BRAC, the 2020 flood in Bangladesh was the longest spell of flooding recorded in last three decades. A total of 30 districts were impacted including 15 moderately to severely. Out of 15 moderately to severely affected districts, three districts were affected most, namely Jamalpur, Kurigram, and Gaibandha. Among these three districts 90%, 84%, and 66% of the respective areas were affected. In this study, only these three severely affected districts are considered for flood mapping and damage assessment. The location map of these three districts is shown in Fig. 3.

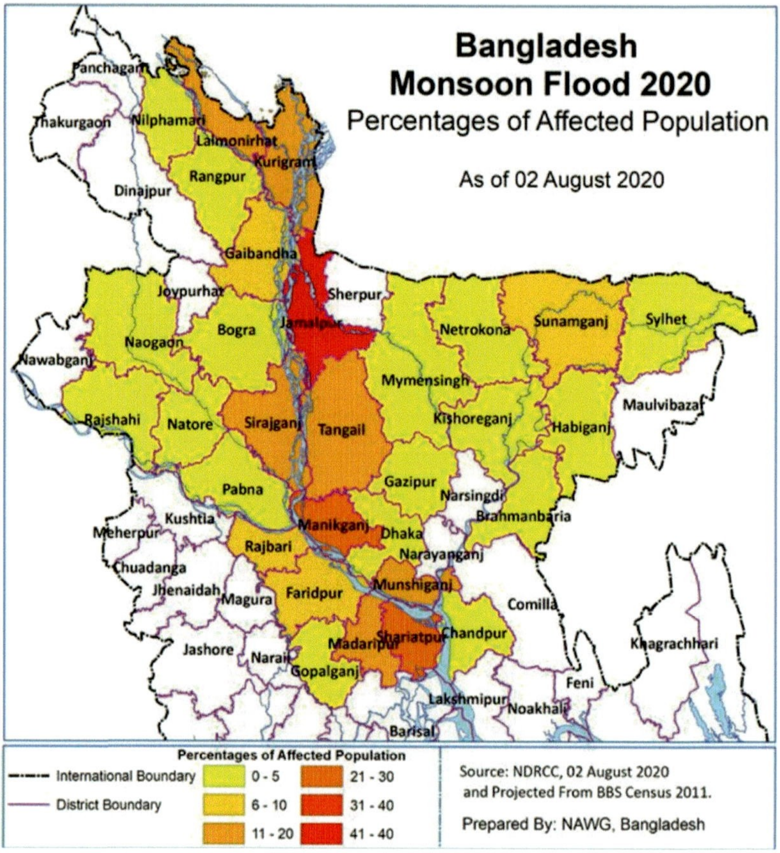

Fig. 2 Affected population by 2020 flood in Bangladesh

Step-2: Delineate Boundary or Scope of the Study

Flood damages can be assessed across several dimensions including population, economic, resources, food security, agricultural, environmental, ecological, biodiversity, infrastructure, etc. Each of these aspects has different dimensions, characteristics, and vulnerabilities. Due to study time constraints, the author has considered only the number of populations affected by the flood. To identify the affected population more specifically data on different age groups, gender, persons with disabilities, poor households, extremely poor households, and agricultural labour-dependent households were considered during damage analysis.

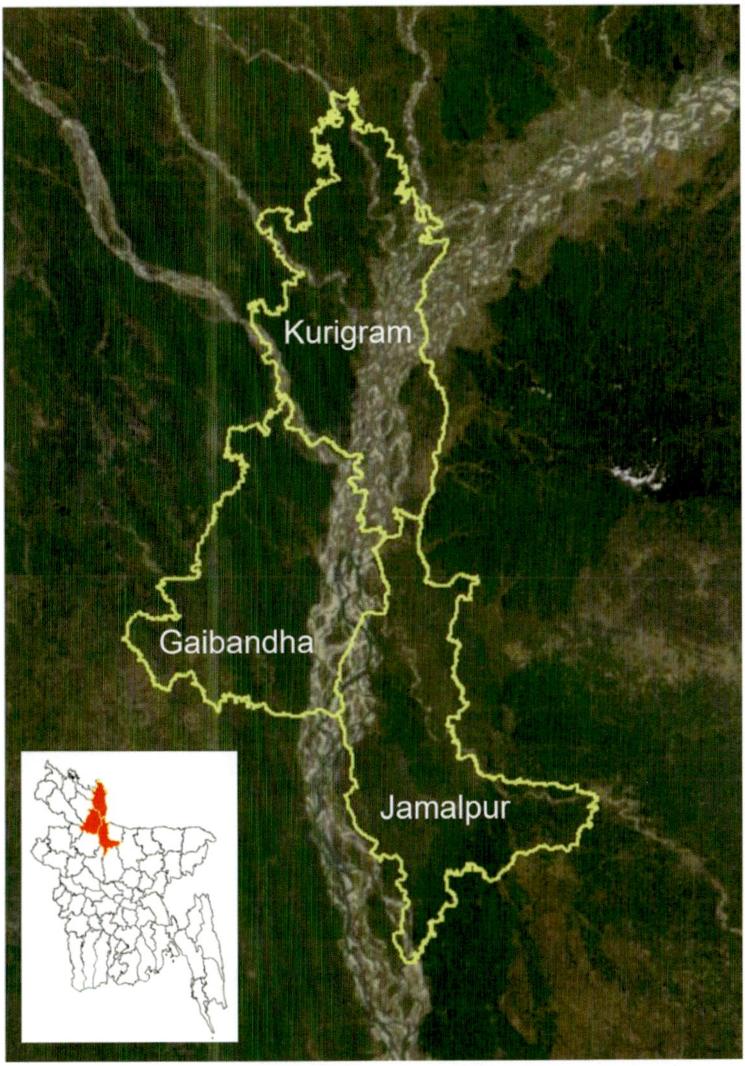

4.2 Step-2: Delineate boundary or scope of the study

Fig. 3 Location map of study area

Step-3: Describe Site Profile

Demographic Characteristics

Around 6,829,080 people live in the study area with a projected population density of 927, 1114, and 1126 per km^2 in Kurigram, Gaibandha, and Jamalpur districts respectively. Data on census population and projected population distribution in the three districts with the projected density is provided in Table 1.

Rationale of This Study

Flooding is one of the most frequent disasters in Bangladesh. The country experiences floods every year and the three selected districts in this study are the most flood-prone districts in Bangladesh. In contrast, there is an acute shortage of skilled manpower for flood damage assessment. Damage assessment in a timely manner is essential for post-flood management and recovery works. In this study, Sentinel-1 SAR data is used which has high spatial and temporal resolution, and it is freely available. UN-SPIDER is the recommended method used in this study which uses the free Google Earth Engine tool.

Research Gap Analysis: Upon searching Google Scholar, Springer, and Elsevier site, no study has been found that was conducted for this study area using GIS and Google Earth Engine. Google Earth Engine is a powerful tool to perform satellite data analysis in time-effective way. As the satellite data are preprocessed here, a less technical knowledge is required to use this tool. Hence, it can be one of the most useful tools for developing countries and in remote areas.

Step-4: Describe Historical Events

Bangladesh Bureau of Statistics showed that between 2009 and 2014, 56.62% of households have been affected by disasters at least once. According to the Flood Forecasting and Warning Centre (FFWC) since 1998 the 2007 and 2017 floods had a similar magnitude causing about 40% inundation of the country.

Step-5: Analyze Flood Inundation and Damages

These three districts were severely affected for the flood. The duration of 2020 flood was from 28th June to 15th August 2020. A picture of the study area published in one of a leading national newspaper is shown below.

Characteristics of Sentinel-1 Data Used in This Study

Sentinel-1 is the first of the Copernicus Programme satellite constellation conducted by the European Space Agency. This mission is composed of a constellation of two satellites, Sentinel-1A and Sentinel-1B, which share the same orbital plane. They carry a C-band (5 cm wavelength) Synthetic Aperture Radar instrument which

Table 1 Projected population within the study area

District	Pop. 2011 (census)	Pop. 2020 (projected)	Density 2020 (per km^2)
Kurigram	2,069,270	2,096,250	927
Gaibandha	2,379,260	2,410,270	1114
Jamalpur	2,292,670	2,322,560	1126
Total	6,741,200	6,829,080	

provides a collection of data in all-weather, day or night. This instrument has a spatial resolution of down to 5 m and a swath of up to 400 km. A 10 m spatial resolution database is freely available. The constellation is on a sun synchronous, near-polar (98.18°) orbit. The orbit has a 12-day repeat cycle and completes 175 orbits per cycle. Therefore, these satellites can give a temporal resolution of 6–12 days.

Methodology for Flood Inundation Map and Damage Assessment

For the SAR data analysis 'VH' band polarization mode was selected. Sentinel-1 operates in four exclusive acquisition modes as stripmap (SM), interferometric wide swath (IW), extra-wide swath (EW), and wave mode (WV). Stripmap (SM) has the highest spatial resolution (5 m) but it is only commercially available. The IW has a 10 m spatial resolution and it is freely available. This is also recommended by the UN-SPIDER method. Therefore, IW acquisition mode is selected for this study. Methodological flowchart is given in Fig. 4.

Step-6: Assess the Risk

To perform preprocessed SAR data analysis, Code Editor from Google Earth Engine was used and all necessary codes were written there. The study area was inundated by flood water from 28 June to 15 August of 2020. To identify how much area was inundated by flooding, two images are required—one is before the starting of the flood and the other is during the flood. The first image was selected between 10 and 25 June of 2020. The second image was selected between 29 July and 15 August of 2020. After creating mosaic for both images, speckle-filter is performed for the radiometric correction (Fig. 5).

After radiometric correction, both layers are compared by applying a threshold value of 1.25. The threshold of 1.25 has been recommended by UN-SPIDER method through trial-and-error process. The predefined threshold of 1.25 is applied assigning 1 to all values greater than 1.25 and 0 to all values less than 1.25. These processes will return binary layers for both before and after flood for the study areas.

From these binary layers, natural slope, permanent and semi-permanent water, and scattered pixels are needed to exclude. All areas with a slope higher than 5% are excluded. Areas that contain water for at least 10 months in a year are considered as permanent waterbody and areas that contain water for at least 4 months in a year are considered as semi-permanent waterbody. These two types of waterbodies are excluded from the binary layers. Finally, all the pixels having less than 4 neighbourhood pixels are considered as scattered pixels and are excluded from the flood layers. Figure 6 shows binary layers of the study area before flood. The figure shows the binary layer after the flood.

After applying all the conditions and comparing both binary layers, potentially flooded areas identified from the image analysis are found. Figure 7 shows the potentially flooded areas within these three districts.

After that, the binary layer for potentially flooded area is exported as an ESRI shapefile. Flood inundation map and the number of affected populations are generated using ESRI ArcGIS application. In Kurigram district 557,243 population, in Gaibandha district 692,237 population, and in Jamalpur district 929,410 population

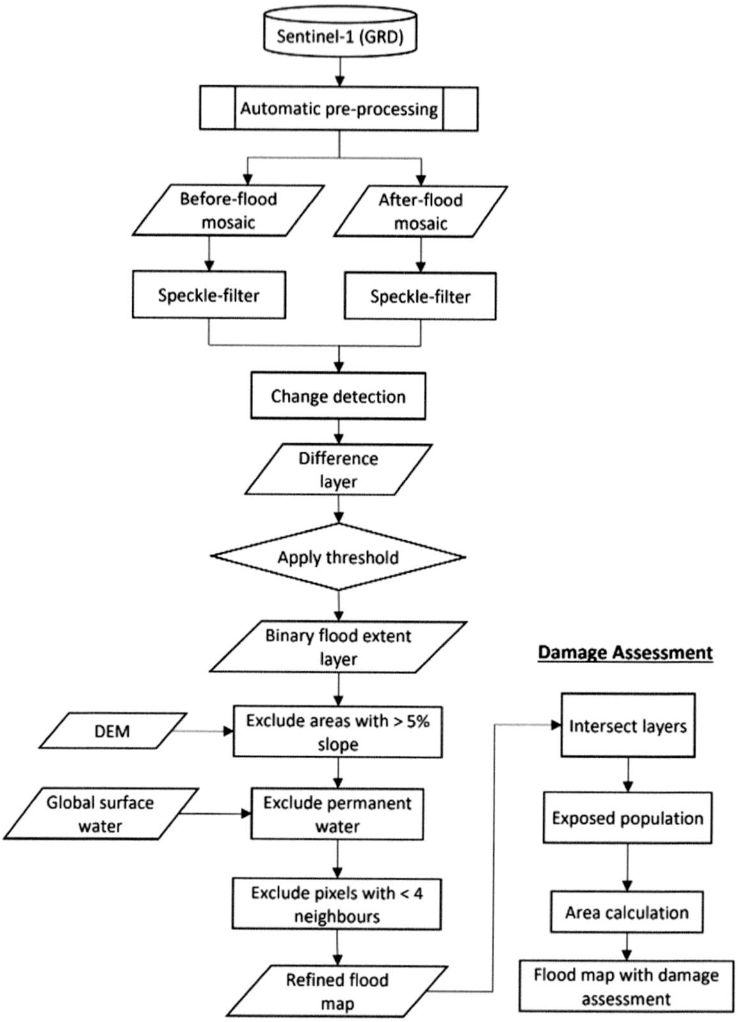

Fig. 4 Methodology for flood inundation map and damage assessment

were affected due to the riverine flood of 2020. Final flood inundation map including number of affected population is shown in Fig. 8.

The smallest administrative boundary are union councils. Classified population data for unions are used for vulnerability analysis. Table 2 shows the number of affected populations from different groups. Population groups in this study are considered as women, senior citizens (60+ years age), child (0–59 months age), person with disability, poor household (HH), extreme poor household, women-headed household, and agricultural labour-dependent household. Across all considered vulnerability indicators, Jamalpur district is the mostly affected area. For

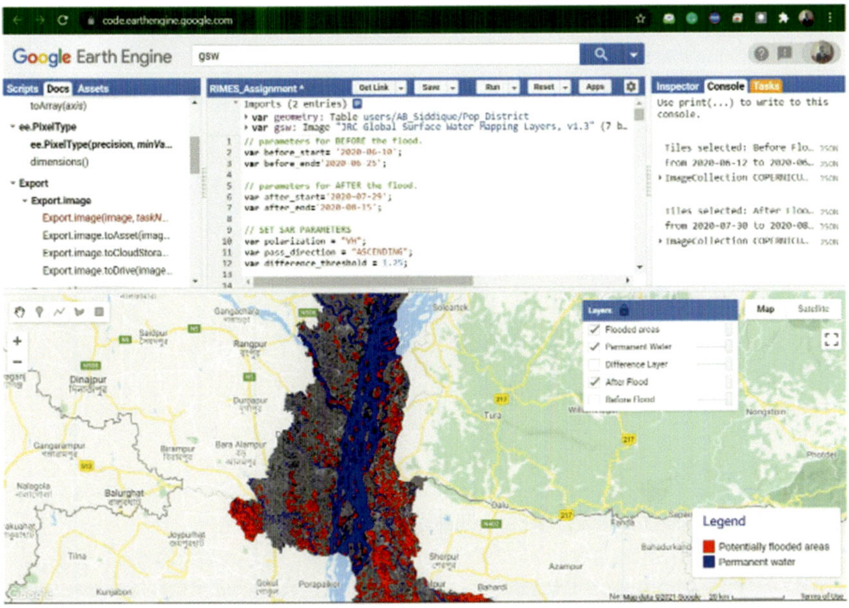

Fig. 5 Code Editor of Google Earth Engine

instance 505,013 women were affected in Jamalpur district, whereas 128,367 and 128,191 women were affected, respectively, from the Gaibandha and Kurigram districts. This also revealed that 48%, 77%, and 42% areas were flooded, respectively, from Gaibandha, Jamalpur, and Kurigram districts.

To justify the validity of this study, the result is compared with the joint need assessment report by CARE, UNOPS, and UK Aid. The comparison is shown in Fig. 9.

Very little variation was found from the comparison. This study has found 77% inundated area in Jamalpur district whereas joint need assessment report found 73% inundation in that district. In Gaibandha district, this study found 3% more inundation and in Kurigram district, this study found 6% less inundation in comparison with the joint need assessment report. This variation may occur due to difference in data collection time. Besides, flood extent may be changed from time-to-time due to prolonged flood.

Step-7: Prepare Recovery Plan for Future Flood Events

Flooding occurs in these areas every year. The result of this study can used to prepare recovery plans for future flood events. Considering the temporal resolution of Sentinel-1 SAR data, flood inundation map and damage assessment can be done within 2 weeks of flooding. It also can be updated time-to-time if the flood is prolonged. Public officials in charge of post-disaster recovery will be able to apply

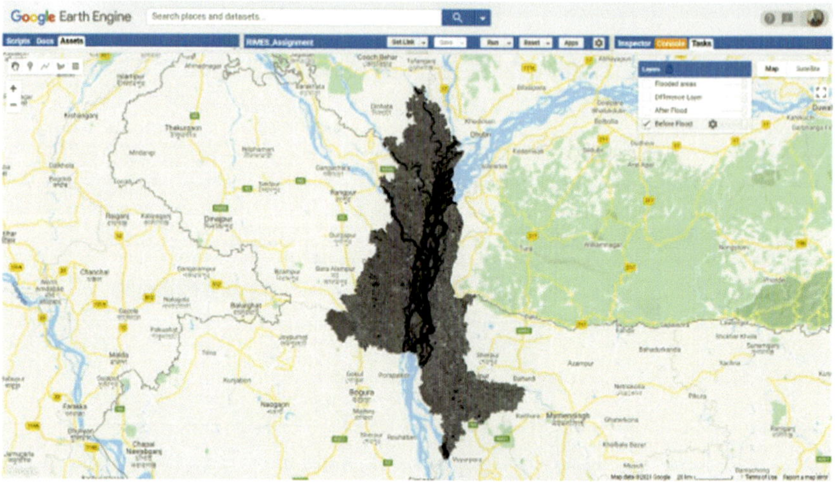

The figure below shows the binary layer after the flood

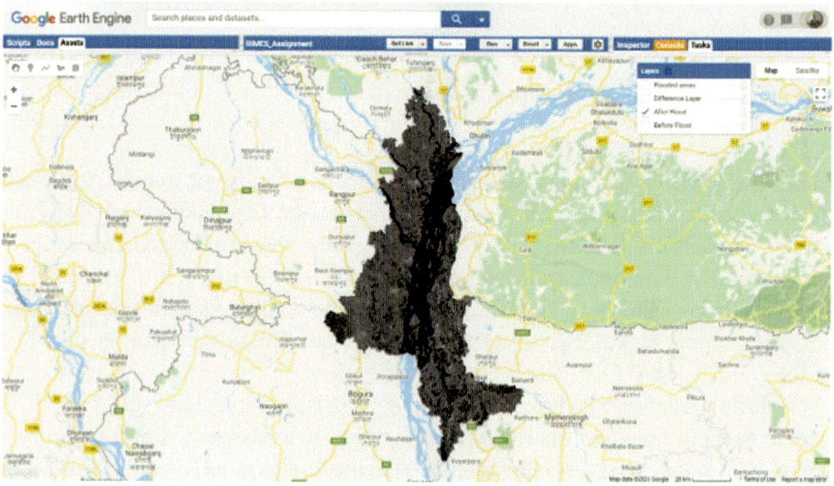

Fig. 6 Potentially flooded areas identified from image analysis

this method with a few days training. Thus, the findings of the study can be helpful in efficient recovery plan and resource distribution during emergency flood situation.

Step-8: ICT Application

Method used in this study is a good example of use of ICT in post-disaster recovery activities. Google Earth Engine is a cloud-based platform. Therefore, satellite data for flooded area can be accessible from anywhere in anytime. An application about flood analytics and management system can be developed to disseminate flood damage information and store data for future analysis. Mobile and other ICT devices can be

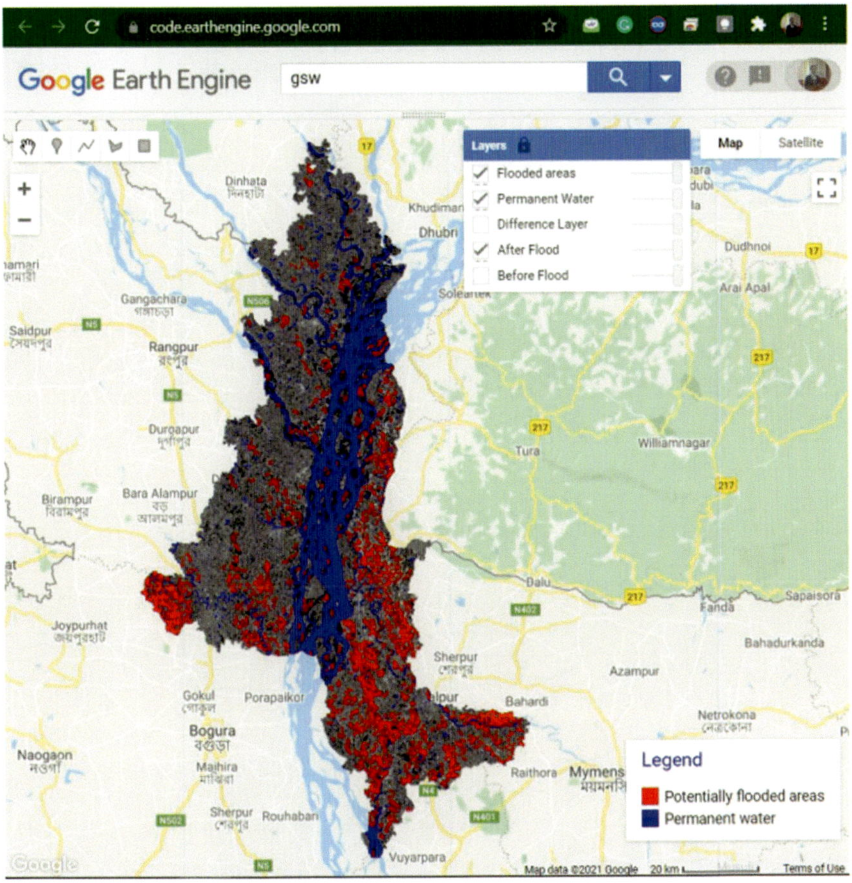

Fig. 7 Potentially flooded areas identified from image analysis

integrated for post-disaster recovery activities in remote areas. A schematic diagram of ICT application in post-flood recovery is shown in Fig. 10.

5 SWOT Analysis

SWOT analysis helps to identify your organization's strengths, weaknesses, opportunities, and threats. SWOT analysis will generate internal factors (strengths and weakness) and external factors (opportunities and threats). The result of internal and external factors will be used as tools and parameters to determine the strategic planning of the community-based disaster management. Therefore, the author has done SWOT analysis for this study which is elaborated in Table 3.

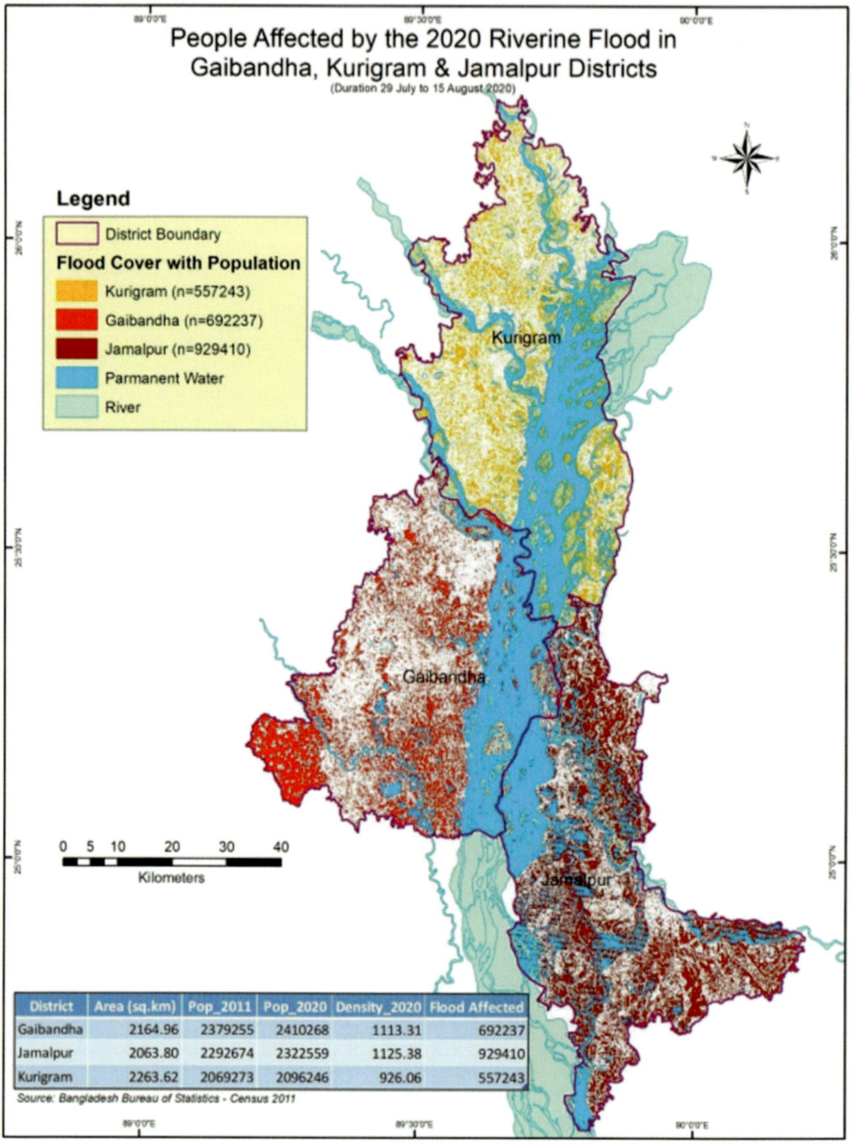

District	Area (sq.km)	Pop_2011	Pop_2020	Density_2020	Flood Affected
Gaibandha	2164.96	2379255	2410268	1113.31	692237
Jamalpur	2063.80	2292674	2322559	1125.38	929410
Kurigram	2263.62	2069273	2096246	926.06	557243

Fig. 8 Flood inundation map with the number of affected populations

6 Recommendations and Conclusion

This study suggests the following recommendations for post-flood recovery in Bangladesh.

Table 2 Vulnerability indicators for the affected community

District	Gaibandha	Jamalpur	Kurigram
Number of affected women	128,376	505,013	128,191
Number of affected (60+ years) age	27,765	110,412	28,058
Number of affected child (0–59 months) age	19,940	81,566	20,042
Number of person with disability affected	4846	14,125	3858
Number of poor HH affected	30,308	128,372	43,559
Number of extreme poor HH affected	18,772	86,062	33,165
Number of women-headed HH affected	8574	29,331	6704
Number of affected agri labour-dependent HH	27,741	82,659	25,465
% of total flooded area	48	77	42

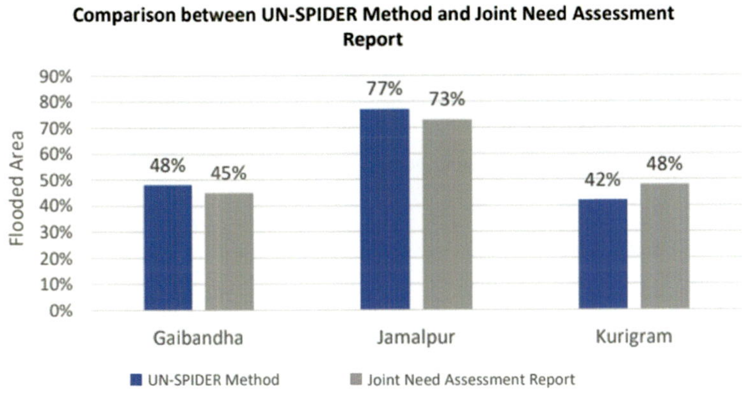

Fig. 9 Comparison between UN-SPIDER method and joint need assessment report

I. Google Earth Engine is a free and cloud-based tool. Hence it is very useful for developing countries like Bangladesh to assess damages in emergency situations in a cost-effective way.
II. Operation and analysis of flood damages in Google Earth Engine requires less technical knowledge as the images are preprocessed. Therefore, government departments can deploy this tool with its existing manpower.
III. Sentinel SAR data has spatial resolution of 10 m and temporal resolution is 6–12 days. So, it can be used in emergency flood period in Bangladesh.

Given the weather conditions during flooding the availability of cloud free optical data is not so easy and SAR data can be utilized for estimating the flooded area. Cloud-based platform, i.e. GEE has been shown to be very useful for the users in preparing an emergency response related to flood and evaluating the damaged area by generating land use land cover maps. Using SAR image by Google Earth Engine can result in precise outcomes that help disaster managers and policymakers towards a better post-disaster management.

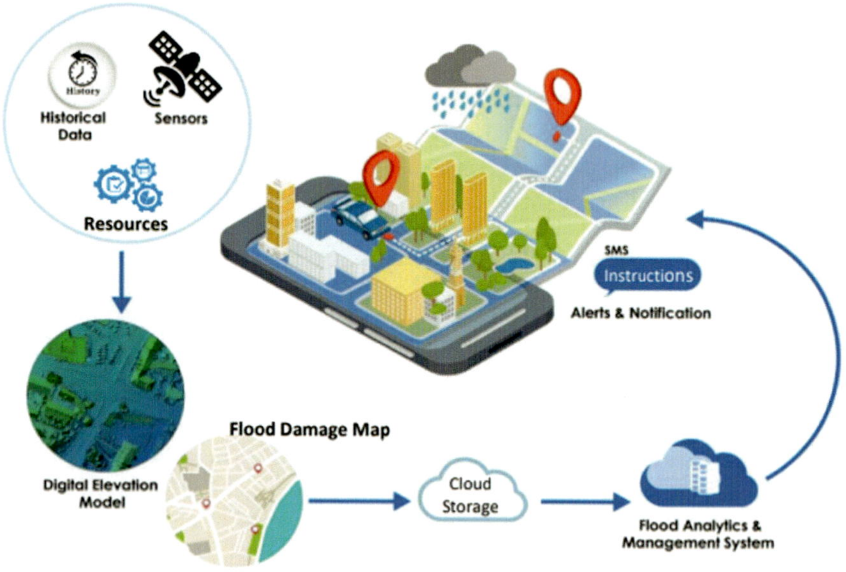

Fig. 10 Use of ICT for flood damage management

Table 3 SWOT analysis for post-flood recovery in Bangladesh

Internal origin (attributes of the organization)	Strength	Weakness
	• Previous experience • Manpower • Local knowledge • Use of technologies • Sufficient ICT devices • Community-based organization • Public representatives	• Shortage of skilled manpower • Lack of information • Lack of coordination • Lack of transparency • Lack of accountability • Lack of fund (internal)
External origin (attributes of the environment)	Opportunities	Threats
	• Free high resolution SAR data • High temporal resolution data • Google Earth Engine • Pre-processed data • Technological advancement	• Upstream river flow • Low elevation • Riverbed siltation • Insufficient fund (external)

References

Amarnath G (2013) An algorithm for rapid flood inundation mapping from optical data using a reflectance differencing technique. J Flood Risk Manag

Brammer H (2016) Floods, cyclones, drought and climate change in Bangladesh: a reality check. Int J Environ Stud 73

CARE (2020) Coordinated preliminary impact and needs assessment, monsoon floods 2020, needs assessment working group. CARE, Bangladesh

Cohen S, Raney A, Munasinghe D, Loftis JD, Molthan A, Bell J, Rogers L, Galantowicz J, Brakenridge GR, Kettner AJ (2019) The Floodwater Depth Estimation Tool (FwDET v2.0) for improved remote sensing analysis of coastal flooding

Demir V, Kisi O (2016) Flood hazard mapping by using geographic information system and hydraulic model: Mert River, Samsun, Turkey. Adv Meteorol 2016:4891015

EM-DAT (2019) Centre for research on the epidemiology of disasters (CRED). Université Catholique de Louvain, Emergency Events Database, Belgium

https://developers.google.com/earth-engine/

https://un-spider.org/advisory-support/recommended-practices/recommended-practice-google-earth-engine-flood-mapping

Islam AS, Bala S, Haque M (2010) Flood inundation map of Bangladesh using MODIS time-series images. J Flood Risk Manag

Jeyaseelan AT (2003) Droughts & floods assessment and monitoring using remote sensing and GIS. In: Satellite remote sensing and GIS applications in agricultural meteorology. World Meteorological Organisation, Geneva

Matgen P, Hostache R, Schumann G, Pfister L, Hoffmann L, Savenije H (2011) Towards an automated SAR-based flood monitoring system: lessons learned from two case studies. Phys Chem Earth Parts A/B/C

Philip S, Sparrow S, Kew S, Van Der Weil K, Wanders N, Singh R, Hassan A, Mohammed K, Javid H, Haustein K (2019) Attributing the 2017 Bangladesh floods from meteorological and hydrological perspectives. Hydrol Earth Syst Sci 23

Sahoo SN, Sreeja P (2017) Development of flood inundation maps and quantification of flood risk in an urban catchment of Brahmaputra River. ASCE-ASME J Risk Uncertain Eng Syst Part A Civ Eng

Talukdar S, Ghose B, Shahfahad et al (2020) Flood susceptibility modeling in Teesta River basin, Bangladesh using novel ensembles of bagging algorithms. Stoch Environ Res Risk Assess 34:2277–2300

Tiwari V, Kumar V, Matin MA, Thapa A, Ellenburg WL, Gupta N, Thapa S (2020) Flood inundation mapping—Kerala 2018; harnessing the power of SAR, automatic threshold detection method and Google Earth Engine

Nature-Based Flood Risk Mitigation for Food Security: The Case of the Mekong River Basin Floodplain, Cambodia

Navin Chea

Abstract The Kingdom of Cambodia is vulnerable and increasingly exposed" to climate change impacts and Kratie Province is considered one of the provinces most vulnerable to flood as it is located along the Mekong River. This case study includes an overview of flood disaster in Cambodia, a situation analysis of flooding in Kratie Province and identifies adaptation strategies and evaluates current efforts at the national, regional, and local levels. The case study author collected data on the ground with farmers, local authorities, and agricultural officials through face-to-face interviews, participatory workshops with farmers, local authorities, and agricultural officials. To cope with floods in the area the local agriculture agency engaged multiple stakeholders to support effective implementation of flood management strategies. To cope with flooding, the local communities engage in a number of hazard reduction measures and strategies including the diversification crops, the use of seed varietals tolerant to flood and the cultivation of short-term crops and improving water management.

Keywords Cambodia and flood risk management · Mekong river and flood disaster · Climate risk assessment · Climate risk management · Disaster risk management · Community engagement and adaptation · Participatory approaches and resilience · Vulnerability · Adaptive capacity · Hazard reduction measures.

1 Adaptation for Cambodia's Agriculture at the National Level

Adaptation for Cambodia's agriculture sector includes:

- Increased support for and reach of existing agricultural research and extension services;

N. Chea (✉)
Royal University of Agriculture (RUA), Phnom Penh, Cambodia
e-mail: Chea_navin@yahoo.com

K. Kim and C. Atkin (eds.), *Cross-Cultural Perspectives on Climate Change Adaptation*, Cities and Nature, https://doi.org/10.1007/978-3-031-50365-8_13

- Implementation of improved water use (water harvesting small-scale irrigation etc.) in drought-prone areas to alleviate rain shortages that cause crop failure—improving farmers' knowledge and access to weather information in carrying out agricultural activities;
- Livelihood diversification to mitigate the reliance on agriculture as the sole livelihood source;
- Restoring and rehabilitating flood protection infrastructures such as dikes.

2 Adaptation for Cambodia's Agriculure at the Sub-National Level

Based on the interviewing with officials at Provincial Department of Agriculture, Forestry and Fisheries (PDAFF), Kratie Province. The result found that the strategy to support farmers better adapts to climate change, mainly flood that was as below:

- Provide technical support to farmers on select seed variety that is tolerant to flood—Meeting with relevant stakeholders include different NGOs, local authorities to evacuate farmers and animals in advance to the safer place.
- Provide seed support after water receding or when farmers' crop damage—Early information to farmers to prepare in advance and follow the information released by meteorology.
- Provide training to official at the commune or village level to collect data of damage and lost.
- Improve the irrigation structure that includes to prevent and store water.
- Work and cooperate closely with stakeholders.

3 Adaptation to Floods by Farmers at the Community Level

Base on the result from the interviewing, participatory workshop the adaptation by farmers include:

- Diversity of crops
- Select seed variety tolerant to flood
- Farmers adjust the crop calendar to escape
- Cultivate short-term varieties of rice
- Improve water management.
- Wet season rice is rainfed and grown at higher elevations to reduce the risk of flood damage, while dry season rice is grown in lower elevation areas as floods recede, and may be supplemented by irrigation. Important tree crops include cashew and mango, with income from higher-priced off-season mango available to some communities.

4 Information Gaps for Writing

- In Kratie Province, there is not available data source that could be found on the historical data of flood disaster that include the loss and damage.
- Lack of historical rainfall data in the past to be available free access. If required data, the official letter must be sent to meteorology at provincial level.
- There is limitation of database of disaster flood history in Cambodia—At district level, the information of flood was not well-documented or recording. Limitation of research study on flood in target province.

5 Introduction

Cambodia is a country vulnerable to the impact of climate change. The Royal Government of Cambodia (RGC) recognized that agriculture has an important role in contributing to the attainment of the Sustainable Development Goals (SDGs), many provinces in Cambodia are impacted by climate change in a different way. Kratie Province is considered one of the provinces most vulnerable to flood as it is located along the Mekong River. In this research paper, the overall objectives were (i) overview of flood disaster in Cambodia, (ii) situation analysis of flood in Kratie Province (iii) identify adaptation by farmers to cope with in Kratie in agriculture. According to flood impact difference sectors in Cambodia, NDC pointed out five vulnerable sectors that include agriculture, infrastructure, forestry, human health, and coastal zones. However, this research paper focused on the agriculture sector only. To meet the above research objectives, the author conducted desk review, field data collection on the ground with farmers, local authorities, agricultural officials through face-to-face interviews[1], participatory workshops with farmers, local authorities, and agricultural officials. In this research study, there were only two districts, namely Chetr Borei and Preak Prasap districts were selected based for this study. The finding revealed that floods occurred frequently and in intensity in Cambodia. Agriculture was recorded as damaged. For example, from 2000 to 2019, there were 113,000 ha of crops that were destroyed. At the provincial level for example in Kratie Province, the flood was recorded to have occurred frequently and in intensity. There were two types of floods that happened, flash floods and riverine floods. The flash flood normally lasted from 1 to 5 days and on average occurred 5 times during the rainy season. For riverine floods normally lasts from 20 to 50 days. Farmers experience flood damage to their crops.

To cope with flood, the local community applied to diversify adaptation measures include cultivating diversity of crops, selecting seed variety tolerant to flood, changing crop calendar, cultivating short-term varieties of rice/crops, improving water management after water receding, adjusting crop cultivation base on the elevation. The coping also combines with fruit trees that provide long-term benefits and could be tolerant with flood. To cope with floods, different stakeholders were jointly together based on their role and responsibility. Agriculture department alone could not be effective in flood management, it requires multi-stakeholders to cooperate.

6 National Context

Geography

The Kingdom of Cambodia is located in mainland Southeast Asia between 10° and 15° N latitudes and longitudes 102° and 108° E that cover an area of 181,035 km^2(RGC 2006). It shares a border with Thailand, Lao PDR, and Vietnam. The country's topography broadly consists of the central plains surrounded by mountainous and highland regions, and a 435 km coastline to the south. Phnom Penh, the capital city, is located in south-central Cambodia, at the confluence of the Mekong, Tonle Sap and Bassac Rivers. The Mekong River and its tributaries dominate the hydrology. The Tonle Sap Lake, an outlet of the Mekong during the rainy season, covers an area of up to 10,400 km^2 in the northwest (GSSD 2015). The country comprises of four regions: Central Plain, Tonle Sap, Coastal and Sea and Plateau and Mountains (NIS 2019). Administratively, the country is divided into 20 provinces and 4 municipalities with a total of 183 districts and 1609 communes (RGC 2006) (Fig. 1).

Population

According to the General Population Census of Cambodia 2019, the population of Cambodia stood at 15,288,489 excluding those that were abroad. The total population has increased from 13,395,682 in the 2008 Census. So, the population has grown by 1,892,807 persons, which represents 14.1%, over the period of 11 years from 2008 to 2019. The male population was 7,418,577 (48.5%) and the female population stood at 7,869,912 (51.5%).

Phnom Penh City of Cambodia has the highest population at 2,129,371 (NIS 2019). Cambodia is one of the smaller nations in the region in terms of populations compared with the larger neighbors Vietnam and Thailand but exceeds that of Laos. More than 80% of Cambodians live in rural areas and poverty remains a serious social issue in Cambodia. Although poverty reduction continues to be a national imperative, Cambodia has made remarkable progress in the last two decades (GSSD 2015) (Fig. 2).

Economy and Development

Cambodia's economy has grown steadily since the liberalization of its markets; the annual GDP growth rate from 1993 to 2004 averaged 9.7%, while per capita growth GDP averaged 6.5%. Based on a 2011 Economic Review of Cambodia, agriculture, fisheries and forestry accounted for 32% of GDP, industry 22%, services 38%, and taxes on product 8%. Industry has been the fastest growing sector in Cambodia's economy since the early 1990s, increasing from 12.6% of GDP in 1993 to 26.2% in

Fig. 1 Map of Cambodia source. *Source* World Atlas

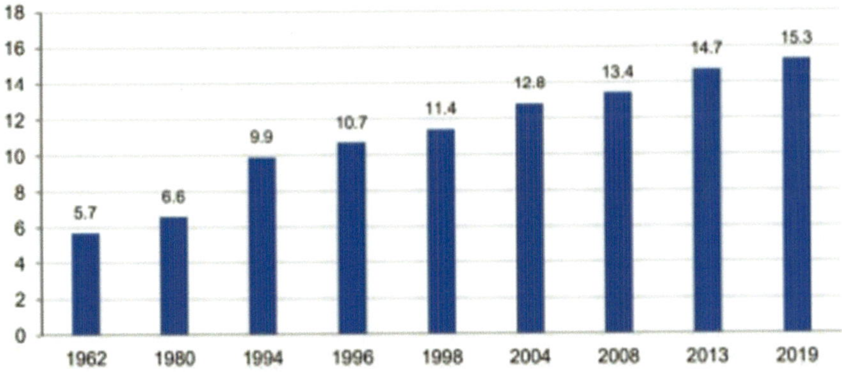

Fig. 2 Cambodia's total populations in millions, trend 1962–2019. *Source* NIS (2019)

2006, with an average annual growth rate of more than 15% over this period. Though suffering a short decline in the late 2000s, industry growth rebounded and remained steady at approximately 22% of GDP. In terms of gross value added by sector in 2004 the textiles and manufacturing sub-sectors made significant contributions. Industry employs 8% of the labor force, far below agriculture, which employs 75%. The challenge for Cambodia's economy is to broaden the opportunities available to its increasing population, while diversifying beyond agriculture and the garment sector. Tourism is the fastest growing sub-sector in services and accounted for about 12% of GDP in 2011 (GSSD 2015).

The Royal Government of Cambodia (RGC) recognizes that agriculture has important roles in contributing to the attainment of the Sustainable Development Goals (SDGs) as indicated in a number of country's strategies and policies, such as the Rectangular Strategy IV, the National Strategic Development Plan (NSDP) 2019–2023, and the Zero Hunger Action Plan, 2016–2025 (FAO in Cambodia 2020). About 80% of the population lives in rural areas and agriculture provides food for daily living, raw materials for agro-industry, and over 30% of GDP from agricultural production and export. The Royal Government of Cambodia (RGC) gives agriculture an important position in its Rectangular Strategy and a high priority in its national development agenda. The RGC expects agricultural development to continue to reduce poverty, through food security, improved household incomes and job creation, and to contribute to economic growth and macroeconomic stability.

7 Cambodia Climate

The tropical monsoon climate of Cambodia is characterized by a rainy season and a dry season. The rainy season, which lasts from May to early October, accounts for 90% of annual precipitation. The dry season, from November to April, brings drier and cooler air from November to March, and then hotter air in April and early May. The maximum mean temperature is about 28 °C and the minimum mean temperature about 22 °C. Maximum temperatures are common before the start of the rainy season and may rise to more than 38 °C. The average annual rainfall from 1994 to 2004 varied between 1400 and 1970 mm. In lowland areas, annual rainfall ranges from 1000 to 1700 mm, while in the highlands it ranges from 1000 to 2700 mm, and for coastal areas, the precipitation ranges from 1000 to 3000 mm. However, over the past decade, some inland provinces have experienced less than 600 mm of rainfall annually, while precipitation has reached 3800 mm in coastal areas.

The geographical occurrence of extreme weather events such as droughts and floods vary, and while floods affect lowland areas, the geographical distribution of droughts is widespread. Though there are some actual benefits from the seasonal flooding experienced in the central plains, providing fish as well as nutrients to the soil, the frequency of severe floods has increased over the last decade. Storms occur most frequently between August and November, with the highest frequency in October. The country is rarely exposed to the full force of tropical cyclones and

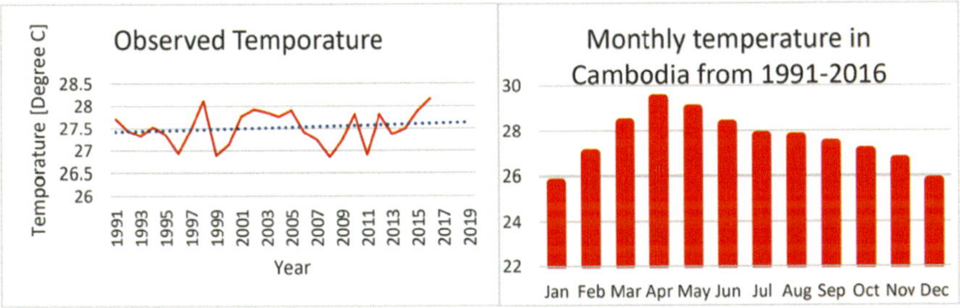

Fig. 3 Observed temperature from 1991 to 2016. *Source* World Bank, FAO Cambodia (2021)

typhoons as it is surrounded by mountain chains, which dissipate a typhoon's force (GSSD 2015). Agricultural production is dependent on the annual flooding and recession of the Tonle Sap Lake and the Mekong River, which brings fertile alluviums to the central plains (RGC 2006).

Observed Temperature

Cambodia's average temperature has increased from 26.85 to 28.16 °C between 1991 and 2016, meaning an overall increase of 1.30 °C, or a rate of 0.05 °C/year. Temperature is relatively uniform across the country, though seasonal variability is significant with a noticeable drop in temperature during the dry season period between Feb–April. Across years, annual average temperatures remain fairly constant, however experience in 2016, where the t-max was around 40–41 °C at daytime and 29–31 °C during the nighttime, indicates that the El Nino effect can significantly influence inter-annual variability (World Bank Group 2021; FAO Cambodia 2021) (Fig. 3).

Observed Rainfall

Available data across years and locations indicates that rainfall patterns have remained relatively constant, however due to wide spatial differences it is hard to define the national mean for precipitation. However, some locations have experienced increased rainfall while the other locations have seen decreases. At higher temporal resolutions, i.e., looking at intra-annual differences and trends, important changes are becoming evident. For example, the rate of rainfall dramatically increased from March, April, and May and continue to increase in June, July, August, and September and slightly decreased from October and November. Despite these increases, break of season (BOS) events (irregular rainfall/dry spell) occurring early during the monsoon season have caused water shortages leading to insufficient access to water for agriculture production and ultimately resulting of crop failures, environmental damage, and health issues (Fig. 4).

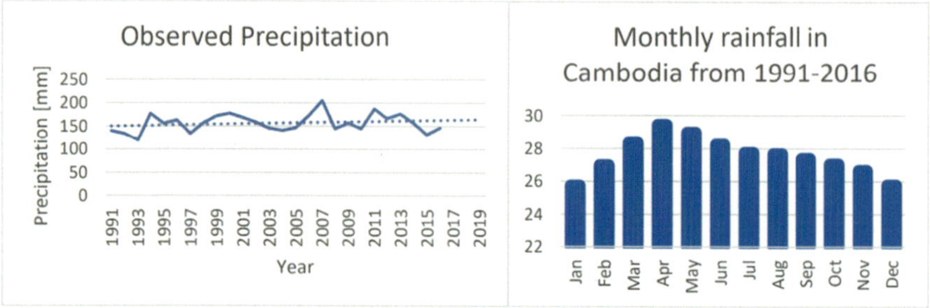

Fig. 4 Observed rainfall in Cambodia from 1991 to 2016. *Source* World Bank, FAO Cambodia (2021)

8 Cambodia's Vulnerable to Climate Change

The Kingdom of Cambodia is vulnerable and increasingly exposed to climate change impacts. In Southeast Asia, the mean annual temperature is expected to increase between 0.8 and 3.2 °C and precipitation between 1 and 8% by 2100 (IPCC 2013). Cambodia has a 70% rural population and the agricultural sector accounts for more than 60% of the employed workforce. With livelihoods largely weather-dependent, a geographic exposure to a range of increasingly erratic hydrometeorological events and low adaptive capacity, the Cambodia is a perfect storm of climate vulnerability (MoE 2021). Cambodia is among the countries most at risk to the impacts of climate change. It ranks 14th in the Global Climate Risk Index (*Global Climate Risk Index 2021—World | Relief Web*) from 2000 to 2019 and 8th in World Risk Index 2016 (http://weltrisikobericht.de/english/). Under future climate conditions, most of Cambodia's agricultural areas will be exposed to higher drought and flood risks (GSSD 2017).

Due to its geography, Cambodia has always been subject to a wide range of hydrometeorological phenomena—flooding, drought, storms, and landslides. Traditionally, the monsoon brings southwest prevailing winds with heavy rains and high humidity from May through October, and a dry season from November through March (MoE 2021) (Fig. 5).

Cambodia exposed to nearly all types of hydrometeorological hazards from floods to droughts, heavy storms, typhoons, and lightning strikes. In addition, the Strategic National Action Plan for Disaster Risk Reduction recognizes disease outbreaks (such as cholera, malaria, and dengue), climate change impacts, fires and hydrological hazards as potential triggers of disasters. However, the geographical characteristics of the country contribute to varying exposure; for example, the floodplains along the Mekong are more exposed to river flooding, whereas the mountain ranges (such as the Cardamom) have a higher risk of localized landslides and flash flooding. Storms and typhoons in Cambodia are often not considered a significant threat as the country is sheltered by mountain ranges which lessen the impacts (UNDRR 2019).

The input–Output approach

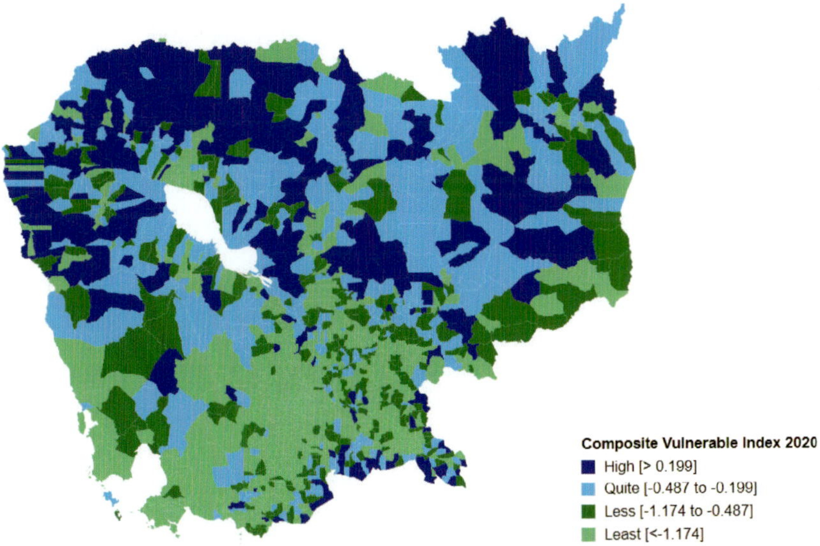

Fig. 5 Climate hazard index (NCSD 2020, CDB 2020)

9 Disaster Impacts and Climate Risk in Cambodia

Flood

Cambodia is a least developing country with a GDP highly dependent on agriculture and it is one among five countries which is located along the Mekong River. In addition, the landscape of the country comprises of rolling plains and lowland. Much of Cambodia's agricultural production (almost rain-fed lowland rice and other perennial crops) is dependent on the Tonle Sap and Mekong river systems' annual flooding and receding, which bring fertile alluvium soils to the central plains. Flooding is caused by two main processes (1) increases in volume of the Mekong and Tonle Sap Rivers resulting in overflow and flooding of the surrounding area and (2) flooding resulting from extreme local rainfall. While these processes are essential for the country's agricultural productivity, they also result in significant negative impacts to infrastructure throughout provinces within the floodplain areas, amount of significant losses of between $100–170 million USD each year (World Bank Group 2011) (Photo 1).

During the monsoon season, Cambodia experiences flash floods normally after heavy rainfall. Some provinces such as Battambang, Kampong Chnang, Kampong Speu, Kampong Thom, Kampot, Kandal, Pursat, and Rattanakiri are usually hit by flash flooding. The second type of flood, the much slower but prolonged flooding, this flood is caused by the overflow of Tonle Sap River and Mekong tributaries, inundating the provinces of Kampong Cham, Kratie, Kandal, Prey Veng, Stung Treng, Svay Rieng, and Takeo (An 2014) (Fig. 6).

Photo 1 Flood in Cambodia. The infrastructure was damaged and there were negative impacts to many people

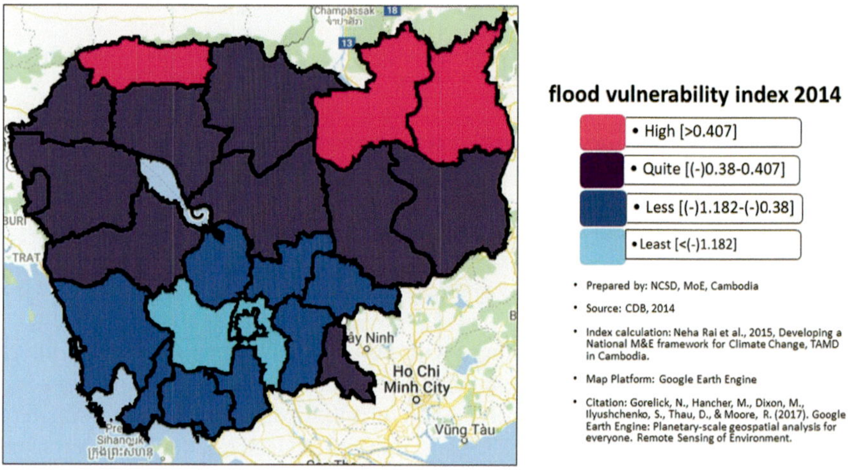

Fig. 6 Provincial Level Flood vulnerability index 2014. *Source* NCSD (2014)

Some examples of extreme flood impacted in Cambodia could be observed. In 2011, floods affected 350,000 households which (over 1.5 million people) and 52,000 households were evacuated. There were 18 provinces in Cambodia were impacted (Fig. 7); which have 4 provinces along Mekong River and Tonle Sap that were considered the worst hit. There were 250 people who were dead and 23 people sustained injuries in the floods in 2011. In addition, there were 431,000 ha of transplanted rice fields that were impacted and 267,000 ha of rice fields were damaged. The infrastructure also seriously impacted from this severe flood, 925 kg of the national, provincial, and urban roads were affected and 360 km of the roads were damaged and there were 1360 schools submerged during the flood. The 2011 floods affected were an estimated loss at 630 million USD (NCDM 2013; An 2014).

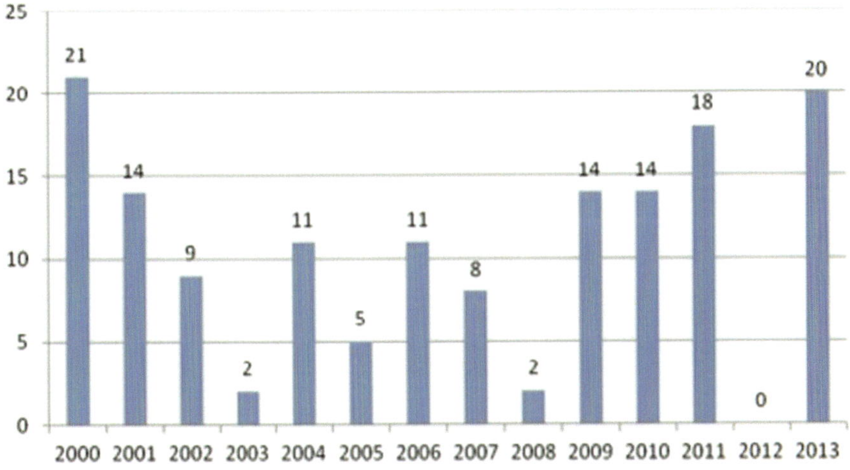

Fig. 7 Number of provinces affected by flood from 2000 to 2013. *Source* An (2014)

In 2013, floods impacted 20 out of 24 provinces (Fig. 7), which had 377,354 households and claimed 168 lives and forced 31,314 households to evacuate themselves to safe places. If compared to floods in 2011, floods in 2013 appear to have been less extensive in scale although in some provinces the impact such as number of evacuated families, damaged crops, and damaged infrastructure were more significant due to a combination of factors that include unexpected gravity of the floods, both in extent and intensity, longer time for waters to recede, repeated floods and flash floods, limitation of preparedness undertaken in advance, and the limitation of early warning (An 2014) (Fig. 8).

The input–output approach (Fig. 9)

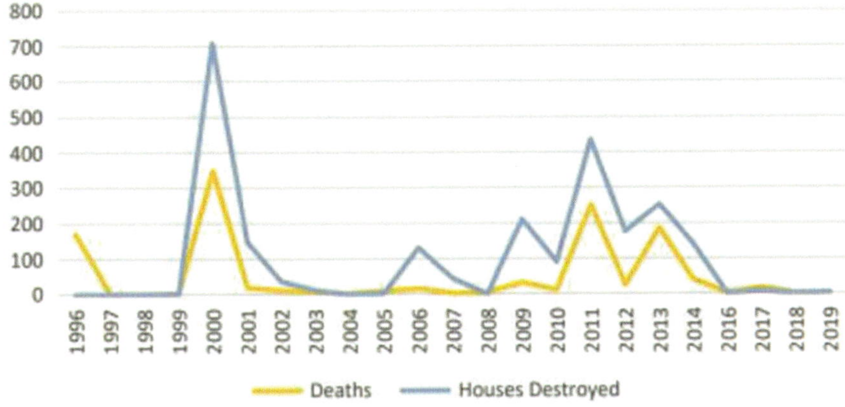

Fig. 8 Flood impacts. *Source* MoE (2021)

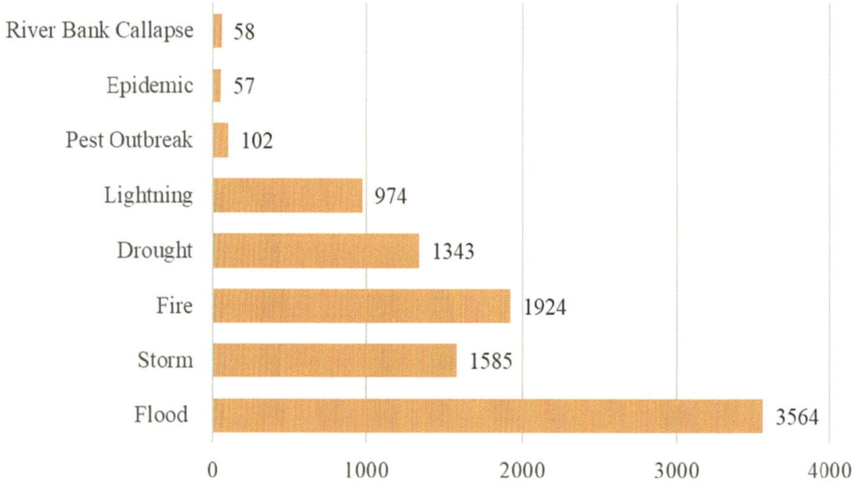

Fig. 9 Disaster occurrence between 1993 and 2013. *Source* UNDRR (2019)

Storm Events

Tropical cyclones are the most costly meteorological disasters which affected East Asia and the Pacific with, on average, 27 tropical cyclones affecting some parts of the region each year. Some typhoons and tropical depressions that reached Indochina do not weaken over the land and produce torrential rainfall and extensive flooding in Cambodia. Typhoon becomes most damaging when it hits during the flooding season (September–October) as it causes heavy precipitation events (An 2014). Cambodia experiences considerable threat and impacts from typhoon events. From 2000 to 2019, almost 30,000 homes were damaged and 113,000 ha of crops destroyed. The following figure indicates the extent of impact (on agriculture) of the three main types of natural disaster in Cambodia between 1996 and 2019 (FAO Cambodia 2021).

The input–output approach (Fig. 10)

The evident showed that on September 29, 2009, Cambodia was affected by Typhoon Ketsana, in which14 out of 25 provinces were hit by the typhoon, and it affected 180,000 households, 43 people were dead and injured 67 people (NCDM, An 2014). Totally, damages were estimated at US$131,996,415 (NCDM 2013) (Photo 2; Fig. 11).

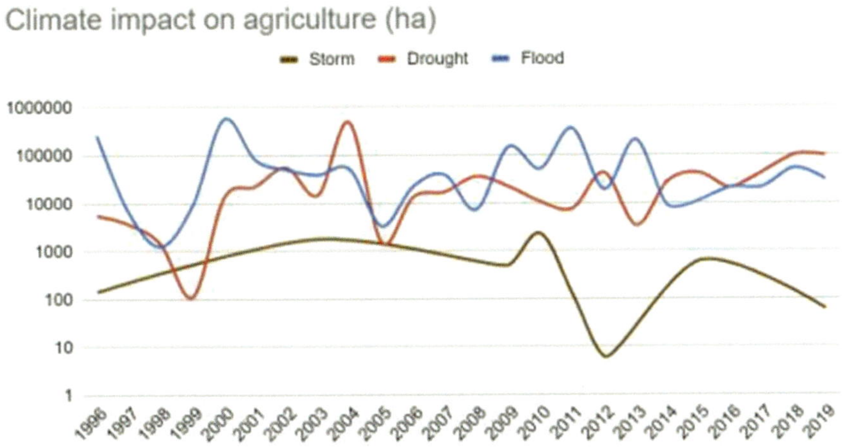

Fig. 10 Climate-related disaster impacts to agriculture production. *Source* NCDM and FAO Cambodia (2021)

Photo 2 Houses were damaged by storm in Cambodia

10 Kratie Province—Local Context

10.1 Flood Situation

The province of Kratie covers an area of approximately 11,094 km^2, with an estimated population of 374,755 (approx. 50% female) based on the national census of 2013. About 70% of the population live along the Mekong River, in which the majority depend on as the key source of livelihoods; and approx. 30% of the residents live in remote mountainous areas with sparse population density. Fisheries and agriculture are the key sources of livelihoods for more than 80% of the population, especially those who live in remote areas (Ngin et al. 2020). The province of Kratie

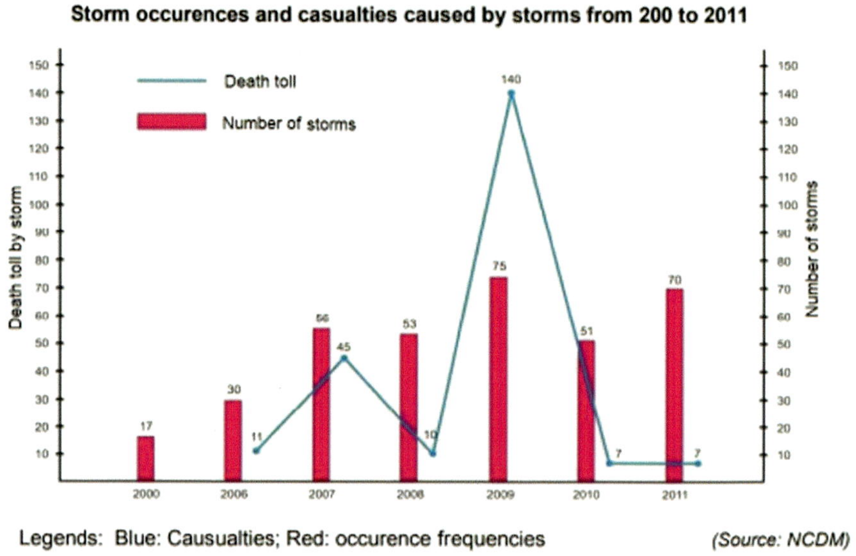

Fig. 11 Storm occurrences and casualties caused by storms from 2000 to 2011

is one of the most vulnerable to climate change in Cambodia. Climate change affects livelihoods in many ways in Kratie. Different disasters have struck Kratie Province, such as floods, droughts, storms, lightning, fire, pest outbreak, riverbank collapse, and epidemics (Ngin et al. 2020). Among those, floods are considered the most devastating hazard, accounting for most destruction and casualties, with droughts (NCDM 2018a). From 1996 to 2017, the floods in Kratie Province adversely affected 558,984 people, damaging 1715 houses and 164,896.50 ha of crops (Neef and Ngin 2019), and storms ranked the second- and third-most destructive disasters in the province (NCDM 2018a, b; Ngin et al. 2020) (Figs. 12, 13; Photo 3).

Based on Fig. 14 floods in Kratie occur from July to September. The mechanism to cope with it should be well-prepared during that period. Most of the people in Kratie's livelihoods focus on farming and fishing. Besides floods, Kratie also faces drought. Drought normally happens in the period of January to April and during November to December.

10.2 Districts Vulnerable to Flood

Preaek Prasap and Chetr Borei consider the flood-prone and drought in Kratie Province. Along Mekong River, people's community experience of flood in target area. The vulnerable of the districts were defined by joining with stakeholder during the workshop in Kratie Province. The difference of stakeholder includes an official from provincial department of agriculture, forestry and fisheries Kratie Province,

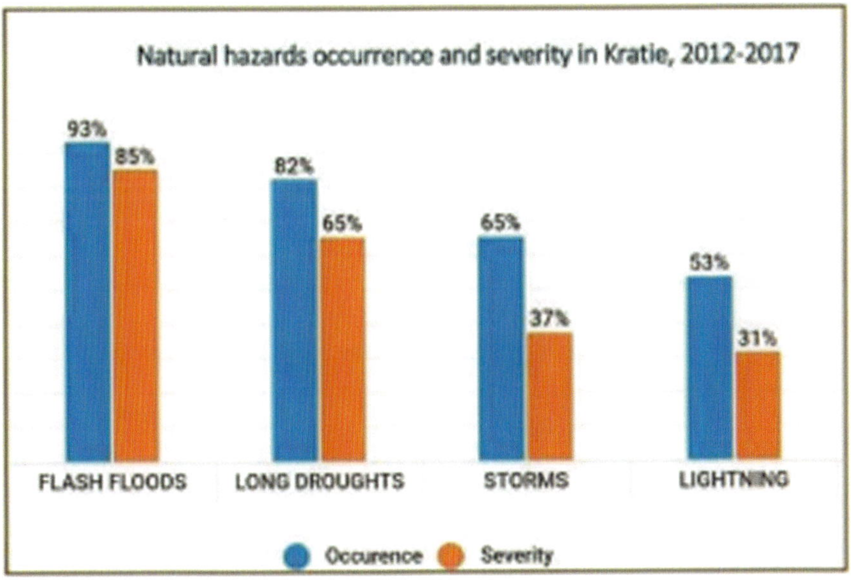

Fig. 12 Severe natural hazards occurrence frequency in Kratie. *Source* Neef and Ngin (2019)

Photo 3 Flood in 2018 in Kratie, land slide along Mekong River. The vegetative strips along the river bank should be made to minimize risk *Source* Provincial department of agriculture, forestry and fishiereis (PDAFF)

official from provincial department of environment, commune extension workers, etc. (Fig. 15).

10.3 Types of Floods and Flood Duration and Intensity

Accordingly, to the interview with official of provincial department of agriculture, forestry and fisheries, local authorities and people in the Kratie province. There are

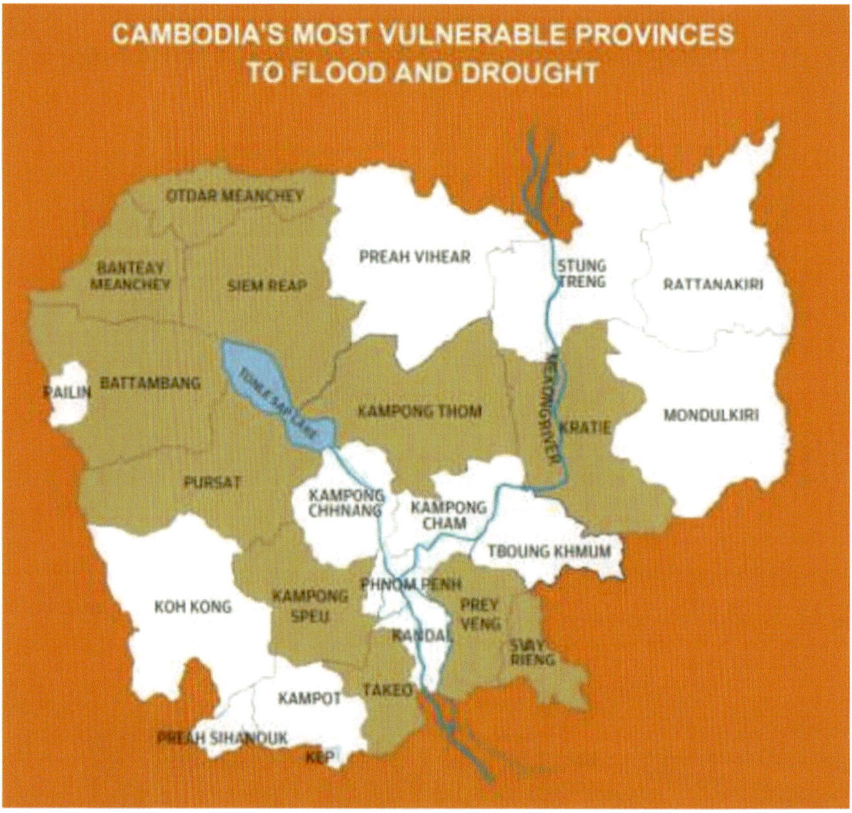

Fig. 13 Kratie vulnerable to flood and drought

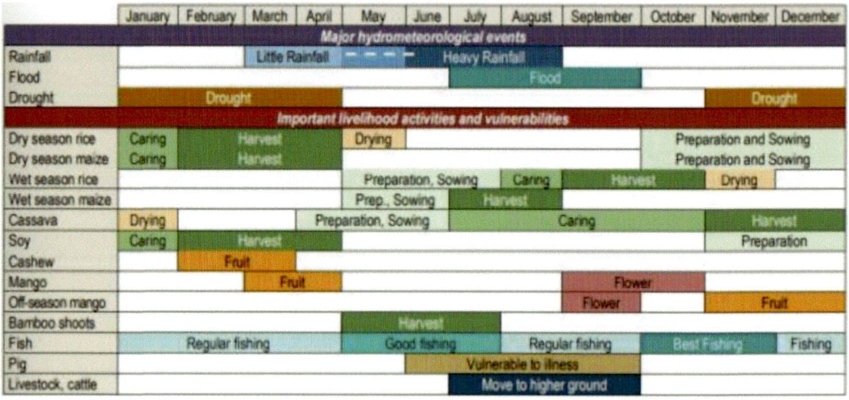

Fig. 14 Seasonal calendar of flood in Kratie. *Source* Neef and Ngin (2019)

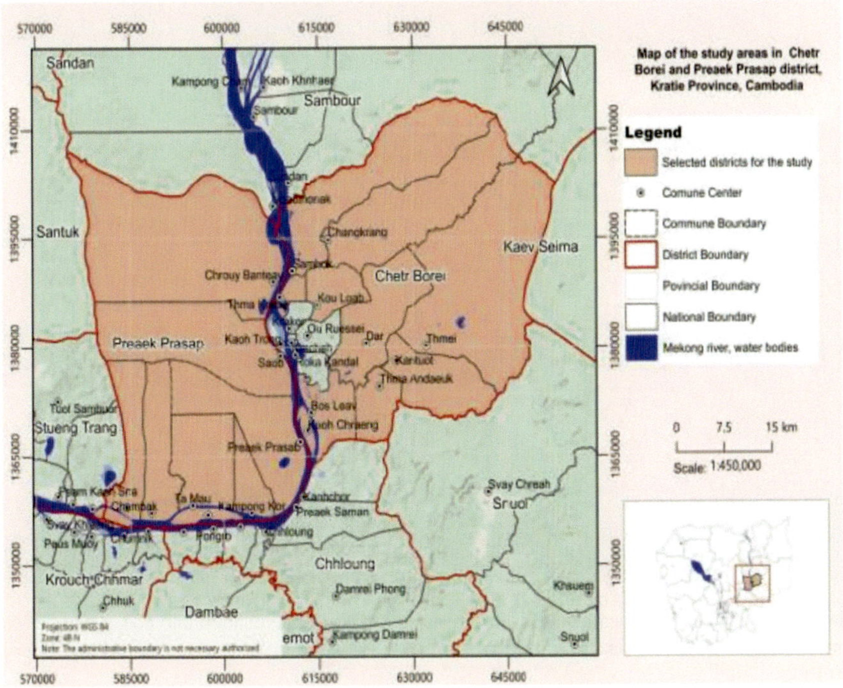

Fig. 15 Districts considered flood and drought prone in Kratie. *Source* Navin Chea

two type of flood in Kratie include flash flood which cause by rainfall and flood cause by overflow of the Mekong River.

Flash Flood: With the intensity of the rainfall, flash flood in Kratie is in especially Chetre Borie and Preak Prasap district. It normally lasts from 1 to 5 days. However, it happened frequently which also affects to livelihood of the farmers and damages crop. It occurs every year. In rainy season, it happened in average 5 times, which once time lasts maximum 5 days.

Riverine Flood: Caused by the overflow of the Mekong River. In general, the duration of the flood is from 20 to 50 days. For the last 5 years from 2015 till 2020, Mekong flood occurred three times based on interview with local authority and people.

Cumulative rainfall in the upper catchments throughout the rainy season causes a slow but steady rise in water levels lasting for several days. This can be aggravated by two factors. First, when this combines with heavy rains around the Tonle Sap Lake, which affects the provinces around the lake and the southern provinces. Second, the most severe floods occur when heavy rains coincide with the arrival of tropical depressions and storms. Mekong River floods are common occurrences in the provinces of Stung Treng, Kratie, Kampong Cham, Prey Veng, Svay Rieng, and Kandal (RGC 2014) (Photo 4, 5 and 5).

Photo 4 Interview with local authority in Preak Prasap on flood situation and how people in the community deal with the issues, as well as how the authority plays role to support and inform farmers

10.4 Land Use Type in Vulnerable Districts

Difference land use type showed in Preak Prasap and Chetr Borei Districts. Crop land and paddy rice field are dominant in target's areas. This is because people in the target's areas rely mainly on the agricultural activities for their livelihood. The crop land and paddy fields observe to locate along the Mekong River or low land areas (Fig. 17).

11 Adaptation to Flood in Agriculture

11.1 Adaptation for Cambodia's Agriculture at the National Level

Adaptation for Cambodia's agriculture sector includes:

- Increased support for and reach of existing agricultural research and extension services;

Photo 5 Participatory methods with farmers, key persons to identify climate risk in the target area, and how farmers cope with those changing mainly for farming

- Implementation of improved water use (water harvesting, small-scale irrigation, etc.) in drought-prone areas to alleviate rain shortages that cause crop failure. Improving farmers' knowledge and access to weather information in carrying out agricultural activities;
- Livelihood diversification to mitigate the reliance on agriculture as the sole livelihood source;
- Restoring and rehabilitating flood protection infrastructures, such as dikes.

11.2 Adaptation for Cambodia's Agriculture at the Sub-National Level

Based on the interviewing with officials at Provincial Department of Agriculture, Forestry and Fisheries (PDAFF), Kratie Province. The result found that the strategy to support farmers better adapt to climate change, mainly flood as below:

Photo 6 Interview with farmers on the ground to provide perception of climate change and adaptation measure in Preak Prasap District

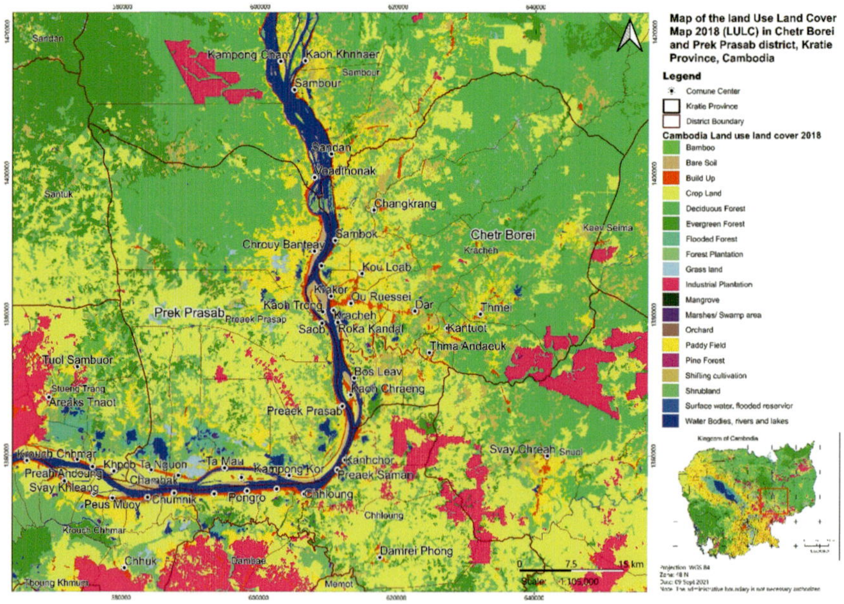

Fig. 16 Land use cover in 2018 in target districts in Kratie. *Source* Navin Chea

- Provide technical support to farmers on select seed variety that is tolerant to flood. Meeting with relevant stakeholders include different NGOs, local authority to evacuate farmers and animals in advance to the safer place.
- Provide seed support after water receding or when farmers' crop damage. Early information to farmers to prepare in advance and follow the information release by meteorology.
- Provide training to official at the commune or village level to collect data of damage and lost.
- Improving the irrigation structure includes to prevent and store water.
- Work and cooperate closely with stakeholders.

11.3 Adaptation to Floods by Farmers at the Community Level

Based on the result from the interviewing and participatory workshop, the adaptation by farmers includes:

- Diversity of crops
- Select seed variety tolerant to flood
- Farmers adjust the crop calendar to escape
- Cultivate short-term varieties of rice
- Improve water management.
- Wet season rice is rainfed and grown at higher elevations to reduce the risk of flood damage, while dry season rice is grown in lower elevation areas as floods recede, and may be supplemented by irrigation. Important tree crops include cashew and mango, with income from higher-priced off-season mango available to some communities.

12 Information Gaps for Writing

- In Kratie Province, there is no available data source that could be found on the historical data of flood disaster which includes the loss and damage.
- Lack of historical rainfall data in the past to be available free access. If required data, the official letter must be sent to meteorology at provincial level.
- There is limitation of database of disaster flood history in Cambodia. At district level, the information of flood was not well-documented or recorded and limitation of research study on flood in target province.

References

An LH (2014) Disaster management country report of Cambodia disaster management

FAO Cambodia (2021) Strengthening capacity in the agricultural and land use sectors for enhanced transparency in implementation and monitoring of Cambodia's Nationally Determined Contribution (NDC): In: Draft report. Food and agriculture organization of the United Nations, Phnom Penh, Cambodia

FAO in Cambodia (2020) The agriculture sector master plan 2030 (ASMP 2030), a promising plan to modernize agriculture in Cambodia: workshop holds for inputs to complete the plan. Retrieved July 23, 2020, from FAO website: fao.org/cambodia/news/rss/detail-events/en/c/1262820/

GSSD (2015) GSSD 2015. Cambodia's second national communication under the United Nations framework convention on climate change. General secretariat, national council for sustainable development/ministry of environment, Kingdom of Cambodia, Phnom Penh. Phnom Penh

GSSD (2017) National adaptation plan process in Cambodia. Retrieved from https://www4.unfccc.int/sites/NAPC/Documents/Parties/CambodiaNAPprocessdocument.pdf

IPCC (2013) The physical science basis. Contribution of working group I to the Fifth assessment report of the intergovernmental panel on climate change. Intergovernmental Panel on Climate Change (IPCC), Cambridge, UK; New York

MoE (2021) Fourth state of environment report: Ministry of Environment of Cambodia, Phnom Penh

NCDM (2013) Climate change strategic plan for disaster management sector. National Committee for Disaster Management, Phnom Penh, Cambodia

NCDM (2018a) Disaster loss database (CamDi). Retrieved from http://camdi.ncdm.gov.kh/DesInventar/%0Aprofiletab.jsp?countrycode=kh855&continue=y

NCDM (2018b) Disaster profile of Cambodia (2005–2017). Consolidated by: H.E. Ma Norith, Deputy Secretrary-General (NCDM)

Neef A, Ngin C (2019) Climate change adaptation in post-disaster recovery-policy brief2 climate change adaptation: inside the flood-prone communities in Kratie Province, Cambodia. ResearchGate. Retrieved from https://www.researchgate.net/publication/331977765

Ngin C, Chhom C, Neef A (2020) Climate change impacts and disaster resilience among micro businesses in the tourism and hospitality sector : the case of Kratie, Cambodia. Environmental Res 186, 109557. Retrieved from https://doi.org/10.1016/j.envres.2020.109557

NIS (2019) General population census of the kingdom of Cambodia 2019. Ministry of Planning, Phnom Penh

RGC (2006) National adaptation programme of action to climate change (NAPA). Ministry of Environment of Cambodia, Phnom Penh

RGC (2014) Post-flood early recovery need assessment report. Royal Government of Cambodia

UNDRR (2019) Disaster risk reduction in Cambodia. Status report 2019. Bangkok, Thailand, United Nations office for disaster risk reduction (UNDRR), Regional Office for Asia and the Pacific

World Bank Group (2011) Climate risks and adaptation country profile. https://climateknowledgeportal.worldbank.org/sites/default/files/2018-10/wb_gfdrr_climate_change_country_profile_for_KHM.pdf

World Bank Group (2021) Climate change knowledge portal: climate change by sector of Cambodia. Retrieved from https://climateknowledgeportal.worldbank.org/country/cambodia/climate-sector

Flood Risk Management in Phnom Penh, Cambodia

Keo Hak

Abstract Cambodia is exposed to nearly all types of hydro-meteorological hazards from floods to droughts, heavy storms, typhoons, and lightning strikes. Since 1970, the country has experienced an economic loss of US $170 million per year due to disasters, and in 2011 alone, disasters caused a 4.3% loss to the country's GDP. The author studies the Phnom Penh floodplain and Mekong Delta and explores the features and importance of floodplain resilience in terms of conveying and attenuating flood flows. The author also undertakes a flood risk assessment and identifies critical elements of flood risk management that are contextualized to local conditions. After a systematic assessment of the national and sub-national enabling environment, the case study author recommends reform of Disaster Risk Management and calls on the government to provide greater funding focusing on communities that are most vulnerable to climate-related disasters.

Keywords Cambodia and flood risk management · Phnom Penh and flood disaster · Climate risk assessment · Climate Risk Management · Disaster Risk Management · National disaster planning · Participatory approaches and resilience · Vulnerability · Adaptive capacity · Hazard reduction measures

1 Introduction

1.1 General Information

Reduction of poverty is a primary national development priority of the Royal Government of Cambodia (RGC) together with sustained and inclusive growth. While the country achieved remarkable human development progress recently, 30% of the population still lives in poverty (National Institute of Statistics 2019).

K. Hak (✉)
Provincial Department of Rural Development Svay Rieng (PDRD-SVR), Ministry of Rural Development (MRD), Svay Rieng, Cambodia
e-mail: keohaksvr168@gmail.com

K. Kim and C. Atkin (eds.), *Cross-Cultural Perspectives on Climate Change Adaptation*, Cities and Nature, https://doi.org/10.1007/978-3-031-50365-8_14

One of the primary causes of persistent poverty is the increasing risk of disaster to life, livelihoods, and economic and environmental assets. Agriculture, employing the majority of the workforce 58% and contributing more than a third to GDP, is regularly affected by drought, floods, and storms. Disasters pose a significant threat to other key development sectors such as infrastructure, industry, tourism, housing, education, health, and environment, which are critical for the country's sustainable development. Since 1970s, the country has experienced an economic loss of US $170 million per year due to disasters, and in 2011 alone, disasters caused a 4.3% loss to the country's GDP. One-third of Cambodia's communes are prone to various types of disasters (RGC 2013).

1.2 Geography and Population

Phnom Penh is in the south-central region of Cambodia and is fully surrounded by Kandal Province. The municipality is on the banks of the Tonlé Sap, Mekong, and Bassac Rivers. These rivers provide freshwater and other natural resources to the city. Phnom Penh and the surrounding areas consist of a typical floodplain area for Cambodia. Although Phnom Penh is 11.89 m above the river, monsoon season flooding is a problem, and the river sometimes overflows its banks. The city, at 11.55° N 104.91667° E (11° 33′ North, 104° 55′ East), covers an area of 678.46 km^2, with some 11,401 ha (28,172 acres) in the municipality and 26,106 ha of roads. The agricultural land in the municipality amounts to 34.685 km^2 with some 1.476 km^2 under irrigation.

Phnom Penh has the highest population among total of the 25 Provinces in Cambodia. Based on data from the National Institutes of Statistics indicated that Phnom Penh has the highest population at 2,129,371 (13.9%), there are 1,039,192 males and 1,090,179 females with a total of 399,203 housholds (National Institute of Statistics 2019).

1.3 Administration

Phnom Penh is a municipality of area 678.46 square kilometers with a government status equal to that of Cambodian provinces. The municipality is divided into 14 administrative divisions called khans (districts). The districts are subdivided into 105 communes and further subdivided into 953 villages. All khans are under the governance of the Phnom Penh Municipality. Dangkao, Meanchey, Porsenchey, Sen Sok, and Russei Keo are considered the outskirts of the city (Fig. 14.1).

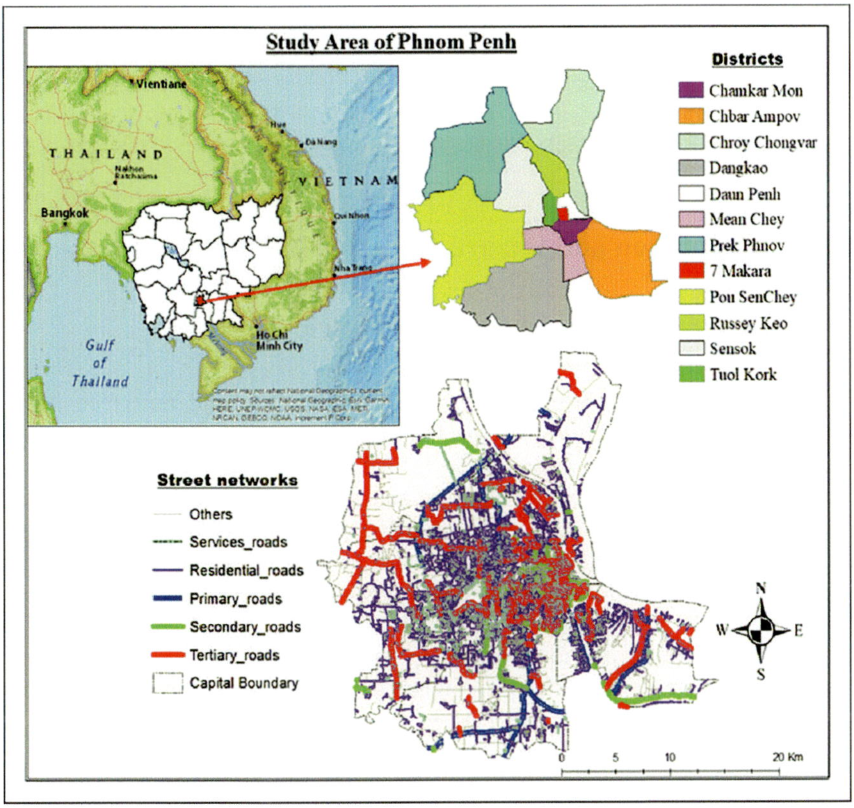

Fig. 14.1 Map of Phnom Penh capital city

2 Research Objectives and Methodology

The purpose of the case study is to (1) conduct a flood situation and hazard analysis focused on Phnom Penh and (2) identify an action plan for flood risk management systems.

2.1 Data Collection and Information

To achieve my study activities, I have conducted my research by desk review from some secondary data and documents that I selected from reports, research papers including some official and non-official papers, and books, such as

- National Action Plan for Disaster Risk Reduction (NAP-DRR) 2014–2018—General Population Census of the Kingdom of Cambodia 2019

- Community Self-Reliance and Flood Risk Reduction
- Disaster Risk Reduction in Cambodia Status Report (July 2019)
- Disaster Risk Management in Cambodia
- Integrated Flood Risk Management Plan for the West Bassac area in Cambodia—Country Report of Cambodia Disaster Management.

3 Flood Situation and Hazard Impacted in Phnom Penh

3.1 Disaster Hazards and Climate Change

Cambodia is exposed to nearly all types of hydro-meteorological hazards from floods to droughts, heavy storms, typhoons, and lightning strikes. Additionally, the Strategic National Action Plan for Disaster Risk Reduction recognizes disease outbreaks (such as cholera, malaria, and dengue), climate change impacts, fires, and technological hazards as potential triggers of disasters (NCDM & Ministry of Planning 2009). However, the geographical characteristics of the country contribute to varying exposure; for example, the floodplains along the Mekong are more exposed to river flooding, whereas the mountain ranges (such as the Cardamom) have a higher risk of localized landslides and flash flooding. Storms and typhoons in Phnom Penh, Cambodia, are often not considered a significant threat as the city and country are sheltered by mountain ranges which lessen the impacts (MRC 2009).

Disasters are inseparable from the economic, social, and environmental features of Phnom Penh and Cambodia country. The country experiences almost all types of hydro-meteorological hazards such as floods, drought, heavy storms (or typhoons), fire incidents, and epidemics. Most geographical regions of the country (i.e., Riverine Central Plains, coastal ecosystems, and Dangrek mountain range in the north and Cardamom mountains in the southwest) are exposed to one or more of these hazards. Additionally, climate change is expected to increase the frequency, intensity, and severity of these extreme natural events. As the majority of Cambodians are farmers and their livelihoods mainly depend upon subsistence agriculture, the vulnerability of people living in rural areas is very high and may continue to rise, requiring improved preparedness and planning (Heng 2014).

3.2 Flood in Situation in Phnom Penh

Phnom Penh, situated on the banks of the Tonle Sap, Mekong, and Bassac rivers, is already highly vulnerable to floods, particularly in the rainy season from June to October. It has more frequency and a strong impact every year. Phnom Penh is one of the capital cities located on the three main rivers in Cambodia Mekong River, Tonle Bassac, and Tonle Sap Lake. Its landscape consists of rolling plains and lowlands. During the monsoon season, Phnom Penh experiences flash floods usually after heavy rainfall, as well as the provinces of Battambang, Kampong Chnang, Kampong Speu, Kampong Thom, Kampot, Kandal, Pursat, and Rattanakiri, are regularly hit by flash flooding. The second type of flood, the much slower but prolonged flooding, is caused by the overflow of the Tonle Sap River and Mekong tributaries, inundating the provinces of Kampong Cham, Kratie, Kandal, Prey Veng, Stung Treng, Svay Rieng, and Takeo (Heng 2014) (Photo 14.1).

3.3 *Floods Caused by Typhoons and Rainfall in Phnom Penh*

Tropical cyclones are the most meteorological disasters affecting East Asia and the Pacific with, on average, 27 tropical cyclones affecting some parts of the region each year (WB 2013). Some typhoons and tropical depressions that reach Indochina do not weaken over the land and produce torrential rainfall and extensive flooding in Phnom Penh, Cambodia. Typhoon becomes most damaging when it hits during the flooding season (September–October) as it causes heavy precipitation events.

Flood events in Phnom Penh are twofold almost daily rainy season flood events and episodic larger-scale flood plain events. During the rainy season (May–October), monsoon rains fill low-lying streets, some to nearly 1.5 m deep. The near-daily rain floods during the rainy season reframe the experience of inhabiting the city, altering its landscape and blurring the distinction between water and land. Roads become waterways and sidewalks disappear beneath the muddy waters. Curbs and tree roots are hidden from Rising view, hindering walking and driving. Businesses unfurl overhangs, open umbrellas, and hang tarps, expanding available dry space. Traffic slows to a near stop as cars, motors, and bicycles navigate the water and intermittently stall out or dip into deep unseen potholes.

The population anticipates the rains and has adapted to the accompanying flooding and its perceived cleansing effects. Nonetheless, the floods disrupt the flow of daily business and activity. Additionally, flooded streets carry potential disease as the stormwater mixes with human waste, and street drains are blocked by municipal trash, slowing drainage and posing a possible public health threat.

As for larger-scale flood events, Phnom Penh was founded in the alluvial plain of the Mekong River, which varies upward of 12 m in depth between the dry and wet seasons. The most devastating flood risk comes from the Mekong River cresting over its natural berm into the city. The volume of water produced by a Mekong flood could take weeks or even months to recede, evaporate, or penetrate the ground.

The factors contributing to the potential for increased flooding in Phnom Penh are deforestation, landfilling lake and rivers, the unknown impacts of climate change, overbuilding in catchment areas, the damming and diversion of natural waterways, and the infill of canals and lakes, combined with no formally accepted or followed master plan (Doyle 2012) (Fig. 14.2).

Photo 14.1 Floods impact in Phnom Penh

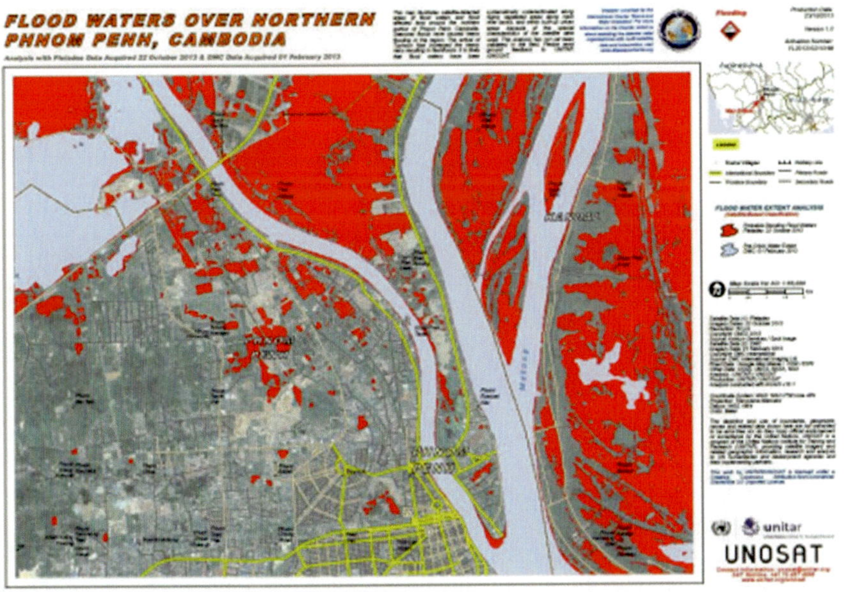

Fig. 14.2 Floods vulnerability area in Phnom Penh. *Source* https://reliefweb.int/map/cambodia/flood-waters-over-northern-phnom-penh-cambodia-23-oct-2013

3.4 *Loss and Damage*

Phnom Penh capital city normally safely situated above the water level; a state of emergency was declared by the authorities because the capital could be inundated in days (BBC 2000). Fortunately, Phnom Penh remained for its greatest part out of the flooded area. In 2000 and 2011, Phnom Penh was affected by floods that caused loss and damage that killed people 253 deaths, destroyed 680 households, affected 358 schools, damaged roads 322 km (Imhoff 2016) and caused GDP estimate loss and damage of about US $52 to US $85 million each year (Photo 14.2).

4 Flood Management Design Process

The study of the Phnom Penh floodplain and Mekong Delta highlighted the role of the floodplain in terms of conveying and attenuating flood flows. Floodplain storage is seen to be critical as a naturally functioning part of the river system that helps to reduce flood levels and downstream effects. Accordingly, changes in the floodplain as well as climate change will have significant implications leading to higher flood water levels, longer duration of flooding, and increased flows.

Photo 14.2 Loss and damage from floods impact

That storage and conveyance features of the floodplain have a direct effect on flood levels, extent, duration, and timing of flooding, a zone approach has been proposed that considers flood behavior based on local natural influences and requirements. The study divided the floodplain into zones as adopted for WUP studies (WUP-JICA) and was modified slightly to provide a more comprehensive coverage of the full extent of the influence of flooding and flood considerations in provinces in Cambodia (KCC 2017).

***Step 1*: Determine Design Flood Requirements**

Identify the flood zone and select design flood event criteria considering the planned development type. It is important to note that the 2000 flood is considered to represent the design flood for the 1:200-year flood event. The design flood event criteria consider risk, vulnerability, and climate change aspects concerning the type of development planned.

***Step 2*: Determine the Level of Flood Hazard**

- Consult the appropriate return period flood hazard map (NFFC 2017) to identify the location of the planned development and determine the expected level of the flood hazard.
- Assess the impact of the planned development's impact on the flood level due to the loss of floodplain storage, flood conveyance, and trans-border flooding considering the flood hazard as defined by the 1:200-year flood event.
- Major works such as roads, levees, irrigation embankments, and water management structures will require the configuration and calibration of a suitable flood hydrodynamic model to assess the effect of the planned development on local flood levels for the required design criteria.
- Propose mitigating measures to ensure there is no net loss of flood storage and flood conveyance.
- Advise the National Mekong Committee of any trans-border flooding concerns.

***Step 3*: Establish Flood Protection Design Level**

- Select the design level for flood protection works considering the maximum water level projected to the HATIEN MSL.
- For semi-critical and non-critical works add a freeboard of 0.3 m to accommodate for waves, localized influences, and uncertainty in the reference level.
- For critical works add a freeboard of 0.6 m to accommodate for waves, localized influences, and uncertainty in the reference level.

5 The Action Plan for Flood Risk Management

5.1 Laws, Policies, and Plans on Disaster Management

See Table 14.1.

Table 14.1 National disaster and climate risk reduction policies, plans, and legislation

Implementation	Policy/Plan	Scope	Purpose
Royal Government of Cambodia	Sub-decree No. 54 ANKR-BK	National	Mandates the establishment of the National Committee for Disaster Management
National Committee for Disaster Management	Sub-decree No. 30 ANKR-BK	National, Provincial	Supports the decree No. 54 by requiring disaster management committees to be established below the national level
National Committee for Disaster Management, Other Relevant Parties	National Action Plan for Disaster Risk Reduction (2008–2013)	National, Provincial, Districts	Provides focus on vulnerability and poverty reduction, a road map for comprehensive DRR
National Committee for Disaster Management, other Relevant Parties	National Action Plan for Disaster Risk Reduction (NAP-DRR) (2014–2018)	National, Provincial, Districts	Articulates the country's DRR strategic focuses and desired outcomes toward resilience building
National Committee for Disaster Management	Law on Disaster Management (2015)	National	Formalizes the role of the NCDM as the lead administrative organ of disaster management activities. Identifies roles and responsibilities of other institutions as well

5.2 Disaster Management Partners

Phnom Penh is one the 25 provinces of Cambodia that needs the support such as financial and technical from all local NGOs, internal NGOs, and development partners from the UN and countries from over the world. Cambodia has a vast network of UN agencies, and international and national NGOs working on humanitarian, recovery, and development programs.

5.3 Disaster Management System

The National Committee for Disaster Management (NCDM) of the Kingdom of Cambodia was established in 1995 and followed by a Sub-Decree No. 30 ANKR.BK, dated April 09, 2002, to serve as the coordinating body for all disaster management-related activities of the Royal Government of Cambodia. NCDM had established the structures of mechanisms to respond to the disaster from national to local levels. It is also recognized as the primary agency for managing emergencies and building the resilience of communities to natural disasters.

The other key mandates of NCDM include implementing prevention and mitigation strategies to reduce the impact of disasters on life, livelihoods, and infrastructure, and issuing and recommending guidelines and policies for legislative and budgetary support for emergency and disaster management. As a coordinating and policy-recommending body, NCDM's goal is to build an effective system for disaster prevention, effective disaster response, and rapid disaster recovery (NCDM 2014).

5.4 Roles and Responsibilities

To ensure the continuity and functioning of the national and sub-national committees for disaster management administration:

- To conduct research into flood, drought, storm, wildfire, epidemics-prone areas, and other hazards in developing preparedness and emergency response plans.
- To instruct the provincial, municipal, and district committees for disaster management and relief communities about work and technical skills that are the basis for collection of disaster data for damage and need assessment and prepare rehabilitation and reconstruction programs of damaged infrastructure in coordination with institutions UN agencies, IOs, and NGOs concerned.
- To formulate a technical skill training program for officials who serve disaster management functions in provinces, municipalities, districts, precincts, and relief communities within the framework of training in and out of the country.
- To coordinate work with ministries/institutions concerned, local authorities, UN agencies, IOs, and NGOs to evacuate vulnerable people to haven and to provide them with security, public education, emergency response, and other programs (Fig. 14.3).

6 SWOT Analysis

6.1 Strengths

- Develop a strategy for disaster management (floods, drought, and, storms)
- Establish a preparedness action plan for flooding response and recovery
- Floods can provide enough water resources for agricultural production
- Floods can import natural fertilizer for agriculture
- Floods provide fish habitats and biodiversity resources.

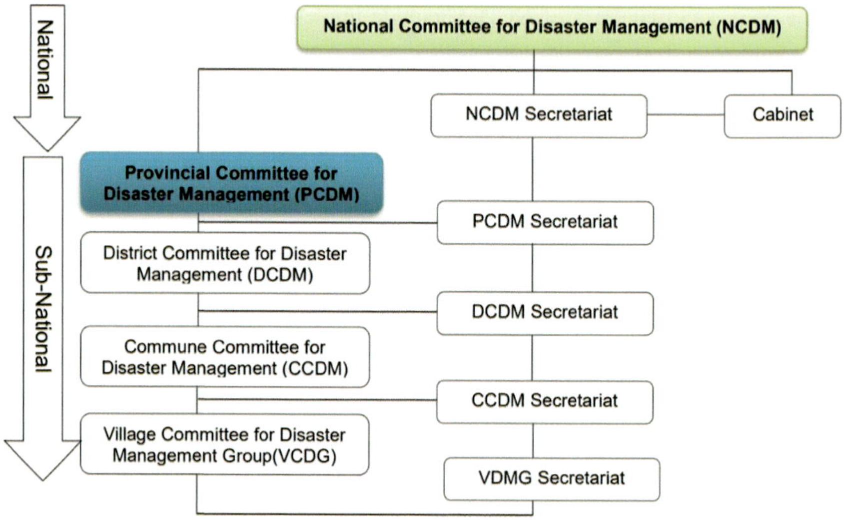

Fig. 14.3 Structure of NCDM mechanism

6.2 Weakness

- Low strategy of Disaster Risk Management caused lack of preparedness and recovery—loss and damaged properties kill people and public infrastructures
- Destroy households and agricultural productions
- Shortage of safe water resources for drinking during floods
- Waterborne diseases
- Caused landslide and homeless
- Floods Impact to human health, education, environment, and economics.

6.3 Opportunity

- Good strategy and activities for preparedness and recovery after disasters
- Mobilize financial support and resources
- Forecast and analyze the scenario for future flooding risk management
- Develop early warning system and social security funds.

6.4 Threatens

- Lack of finances to implement the strategy and preparedness activities –
- Floods impact human life and ecosystems

- Floods can damage households, roads, and other national properties – Floods impact to national development and economics.

7 Lessons Learned and Recommendations

7.1 *Lessons*

After I conducted a study topic on Flood Risk Management in Phnom Penh , Cambodia, I have learned a lot with good and weak points:

- Climate change become more impact which is caused by human activity and natural activity.
- Climate change caused disasters more serious such as floods, droughts, storms, and heat waves to increase temperatures.
- Disaster caused more frequent floods, droughts, and storms that strongly impacted to human life and the environment.
- Floods have more frequency and impact in Phnom Penh, and they have damage and loss to government and people. For example, floods killed many people and damaged a lot of households, roads, schools, and other infrastructure.
- To respond to the disasters, the government has established the National Committee for Disaster Management (NCDM) for national and sub-national. Also created laws, policies, strategies, and action plans for responding with short and long term.
- UN agencies, internal organizations, local NOGs, and development partners contributed technical and financial to respond to disaster impact.

7.2 *Recommendations*

The following are two proposed disaster management recommendations that can reduce the negative impacts of climate-related disasters on communities:

- Revise the notational and sub-national of Disaster Committee Management and the government should increase the national budget to take action to respond to disaster impact. In this case, the government can update the policy to prioritize activity and identify the vulnerability area and group. For example, sub-national governments may have enough finances to help people who are impacted by disasters such as floods, droughts, and storms by building new households as well as providing food. Also, the capacity to restore after disasters impacts to repair of roads and public infrastructures, and support farmers' agriculture production after damage from flood impacted.
- UN agencies, internal organizations (IOs), local NOGs, and development partners can be involved to collaborate with the government to take action and implement

action plans to reduce disaster impact. In addition to responding and reducing the effects of disasters not only the government but also UN agencies, IOs, NGOs, and development partners should cooperate to implement action plans and support technology and finance in local areas.

References

Doyle SE (2012) City of water: architecture, urbanism and the floods of Phnom Penh. Nakhara: J Environ Design Plan 8(0):18–19. http://www.aj.arch.chula.ac.th/en/index.php/CU/article/view/120%5Cninternalpdf://171/120.html

Imhoff R (2016) Analyzing future flood risks in Cambodia. Deltares Enabl Delta Life 64

KCC, N. C. T. (2017) Climate resilient design guidelines for structural flood and drought control measures. Solid State Ionics 2(1). http://linkinghub.elsevier.com/retrieve/pii/S0167273817305726%0Ahttp://dx.doi.org/ 10.1038/s41467-017-01772-1%0Ahttp://www.ing.unitn.it/~luttero/laboratoriomateriali/RietveldRefinements.pdf% 0Ahttp://www.intechopen.com/books/spectroscopic-analyses-developme

Leng Heng A (2014) Country report of Cambodia disaster management (Issue February)

MRC (2009) Integrated flood risk management plan for the West Bassac area in Cambodia. Mekong River Commission Secretariat 6:1–500. papers3://publication/uuid/F3F9A5CE-597E-45F8-BC03-876D0722F1E5

National Institute of Statistics (2019) General population census of the Kingdom of Cambodia 2019. In: Ministry of Planning Kingdom of Cambodia (Issue June 2019). http://www.nis.gov.kh/nis/Census2019/ProvisionalPopulationCensus2019_English_FINAL.pdf

NCDM (2014) Disaster management status of the Kingdom of Cambodia 1–12

RGC (2013) National action plan for disaster risk reduction (NAP-DRR) 2014–2018 (Issue November 2013). http://www.rcrc-resilience-southeastasia.org/wpcontent/uploads/2017/12/2014_national_action_plan_for_disaster_risk_reduction_nap-drr_2014-2018_english.pdf

Flood Cause Analysis Using Remote Sensing and Upper-Air Observation Data: Case Study of Jakarta

Azhari Putri Cempaka

Abstract In the past decades, the flooding in Jakarta has become a huge issue with major floods in 2002, 2007, 2013, 2014, and 2020 causing billions of dollars of direct and indirect economic damage, and huge social upheaval. In this context, case study author Azhari Cempaka undertakes a hazard, exposure and vulnerability analysis to estimate a median increase in flood risk of 183% in 2030 compared to baseline conditions and synthesizes data to present a forward look at risk drivers. The author employs a Damagescanner model to assess the impacts of changes in physical and socioeconomic drivers on flood risk. In addition, the author identifies the ways in which land subsidence and land use change impact the likelihood of flooding disasters and explores the potential benefits of an SMS-based flood early warning system as a high-value hazard reduction measure. From this analysis, it was found that an SMS-based flood early warning system could reduce annual flood risk in Jakarta by US$ 3.5 million (1.9%) under a realistic scenario, and up to US$ 22.6 million (12%) under an optimistic scenario.

Keywords Jakarta and flood risk management · Damagescanner and flood disaster · Climate risk assessment · Climate risk management · Disaster risk management · National disaster planning · Participatory approaches and resilience · Vulnerability · Adaptive capacity · Early warning systems

1 Introduction

In the past decades, the flooding in Jakarta has become a huge issue. For example, the major floods in 2002, 2007, 2013, 2014, and 2020 caused billions of dollars of direct and indirect economic damage, and huge social upheaval. Hence, measures are needed to try to reduce the impacts of flood hazards. Traditionally, most planned measures to decrease flood risk in Jakarta have focused on structural approaches such as dikes, canals, and flood gates. This follows the traditional approach to flood

A. P. Cempaka (✉)
Indonesia Agency for Meteorology, Climatology, and Geophysics, Central Jakarta, Indonesia
e-mail: cempakaputria@gmail.com

K. Kim and C. Atkin (eds.), *Cross-Cultural Perspectives on Climate Change Adaptation*, Cities and Nature, https://doi.org/10.1007/978-3-031-50365-8_15

management, whereby limiting the impacts of floods is mainly achieved by reducing the likelihood of the hazard by implementing flood abatement structures.

The flood risk management approach has become more and more mainstream in the last few decades. Here, risk is defined as a function of the hazard, exposure, and vulnerability. In this approach, measures are taken to address all aspects of flooding, not only reducing the magnitude and/or frequency of the hazard, but also measures to decrease the impacts of a flood if it should occur. Such an approach calls for a more integrated approach, combining technical, institutional, socioeconomic, and financial aspects in the planning of flood risk management.

An example of a non-structural measure to decrease flood risk is a flood early warning system (FEWS). FEWS have traditionally been used by inhabitants along rivers in many parts of Indonesia. During the flood season, members of the community serve shifts as guards near the riverbank, monitoring the depth of the river and sending flood alarms to other inhabitants using a traditional instrument, namely the kentongan (wooden or bamboo gong). In this traditional system, a four-knock rhythm means the flood is about to occur.

In Jakarta, with the norm of placing a guard hut in each block, each kentongan alarm covers a radius of 100 m. In Jakarta, a traditional FEWS has been augmented by the implementation of gauges using automatic water level recorder (AWLR). At some posts, the monitoring stations have guards working in 24-hr shifts. The result of this system is that flood waters flowing from the most upstream station in the mountains (50 km upland) can be recognized up to nine hours before reaching the city.

Combining FEWS and the response system, researcher Adi Utarini developed a more advanced approach using the hazard map that appeared in the flood risk study of Budiyono et al. (2016). The system is able to alert people and provide information including the predicted inundation depth and duration of impending floods. With the wide ownership of mobile phones and TV, timely predictions of floods can be broadcasted directly to users around the modeled inundation areas via digital TV or short messaging service (SMS), as has been recognized in the Decree of the Minister of Communication and Information Technology number 32/2013 (Permenkominfo 32, 2013).

2 Objectives and Research Questions

The main objectives of this study are:

- To use Damagescanner model to assess the impacts of changes in physical and socioeconomic drivers on flood risk; and
- To use the model to assess the impacts of various adaptation measures for reducing flood risk.

In order to address these objectives, the following research questions have been formulated:

- What are the possible future changes in river flood risk in Jakarta as a result of climate change, land subsidence, and land use change?
- What is the potential reduction in flood risk that could be achieved in Jakarta through the implementation of an SMS-based Flood Early Warning System?

3 Background and Problem Definition

City Profile

Jakarta is located on the north coast of the island of Java in the Indonesian archipelago in Southeast Asia. It is the country's largest city and its political and economic hub. The city's built environment is characterized by numerous skyscrapers, concentrated in the central business district but also built ad hoc throughout the city, especially in the past 20 years. The rest of Jakarta generally comprises low-lying, densely populated neighborhoods, which are highly diverse in terms of income levels and uses, and many of these neighborhoods are home to varied informal economic activities. The population of Jakarta is considered wealthy relative to neighboring provinces and other islands, and indeed its GDP per capita is more than four times the national average. Jakarta is located in a deltaic plain crisscrossed by 13 natural rivers and more than 1400 km of man-made waterways. About 40 percent of the city, mainly the area farthest north near the Java Sea, is below sea level. Jakarta is prone to flooding from water draining through the city from the hills in the south, and also from coastal tidal flooding (Table 1).

History of Flooding in Jakarta

Flooding has been a problem in Jakarta for centuries. Since the era of kingdoms in Java, and during the history of colonization when the city was named Batavia, floods have caused huge losses to the city. The first record of floods and flood management in what is now Jakarta is inscribed on Prasasti Tugu (the Tugu Monument) from the fifth century. Records on the more recent flood history in Jakarta are available from the colonial government, including the years 1621, 1654, 1876, 1893, 1895, 1899, 1904, 1918, and 1932.

Table 1 Jakarta profile

Jakarta profile	
Total city population in 2022	10.56 million
Population growth (% annual)	1.02%
Land area (km^2)	651
Population density (per km^2)	13,000
Country's per capita GDP (US$)	3,869.59
% of country's population	3.94
GDP (US$)	11,946

Important floods after the colonial period have been recorded in 1976, 1996, 2002, 2007, and 2013. The last three flood events are better documented and provide good information for risk assessment purposes. While floods are not a new problem in Jakarta, their impacts have increased in recent decades, as a result of rapid changes in both physical and socioeconomic drivers (Budiyono et al. 2016). A few examples of physical drivers include land subsidence due to groundwater extraction, and a low drainage capacity of the waterways clogged with solid waste and sediments. Rapid increases in socioeconomic drivers, like population and wealth, have also led to extensive changes in the land use of the city and its surroundings. These land use changes can affect flooding in Jakarta in two main ways: (a) increasing river discharge and sedimentation in Jakarta's rivers; and (b) increasing the value of assets and number of people potentially exposed to floods.

Drivers of Flood Risk in Jakarta

Physical and socioeconomic drivers both contribute to changes in flood risk in Jakarta. For example, land subsidence is a serious problem due to groundwater extraction, construction loading, natural consolidation of alluvium soil, and geotectonic adjustments. The first three, and especially groundwater extraction, are believed to be the most dominant drivers.

Recent estimates suggest that the northern part of the city experiences an average subsidence rate of 4 cm/year. However, there are also areas where subsidence is occurring more rapidly. Future flood risk in Jakarta may also be influenced by climate change. During the twentieth century, the mean annual temperature in Indonesia as a whole increased by 0.3 °C. Other research indicates that there has been a change in mean annual rainfall totals for Jakarta over the twentieth century, even though the number of days with rainfall has decreased (Fig. 1).

Across the country as a whole, observations suggest that mean annual rainfall over the same period decreased by 2–3%, mainly in the wet season from December to February. Several climate models project a temperature increase of ca. 0.1–0.3 °C per decade over the twenty-first century. The same projections suggest that mean annual rainfall may increase in future across most of Indonesia, although in Java it may decrease (Fig. 2).

Jakarta will also be affected by climate change due to projected sea-level rise. Sea-level rise is currently taking place at a rate of 1–3 mm/year in most parts of coastal Asia (IPCC 2007). In the Bay of Jakarta, observations based on altimetry satellite detection show mean sea-level rise of ca. 2–4 mm p.a. between 1992 and 2005. Detailed projections of climate change's impacts on sea-level rise specific to the Jakarta Bay are not yet available. Nevertheless, these observed changes over recent decades are within the range of global rise reported by IPCC (2007), i.e., a likely minimum and maximum global mean sea-level rise until 2100 of between 18 and 59 cm.

In terms of global socioeconomic developments in Jakarta and the exposure of population and assets to flooding, Jakarta's population has risen rapidly from 2.7 to 9 million between 1960 and 2007. This is projected to increase further in future. At

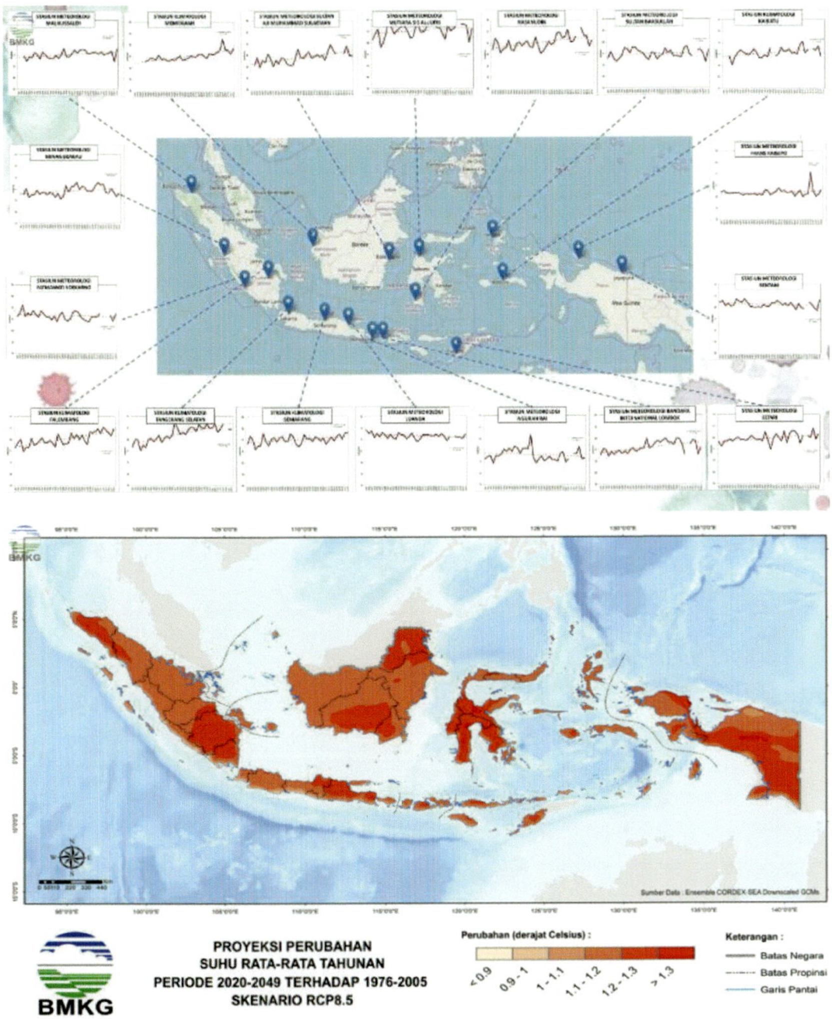

Fig. 1 **a** Temperature trends from BMKG observation data in the period 1981–2018. In general, the average temperature in Indonesia has a positive trend with a magnitude that varies around 0.03 °C each year. **b** Average temperature projection in period 2020–2049 increase 1 to > 1.3 °C. *Source* BMKG

the same time, the GDP of Indonesia has also increased rapidly, and is projected to increase in future, further increasing the exposure of assets and wealth to flooding hazards. The rapid growth of population and economic developments have led to extensive changes in land use in Java, and in Jakarta in particular.

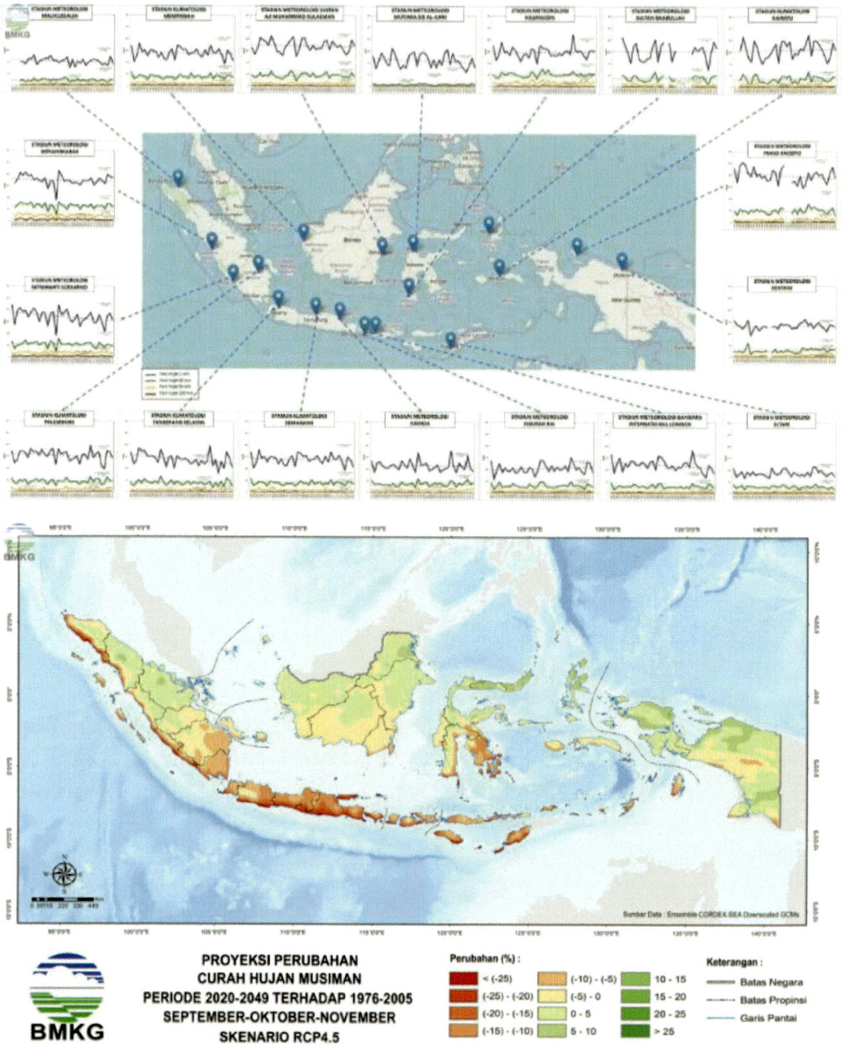

Fig. 2 **a** Rainfall trends from BMKG observation data in the period 1981–2018. In general, Indonesia's rainy day has a positive trend although in some areas like Java island it is negative with varying magnitudes. **b** Rainfall projection in the period 2020–2049. The rainfall projection in most parts of Indonesia is increase 5–10%, while in most parts of Java island is decrease 5–25%. *Source* BMKG

4 Methodology

Damagescanner-Jakarta is a flood risk model for Jakarta developed by Budiyono et al. (2016) in the Python programming language. Damagescanner-Jakarta estimates flood risk as a function of hazard, exposure, and vulnerability. In brief, the model

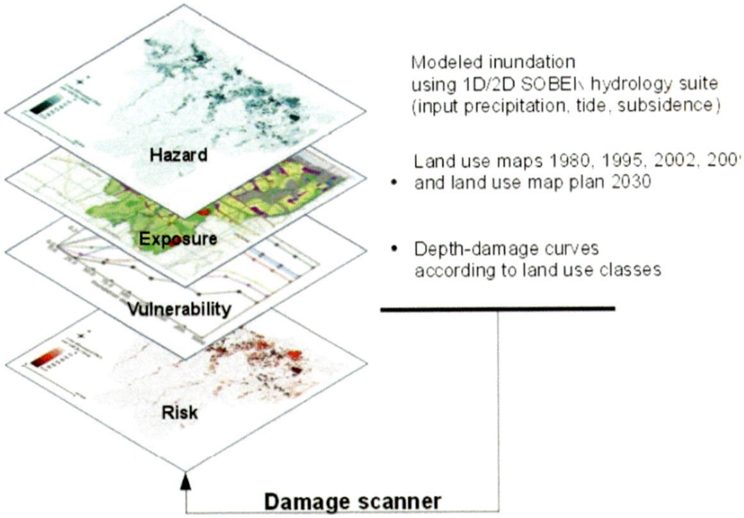

Fig. 3 Flow diagram representation of Damagescanner-Jakarta

has a horizontal resolution of 50 m × 50 m, and works by combining maps of hazard and exposure with a depth–damage function to represent vulnerability. For each grid cell, the model identifies the depth of inundation found in the flood hazard map. For this grid cell, it then identifies the land use class and the associated value of maximum damage for this class. This procedure is carried out for floods of several return periods between 2 and 100 years (Fig. 3).

Hence, change in precipitation intensity was estimated using output data from a large range of global climate models (GCMs) and Representative Concentration Pathways (RCPs). These downscaled future climate data were used for five GCMs, namely GFDL-ESM2M, HadGEM2- ES, IPSLCM5A-LR, MIROC-ESM-CHEM, and NorESM1-M, and for the following Representative Concentration Pathway (RCP) scenarios: RCP2.6, RCP4.5, RCP6.0, and RCP8.5.

5 Results

5.1 Hazard

Flood hazard is represented by maps showing inundation depth and extent for several return periods (1, 2, 5, 10, 25, 50 and 100 years). These inundation maps for Jakarta were produced by the Flood Hazard Mapping (FHM) framework by Deltares in 2007 and 2009, as well as the Flood Management Information System (FMIS) projects by Deltares in 2012 together with the Research & Development Center for Water Resources (Pusair) and the National Office for Climate (BMKG) for the province of DKI Jakarta (Special Capital Region of Jakarta) administration and the national government of Indonesia. The FHM framework includes a hydrological and hydraulic model of the Ciliwung River integrated with an overland flow model of DKI Jakarta. The framework is used by and updated in close communication with stakeholders in Jakarta [i.e., local office of Public Works (PU DKI) and the office for Ciliwung Cisadane river management (BBWSCC)]. As an example, the hazard map for a 50 year return period flood is shown in Fig. 4. The spatial resolution of the original hazard map is 50 m x 50 m, and the depths are given for increments of 1 cm.

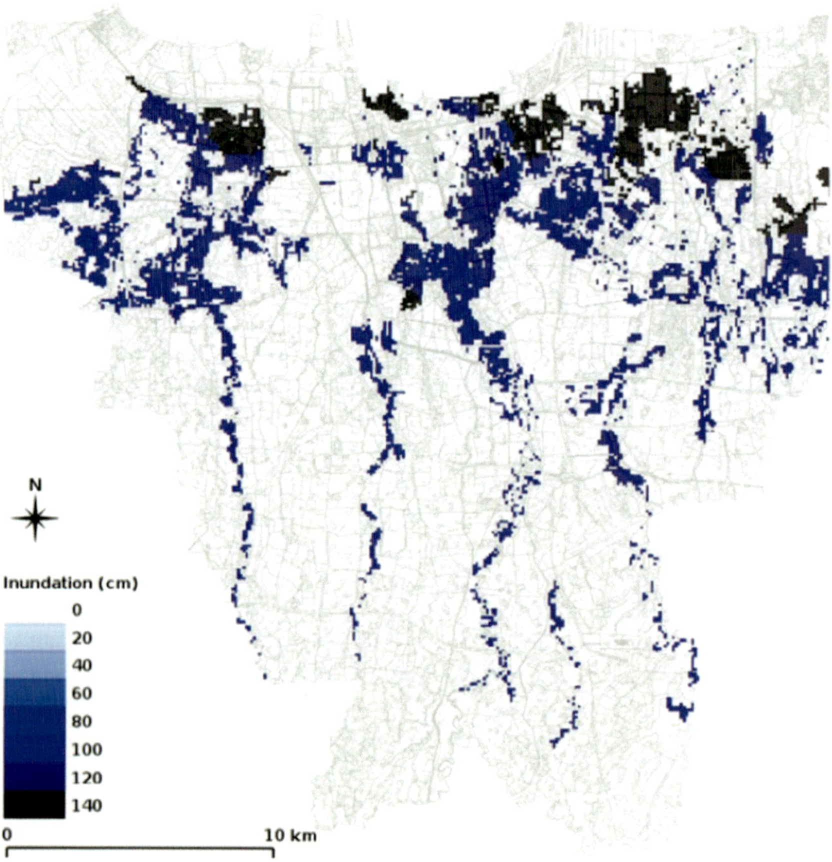

Fig. 4 **a** Map of Jakarta showing the modeled hazard for 50 years flood return period. **b** Change in inundation depth for a return period of 100 years in the flood hazard map

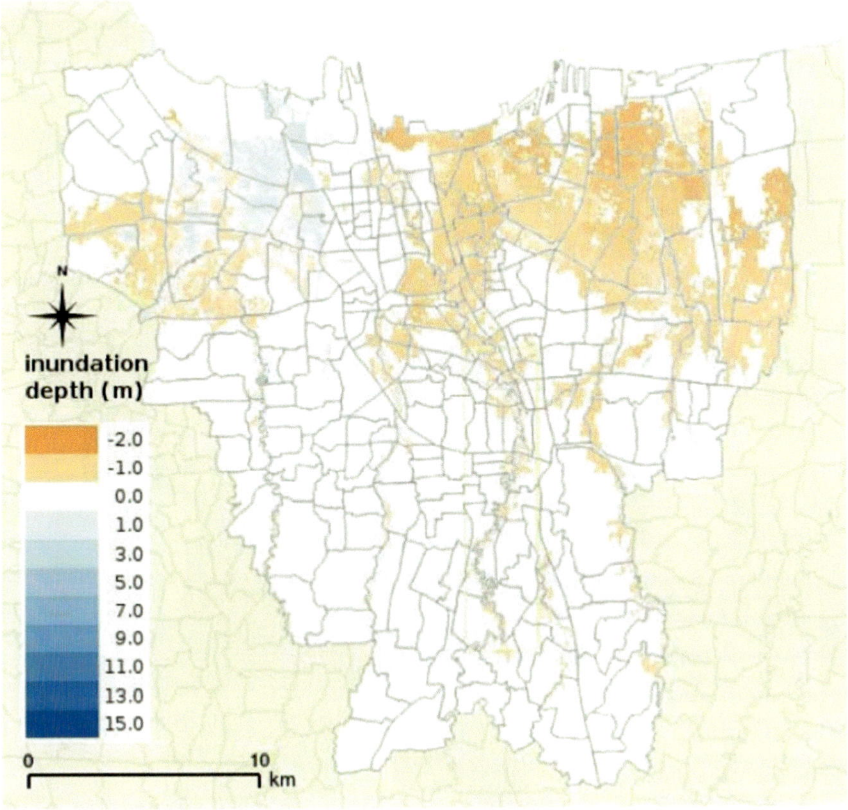

Fig. 4 (continued)

5.2 Exposure

Exposure is represented by a map showing the economic exposure per grid-cell, which is in turn based on a land use map. The map shown in Fig. 5 shows land use at a horizontal resolution of 50 m × 50 m, for 12 land use classes, namely: agriculture and open space; low density urban kampung (self-made houses or residential areas with no site plan); swamp, river and pond; industry and warehouse; commercial and business; planned house; education and public facility; government facility; high density urban kampung; transportation facility; and park and cemetery.

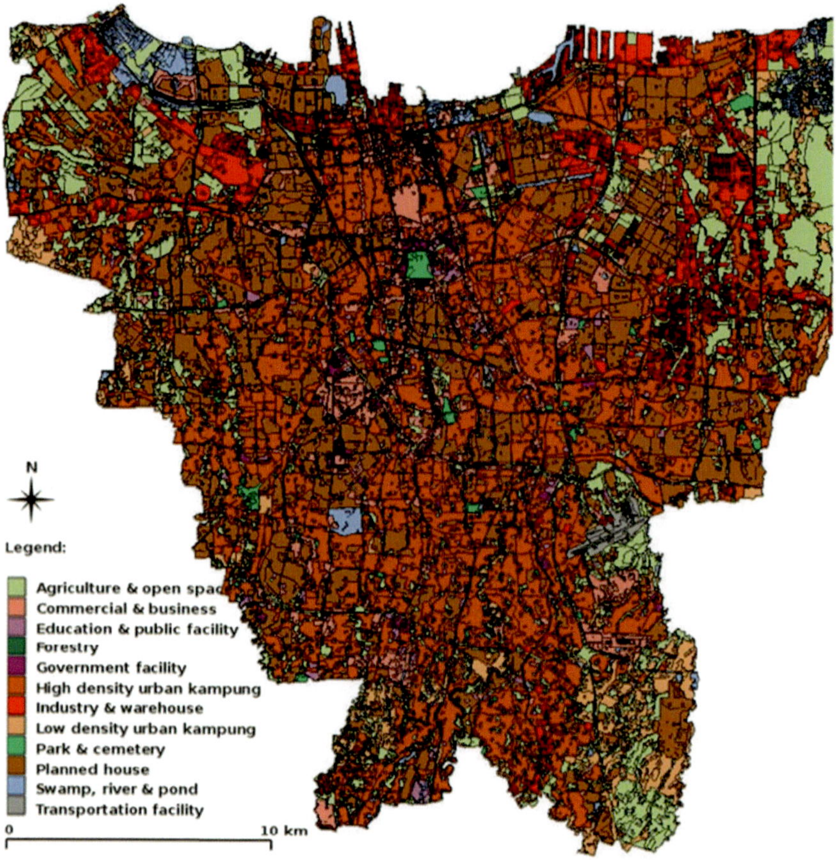

Fig. 5 Jakarta land use map for 2022 from the office city planning, showing twelve land use classes

5.3 *Vulnerability*

The third element of the risk framework is vulnerability. The definition of UNISDR is that vulnerability is "the characteristics and circumstances of a community, system or asset that make it susceptible to the damaging effects of a hazard". In Damagescanner-Jakarta, vulnerability is represented by depth–damage functions, which show the fraction of the maximum economic exposure value per land use type that would actually result in damage for different inundation depths (Fig. 6).

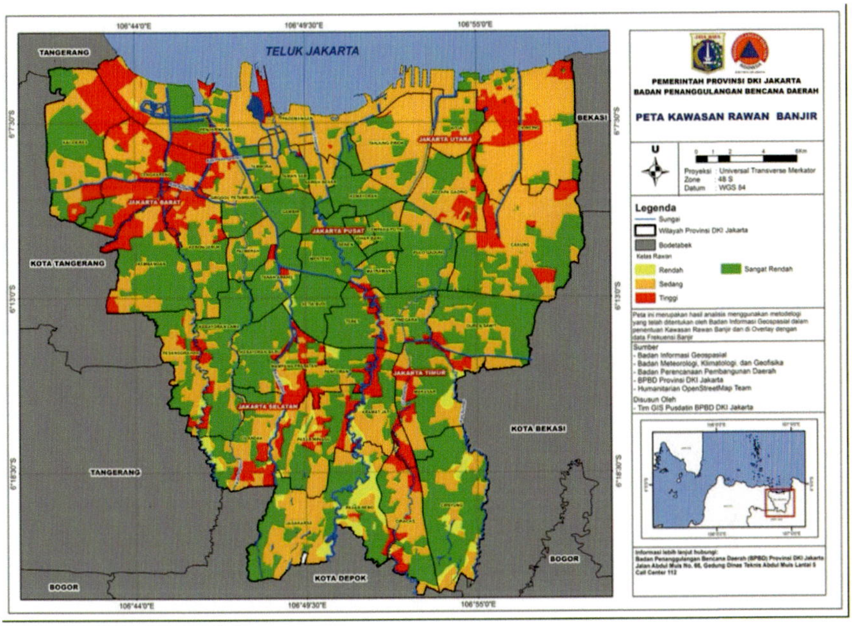

Fig. 6 Jakarta flood vulnerability map

6 Flood Risk Assessment—Impacts of Future Changes in Individual Risk Drivers

6.1 *Climate Change*

The potential influence of climate change on future flood risk compared to baseline flood are shown in Table 2. Here, the future risk (in 2030 and 2050) is shown for each of the different combinations of precipitation intensity (represented by the RCP scenarios) and sea level rise (low and high scenarios). The median and standard deviation of the results across the five GCMs are shown for each combination of RCP and sea level rise scenarios. From these results, there is no clear signal of change in future flood risk as a result of climate change alone.

For 2030, under the low sea level rise scenario, the median risk is in fact lower than for the baseline (US$ 186 million p.a.) for all RCPs. However, the standard deviation is large. Under the high sea level rise scenario, the median risk is higher than baseline for two RCPs, and lower for the other two RCPs; again the standard deviation between GCMs is large. For 2050, the results generally show slightly higher risk compared to 2030, under both sea level rise scenarios.

In 2030, the highest risk values are simulated under RCP6.0 and RCP8.5, while there are only small differences between the other RCPs. According to IPCC (2022), the global radiative forcing by 2030 is the highest under RCP8.5. By 2050, we see an

Table 2 Median and standard deviation of flood risk (million US$) between the 5 GCMs, for each RCP in 2030 and 2050

	Low SLR		High SLR	
	Median	Standard deviation	Median	Standard deviation
2030				
RCP 2.6	118.0	51.8	152.4	47.1
RCP 4.5	112.0	80.6	147.1	75.5
RCP 6.0	118.3	85.1	152.6	79.1
RCP 8.5	127.0	82.6	160.4	77.7
2050				
RCP 2.6	118.1	48.5	125.5	43.9
RCP 4.5	121.8	83.5	155.7	78.6
RCP 6.0	118.6	97.8	152.9	93.7
RCP 8.5	148.0	102.2	179.1	98.5

increase in the difference between the risk estimates under RCP8.5 and those under the other RCPs.

6.2 *Land Subsidence*

Assuming only an increase in land subsidence for 2030 leads to an increase in annual expected damage of 126% between the baseline and 2030, i.e., an increase from US$ 186 million to US$ 421 million. The increase in risk resulting from projected subsidence, however, is not uniform across the city. Note also that the actual influence of subsidence will strongly depend on the changes in other environmental and socioeconomic drivers (Fig. 7).

7 Impacts of Future Changes in Combined Risk Drivers on Flood Risk

In the previous subsections, the change in risk between the baseline situation and the future scenarios has been shown for each risk driver separately. In reality, the future situation will depend on the combined change of all the drivers. Hence, this section shows the impacts of combinations of different risk drivers on future risk. In Fig. 8, probability density functions (PDFs) of the simulated annual expected damage are shown, whereby each PDF is derived from a two-parameter gamma distribution fit to the 20 GCM/RCP combinations. The dotted black vertical line represents baseline flood risk, i.e., US$ 186 million.

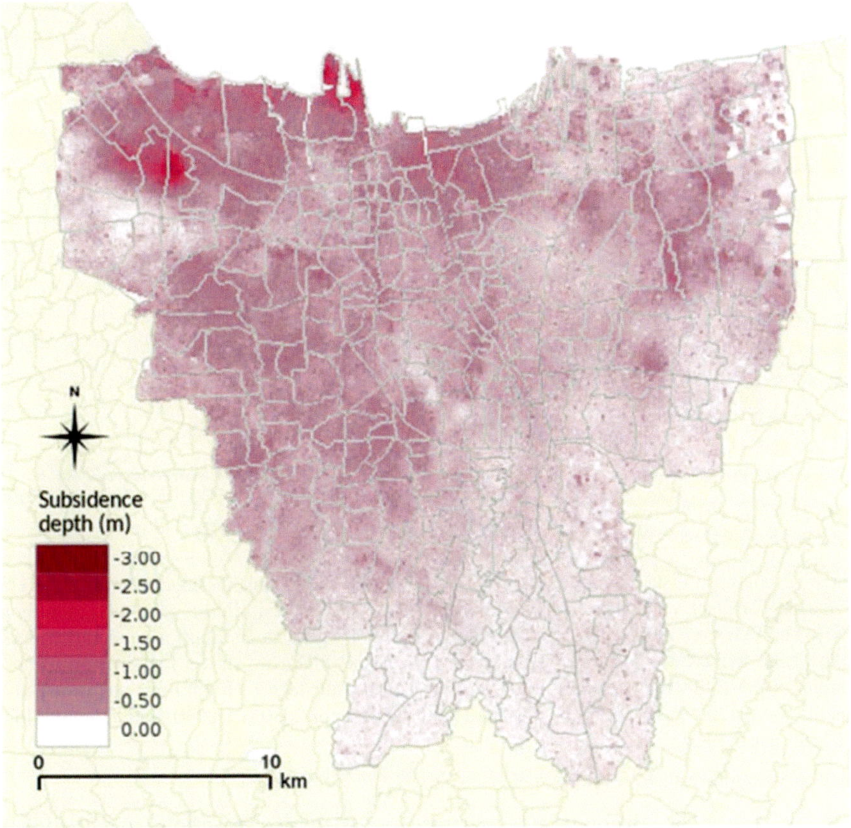

Fig. 7 Spatial distribution of projected total land subsidence over period 2012–2025

Figure 8 clearly shows the strong influence of projected subsidence on the overall change in risk. All of the PDFs representing scenarios with subsidence (shown in red) show much higher annual expected damage than those without subsidence (shown in blue). The PDFs also clearly show the large uncertainty associated with the projected changes in precipitation from the different GCMs and RCPs, which is large under all of the PDFs. However, the results show that if we include land subsidence in future projections, the probability of future flood risk exceeding baseline flood risk exceeds 99.999% when accounting for changes in precipitation. The results also show the importance of the interaction between different drivers. For example, if we examine the difference between the PDFs for low and high sea level rise, we see a small difference under the scenarios with no subsidence and land use change in 2030. However, if we make a similar comparison using the scenarios that include subsidence, we see a larger difference between the low and high sea level rise scenarios. Similar differences can be found when comparing the scenarios with and without projected

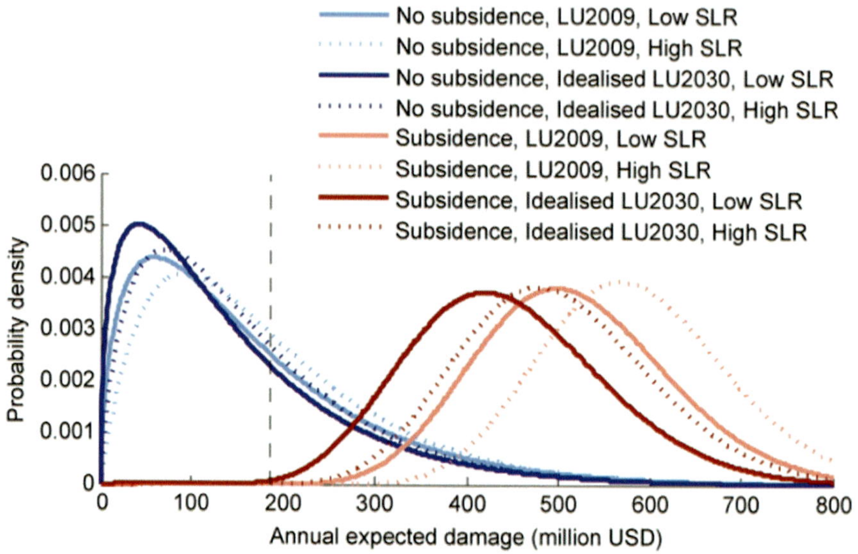

Fig. 8 Probability distribution function (PDFs) of future flood risk in Jakarta under different scenarios. The black vertical dashed line shows risk associated with baseline conditions (186 million US$). The PDFs are obtained by applying a two parameter gamma distribution to simulated risk values from five GCMs and four RCP emission scenarios. PDFs are shown for different combinations of the following scenarios: **a** subsidence and no subsidence; **b** land use under baseline conditions (LU2009) and under the idealized land scenario for 2030 (LU2030); and **c** high or low sea level rise (SLR)

land use change. The differences between the two scenarios are amplified with higher rates of subsidence and/or sea level rise.

8 Potential Flood Risk Reduction in Jakarta Through Vulnerability Reducing Measures Associated with SMS-Based FEWS

Potential economic damage could be avoided in Jakarta through the implementation of an SMS-based flood early warning system measured by a hydrodynamic model. The potential damage was assessed using the Damagescanner-Jakarta flood risk model. An existing survey carried out along the Pesanggrahan River that examined the measures that households took to avoid flood damages during the 2010 flood was used. The survey found that a FEWS system could lead to a decrease in flood risk, as people report taking actions to move assets upwards or to prevent floodwaters entering households, if sufficient warning is available. The survey shows that some households take measures to avoid damages for inundation up to a depth of 125 cm.

By using these findings to amend the depth-damage functions in Damagescanner-Jakarta, first order estimates of flood risk with and without the FEWS were able to be simulated.

SMS-based FEWS can provide a large contribution to reducing flood risk in Jakarta. By moving assets upwards (or outwards in the case of vehicles) and protecting properties with household-level structures, expected annual damage in residential areas in Jakarta could be reduced by up to 84% under an optimistic scenario. This equates to an overall reduction in EAD of 12%. This is a very optimistic value, since it is assumed that all residents receive the early warning, have enough to act on it, and do indeed act on it. Under the 'realistic scenario', the reduction is projected to be much lower, at 2%. This is based on the actions that residents already take to reduce vulnerability due to flood warnings.

9 Conclusion

Probabilistic scenarios allow us to gain a better understanding of the potential future changes than if we only use several climate change scenarios. Combining all of these scenarios, a median increase in flood risk found of 183% in 2030 compared to baseline conditions. This value is based on median projection for the influence of changes in extreme precipitation on flood risk.

The single driver with the largest influence on future flood risk is land subsidence (+ 126%). Clearly, addressing this driver could potentially have a large influence on reducing future flood risk. Under an idealized land use scenario for 2030, whereby the official Jakarta Spatial Plan 2030 is fully implemented, flood risk could be reduced by 12% compared to baseline conditions, if all other driving factors are held at baseline levels. On the other hand, if past trends in risk increase due to land use change continue, flood risk could increase by 45% by 2030 due to land use change alone. This demonstrates the importance of effective land use planning for flood risk reduction.

An advanced system could reduce risk, since the traditional approach is limited. For this, a hydrodynamic model was coupled with an SMS-based flood early warning system (FEWS). Using the results of a survey of inhabitants along the Pesanggrahan river in South and West Jakarta, the vulnerability curves were adjusted to reflect the damage that inhabitants could potentially avoid for different flood depths, for a given warning time from the new SMS based early warning system. From this analysis, it was found that an SMS-based Flood Early Warning System could reduce annual flood risk in Jakarta by US$ 3.5 million (1.9%) under a realistic scenario, and up to US$ 22.6 million (12%) under an optimistic scenario.

References

Budiyono Y, Aerts JCJH, Tollenaar D, Ward PJ (2016) River flood risk in Jakarta under scenarios of future change. Nat Hazards Earth Syst Sci 16:757–774

IPCC (2007) Climate change 2007: the physical science basis. In: Solomon S, Qin D, Manning M, Chen Z, Marquis M, Averyt KB, Tignor M, Miller HL (eds) Contribution of working group I to the fourth assessment report of the intergovernmental panel on climate change. Cambridge University Press, Cambridge

IPCC (2022) Full report. In: Pörtner H-O, Roberts DC, Tignor M, Poloczanska ES, Mintenbeck K, Tignor M, Alegría A, Craig M, Langsdorf S, Löschke S, Möller V, Okem A, Rama B (eds) Climate change 2022: impacts, adaptation, and vulnerability. contribution of working group II to the sixth assessment report of the intergovernmental panel on climate change. Cambridge University Press (in press)

Water-Related Disaster Security in the Coastal Community of Jordan City, Guimaras Island, Philippines Using Climate and Disaster Risk Assessment (CDRA)

Neil Paul Fara-on

Abstract Due to its population's vulnerability and susceptibility to natural hazards, the Philippines is ranked among the most at-risk nations in the world with meteorological and meteorologically induced hazards intensifying over the past decade, resulting in a rise in fatalities and economic devastation. Case study author focuses on Jordan, a municipality in Guimaras Island in the Philippines which spans12,611 ha and uses the government's adopted methodology, the Climate and Disaster Risk Assessment (CDRA). Using this framework, the author evaluates local risk, vulnerability, and exposure to water-related disasters. In addition to sharing flood disaster risk maps, the study ends with policy recommendations to mitigate the flood disaster risks as intensified by climate change including water policy and planning as well as the use of nature-based solutions and the implementation of early warning systems to support effective flood risk management.

Keywords Philippines and flood risk management · Climate and disaster risk assessment and Philippines · Guimaras and disaster response · Climate risk assessment · Climate risk management · Disaster risk management · National disaster planning · Participatory approaches and resilience · Vulnerability · Adaptive capacity · Early warning systems

1 Introduction

Jordan is a 3rd Class Municipality in Guimaras Island with a land classification of agricultural, having an area of 12,611 ha. Composed of 14 barangays, Jordan has a total of nine (9) coastal barangays and five (5) inland/upland barangays. For CY 2010, Jordan has a population of 35,732 with 7464 total households and 8465 total families.

N. P. Fara-on (✉)
Department of the Interior and Local Government, Quezon City, Philippines
e-mail: npfaraon@gmail.com

K. Kim and C. Atkin (eds.), *Cross-Cultural Perspectives on Climate Change Adaptation*, Cities and Nature, https://doi.org/10.1007/978-3-031-50365-8_16

The study adopts the Philippine government's Climate and Disaster Risk Assessment (CDRA). The study focuses on the natural hazard risk of flooding with climate-induced increasea in temperature. Population exposure is the focus of climate change vulnerability assessment and disaster risk assessment.

The purpose of the study is to assess the water-related disaster security with a focus on flood disaster risk in the Jordan municipality on the Island Province of Guimaras, Philippines. Further, it will identify risk, vulnerability, and exposure to water-related flood disaster of the site location. Lastly, it will present flood disaster risk maps developed by the Philippine government.

The study ends with policy recommendations from the Asian water development outlook report 2020 to mitigate the flood disaster risks as intensified by climate change.

2 Background of the Study

Due to its population's vulnerability and susceptibility to natural hazards, the Philippines is ranked among the most at-risk nations in the world. This situation is made worse by threats such as climate change. Meteorological and meteorologically induced hazards have intensified over the past decade, resulting in a rise in fatalities and economic devastation, particularly in regions unprepared for such phenomena (Jordan CDRA 2020).

The Philippines are frequently impacted by numerous climate hazards induced by rainfall variability, increasing temperature, and sea level rise. In recent decades, it has experienced extreme weather, including temperatures above 35 °C, more intense rainfall, and typhoons stronger than the 30-year average for the region. In 2015, the Philippines had the highest sea level rise, three times the global average, recorded by the World Meteorological Organization (WMO) (NEDA 2018).

A more focused intervention prioritizing climate change adaptation and disaster risk reduction in the country's cities and municipalities needs to be put in place, as the community's vulnerabilities are becoming more pronounced.

Climate change adaptation centers on water. Changes in the frequency and intensity of extreme weather events will primarily manifest through water-related events. These impacts include too little water leading to droughts and water shortages, too much water causing floods and typhoons, and contaminated water causing salt water intrusion and pollution. Climate change-related water issues require improvements in water security (Babel et al. 2020).

The Asian Water Development Outlook (AWDO) of the Asian Development Bank (ADB) defines water security as "the availability of adequate water to ensure safe and affordable water supply, inclusive sanitation for all, improved livelihoods, and healthy ecosystems, with reduced water-related risks toward supporting sustainable and resilient rural–urban economies." (ADB 2020).

One of the five key dimensions of water security is the water-related disaster security including the nation's recent exposure to water-related disasters, its susceptibility to such disasters, and its resilience and resiliency. It consists of three water-related indicators of disaster risk:

1. Climatological risk (drought),
2. Hydrological risk (floods and mudslides), and
3. Meteorological risk (storms and storm surges).

The United Nations Office for Disaster Risk Reduction defined *disaster risk* as "the potential loss of life, injury, or destroyed or damaged assets which could occur to a system, society or a community in a specific period of time, determined probabilistically as a function of hazard, exposure, vulnerability and capacity." (ADB 2020).

Disaster risk is the product of three components that can change over time: hazard exposure, vulnerability, and coping capacity. Hazard exposure is exacerbated by climate change, which brings about changes to regional and local climatology and hydrology (ADB 2020).

In the AWDO report 2020 of the ADB, the Philippines had a National Water Security (NWS) Score of 67.8 (out of 100) with the NWS stage of capable. In the key dimension of water-related disaster security, the country has a score of 12.1 (out of 20) which is described as institutional commitment to disaster risk reduction.

The Philippine government conducts Climate and Disaster Risk Assessment (CDRA) in its local communities. This is done primarily by the local government units at the Provincial and Municipal Levels. Climate and Disaster Risk Assessment (CDRA) is a vital process to help the coastal community become disaster risk resilient and climate adaptive. CDRA generates climate data and rationalizing risk and vulnerability information for the welfare of the community. (Jordan CDRA 2020).

The CDRA guidelines are adopted in the Supplemental Guidelines on mainstreaming climate change and disaster risks in the 2015 comprehensive land use plan of the Housing and Land Use Regulatory Board (HLURB) of the Philippine Government. Based on the HLURB supplemental guideline, the systems of interests of CDRA are population exposure, urban use area exposure, natural resources-based production areas, lifeline utilities and critical point facilities.

Guimaras is one of the most vulnerable provinces in the Philippines to natural disasters such as tsunamis, landslides, and drought. Changing weather patterns, including temperature and precipitation, that have been attributed to the negative effects of climate change have exacerbated this situation. Recognizing the importance of a more balanced and systematic approach to planning for natural hazards, the community's disaster risk reduction management should incorporate climate change adaptation (Jordan CDRA 2020).

3 Objectives of the Study

- To assess the water-related disaster security with focus on flood disaster risk of Jordan municipality in the Island Province of Guimaras, Philippines
- To identify risk, vulnerability, and exposure to water-related disaster flooding security of the Jordan municipality in the Island Province of Guimaras, Philippines
- To present flood disaster risk maps developed by the Philippine government

4 Methodology

The study adopts the Philippine government steps on the Climate and Disaster Risk Assessment (CDRA). The study focuses on the natural hazard risk of flood with climate increase in temperature. Population exposure is the focus of the climate change vulnerability assessment and disaster risk assessment (Fig. 1).

5 A Case Study Overview

5.1 Site Selection

Jordan is located in the midwest portion of the island of Guimaras. It is bounded by the Municipality of Buenavista in the northeast; the Iloilo strait in the northwest; Municipality of Sibunag in the southwest and southeast; and Municipality of San Lorenzo in the east.

Jordan is a 3rd Class Municipality with a land classification of agricultural, having an area of 12,611 ha. Composed of 14 barangays, Jordan has a total of nine (9) coastal barangays and five (5) inland/upland barangays. For CY 2010, Jordan has a population of 35,732 with 7464 total households and 8465 total families.

Jordan is also known as the economic hub of the province due to the presence of various economic activities and business establishments such as banks, pawnshops, groceries/convenient stores, lending institutions, and hotels and restaurants (Fig. 2).

5.2 Site Characteristics

Guimaras province is an island ecosystem composed of five coastal municipalities. It is a fourth class province and the smallest and youngest province in the administrative region of Western Visayas in Central Philippines.

The island province covers a land area of 60,547 ha and with a total coastline length of 470.29 km. The topography of Guimaras Island varies from level to steeply

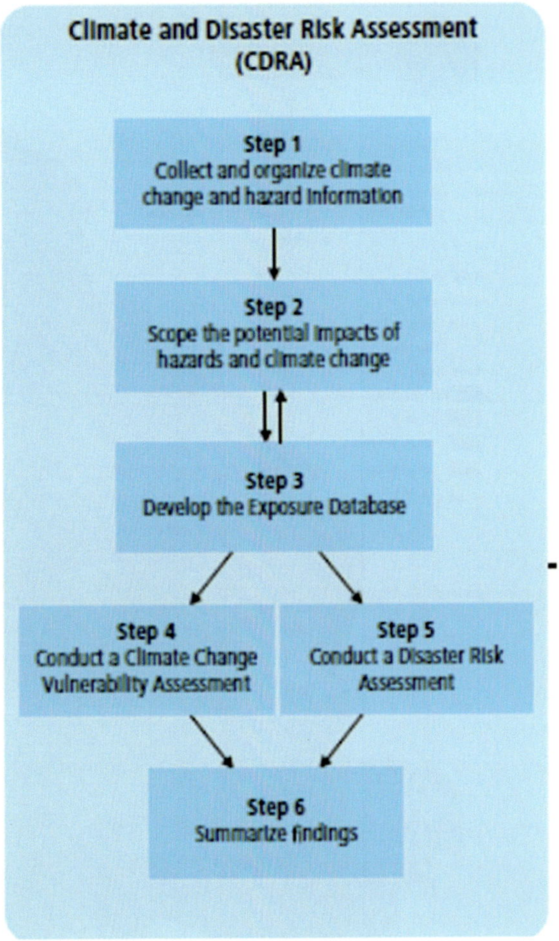

Fig. 1 Mainstreaming framework of CRA into comprehensive land use plan (CLUP) formulation. *Source* HLURB, CLUP Guidebook: Supplemental Guidelines on mainstreaming climate change and disaster risks in the comprehensive land use plan, 2015

sloping with land elevation ranging from 0 to nearly 300 m above sea level. Mt. Dinalman, located in Millan, Sibunag, has the highest elevation of 267 m above mean sea level.

The island's topography shows quite steep slopes on the western side of island with plateaus and peaks above 200 m in the central portion. A large part or 69% of the total land area is within the 0–18% slope, 19.73% is above 18–30% slope 9.42% is above 30–50% slope and 1.74% is above 50% slope (Guimaras Forest Land Use Plan 2017–2026).

There are two principal watersheds in the island, the Eastern watershed covering 387 km^2 or 65% of the total area, and the Western watershed covering 209 km^2 or 35% of the total area including the inhabited islands with surface water sources. The remaining 9 km^2 are small-inhabited islands. In the Eastern watershed, there are 29 river basins, of which Cabano (85.4 km^2), Sibunag (109.5 square kilometers) and

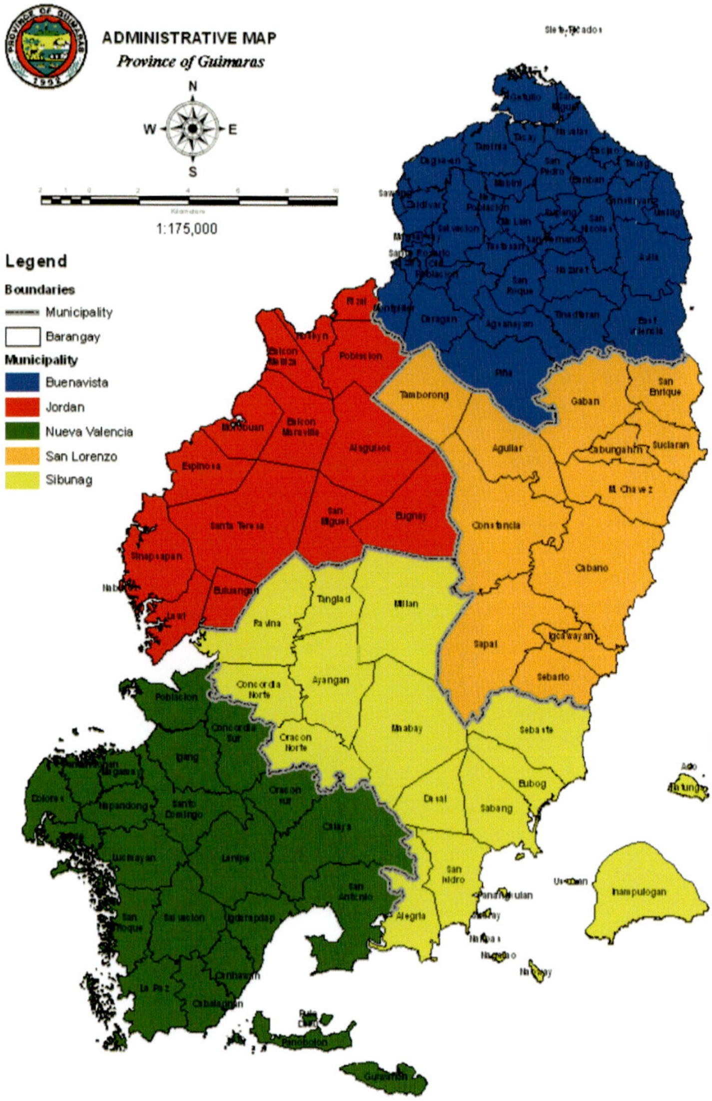

Fig. 2 Administrative map of the Island Province of Guimaras

Mantangingi River (59 km^2) are considered perennial, indicating that the base flow is supported by aquifer drainage. In the Western watershed, on the other hand, there are 10 individual river basins where Igang River is considered perennial (PEMSEA and Guimaras 2018).

Guimaras' climate is tropical with an average temperature of 27.5 °C. The upper part of the province (Buenavista, San Lorenzo and Part of Sibunag and Jordan) has a Type III climate, where seasons are not very pronounced. It is relatively dry from November to April and wet the rest of the year.

6 Climate and Disaster Flood Risk Assessment

While executing the climate and flood risk assessment exercise, the following steps will be followed;

STEP 1: Collect and Analyze Climate and Hazard Information

This step entails collecting data on climate change and identifying potential threats to the local area (Figs. 3 and 4; Tables 1, 2 and 3).

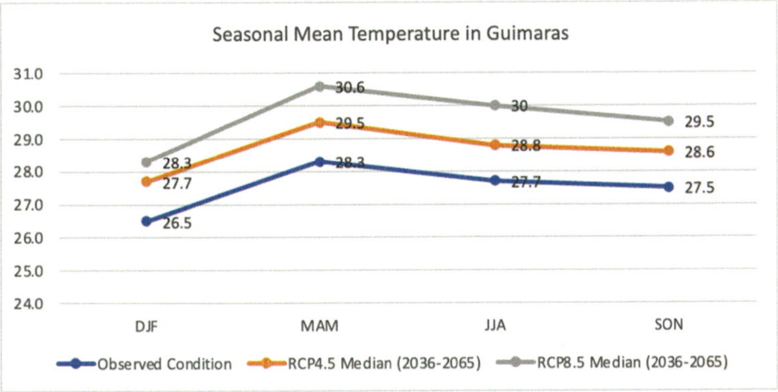

Fig. 3 Seasonal mean temperature in Guimaras. *Source* Observed Climate Trends and projected Climate Change in the Philippines, DOST PAGASA 2018

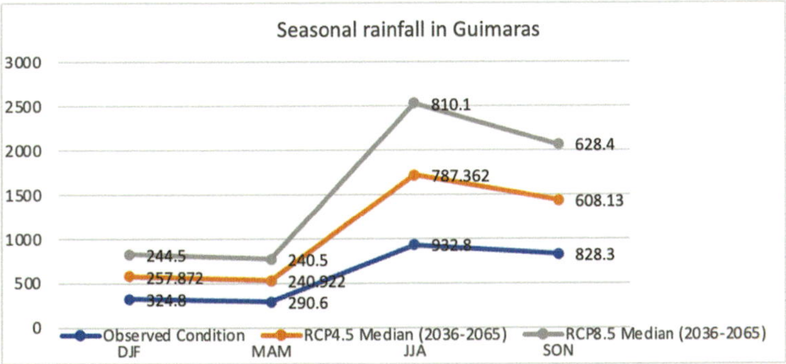

Fig. 4 Seasonal rainfall in Guimaras. *Source:* Observed Climate Trends and projected Climate Change in the Philippines, DOST PAGASA 2018

Table 1 Seasonal mean temperature projected change (in °C) for 2036–2065 under RCP4.5 and RCP8.5 scenarios for the province of Guimaras

Observed (1971–2000)				Projected (2036–2065)									
DJF	MAM	JJA	SON	Scenario	Range*	DJF (Dec—Jan–Feb		MAM (Mar–Apr–May		JJA (Jun–Jul–Aug)		SON (Sep–Oct–Nov)	
						Change	Projected Value	Change	Projected Value	Change	Projected value	Change	Projected Value
26.5	28.3	27.7	27.5	Moderate Emission (RCP4.5)	Lower Bound	1.0	27.5	1.0	29.3	0.9	28.6	1.0	28.4
					Median	1.2	27.7	1.2	28.5	1.1	28.8	1.1	28.5
					Upper Bound	1.5	28.0	1.5	29.8	1.5	29.2	1.6	29.1
				High Emission (RCP8.5)	Lower Bound	1.2	27.8	1.3	29.6	1.3	29.1	1.3	28.8
					Median	1.5	28.1	1.6	29.9	1.4	29.1	1.4	28.9
					Upper Bound	1.8	28.4	2.1	30.4	1.0	29.6	2.0	29.5

Table 2 Seasonal rainfall change (in %) for 2036–2065 under RCP4.5 and RCP8.5 Scenarios in the province of Guimaras

Observed (1971–2000)				Projected (2036–2065)									
DJF	MAM	JJA	SON	Scenario	Range*	DJF (Dec–Jan–Feb		MAM (Mar–Apr–May		JJA (Jun–Jul–Aug)		SON (Sep–Oct–Nov)	
						Change	Projected Value	Change	Projected Value	Change	Projected value	Change	Projected Value
227.0	233.0	843.0	699.0	Moderate Emission (RCP4.5)	Lower Bound	4.7	237.8	−1.0	231.2	−14.6	720.7	−15.1	594.2
					Median	13.6	257.9	3.4	241.4	−6.6	788.2	−13.0	608.5
					Upper Bound	24.6	282.8	12.9	263.7	11.4	939.6	12.2	785.2
				High Emission (RCP8.5)	Lower Bound	−4.2	217.5	−10.8	208.4	−18.7	686.0	−17.9	574.7
					Median	7.7	244.5	3.2	241.0	−3.9	810.8	−10.1	629.0
					Upper Bound	27.9	290.5	18.0	275.5	14.0	961.5	9.3	764.5

Table 3 Climate projections for extreme events

No. of days w/Tmax > 35 °C			No. of dry days			No. of days w/rainfall > 200 mm		
Observed baseline (OBS) (1971–2000)	2006–2035	2036–2065	OBS	2006–2035	2036–2065	OBS	2006–2035	2036–2065
460	1431	3076	7839	5227	5226	4	5	4

Hazard Information and Characterization of Hazards

The municipality of Jordan has experienced natural disasters. Although not major ones, there are indicators of the vulnerabilities and susceptibilities of the municipality. There are available records for the past 10 years that can provide insights on hazards impacts. The municipality had experienced flooding, landslide, and El Nino.

Using available statistical and spatial data from various sources, the hazards were identified and analyzed. Figure 5 presents the hazards and the number of barangays that are susceptible. Most of the barangays in Jordan are susceptible to rain-induced landslide and rain-induced flooding.

Table 4 is a summary of previous hazard events experienced in the Municipality of Jordan. Typhoons that caused heavy rains had resulted to disruptions in transport but overall, the effects on people and households were minimal.

Summary Hazard Matrix Per Barangay

The susceptibility of barangays to the three major hazards are summarized in the Table 5. Rain-induced landslide was identified as the hazard that will affect the greatest number of barangays.

Rain-Induced Flooding

The low elevation areas in the municipality are particularly susceptible to flooding. These are areas that are not well-drained due to several factors such as inadequate or non-existent drainage system and because of road widening projects that expanded the roadways with high embankments resulting to increased runoff (Fig. 6).

Highly susceptible are barangays Morobuan, Balcon Maravilla, Lawi, Balcon Melliza and Buluangan while barangays Rizal, Hoskyn, Bugnay, Sinapsapan, Espinosa, Poblacion, Sta. Teresa and San Miguel are moderately susceptible.

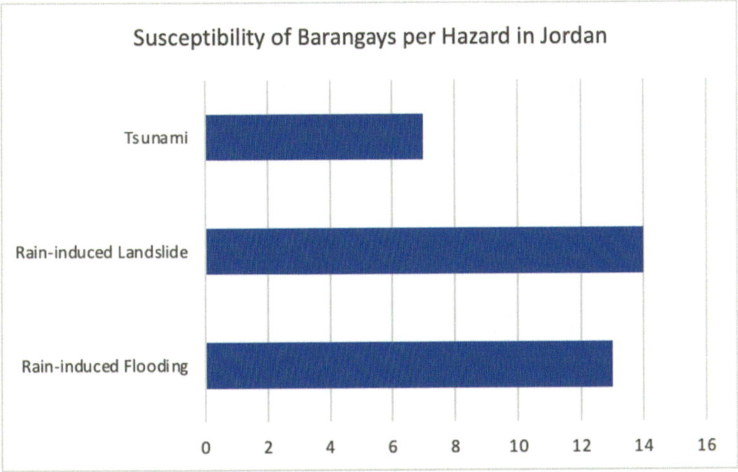

Fig. 5 Susceptibility of Barangays per hazard in Jordan

Table 4 Records of previous disasters, municipality of Jordan, province of Guimaras

Hazard events and description/ Date	Affected Barangays	Number of casualties (persons)			Number of affected (persons)		No. of houses damaged		Damage to properties (Php)					Source of information
		Dead	Injured	Missing	Persons	Families	Totally	Partially	Infra	Agri	Insti	Private / commercial	Total	
Landslide (9/10/2017) due to heavy rain of typhoon Lanie and Maring	1 Barangay Hoskyn	0	0	0	38 22 M/ 16F	15	0	0	1 Unserviceable Road	0	0	0	0	DRRM Report on 9/11/2017
Rain Induced Flood (9/26/2017) typhoon Lanie and Maring	4 Barangays Buluangan Poblacion Morubuan Bunay	0 0 0 0	0 0 0 0	0 0 0 0	0 0 0 0	0 0 0 0	79 106 120 0	0 0 0 0	0 0 0 0	2 Carabaos 5 (Ha) Crops 0 0 1 Carabao	0 0 0 0	0 0 0 0	P188,000	DRRM Report on 9/26/2017
El Nino	14 barangays	0	0	0	0	0	0	0	0					

Source Updating PDPFP and CLUPs: Workshop 2. October 2019

Table 5 Climate Hazard Inventory

Jordan	Flooding			Rain-induced landslide			Tsunami		
	High	Medium	Low	High	Medium	Low	High	Medium	Low
San Miguel		✓			✓				
Santa Teresa		✓			✓				
Poblacion		✓		✓			✓		
Alaguisoc					✓				
Morobuan	✓			✓					
Rizal		✓		✓			✓		
Hoskyn		✓			✓		✓		
Bugnay		✓		✓			✓		
Sinapsapan		✓			✓		✓		
Balcon Maravilla	✓			✓					
Lawi	✓				✓		✓		
Espinosa		✓			✓		✓		
Balcon Melliza	✓			✓			✓		
Buluangan	✓			✓					

Source Updating PDPFP and CLUPs: Workshop 2. October 2019

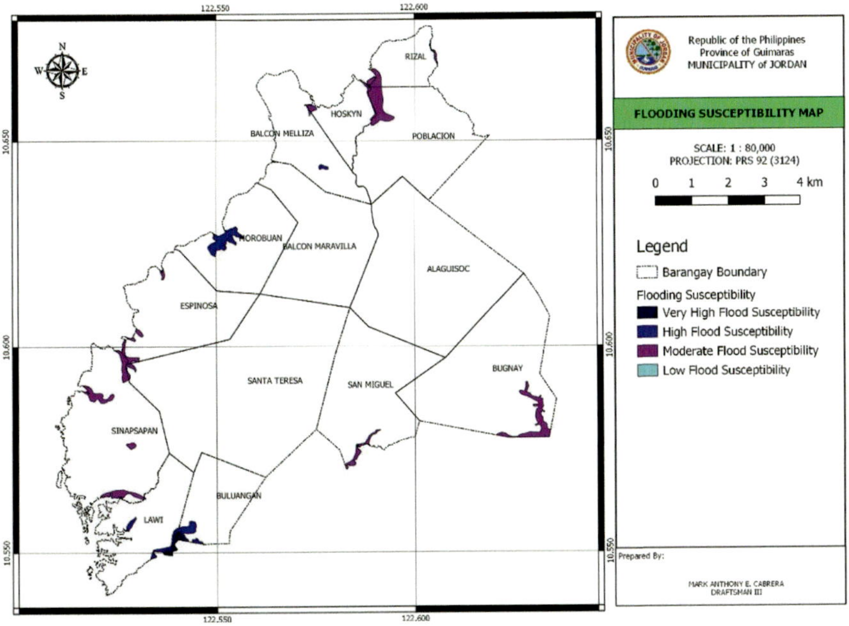

Fig. 6 Flood prone areas in Jordan

STEP 2: Scoping Potential Impacts of Disasters and Climate Change

Scope the potential impacts of hazards and climate change—identifying key areas or sectors that may be affected by climate change and natural hazards and determining likely impacts (direct and indirect).

The impact chain analysis identified increase in precipitation/rainfall and increase in temperatures as the climate stimuli that will create various impacts on the municipality. These stimuli will result in flooding, landslides, and drought and in turn will affect the population. The following are the significant impacts:

- Flooding can cause water-borne diseases and impact key facilities like the municipal government center and schools.
- Landslides can cause damage to properties, suspension of classes/work, loss of livelihood/income, loss of lives, and damage to crops. It can also reduce the delivery of basis services, inaccessibility of areas due to disrupted road networks, power and communication disruption, contamination of potable water, and loss of tourism receipts.
- Droughts will result in increased health risks/hazards, decreased agricultural production, grass/forest fires, soil salinity, coral bleaching, fish kills, potable water shortages, and irrigation water disruption. Based on historical records, Guimaras was among the areas affected by the 1989–1990 drought that resulted in an estimated 5×105 Mt of rice and corn production losses and hydropower generation loss of PhP348M and 10% cutback in water production in Metro Manila (Phil. Climate Assessment 2016).

Impact Chain Analysis for Climate Stimuli

STEP 3 Analysis of Exposure and Sensitivity of Systems to Hazards and Climate Changes

Develop the exposure database—gathering baseline map and attribute data on exposure, vulnerability/ sensitivity and adaptive capacity as basis for the Climate Change Vulnerability Assessment (CCVA) and Disaster Risk Assessment (DRA) (Fig. 7).

The systems of interests is the population exposure and sensitivity to hazards and climate system. The exposure database was established using the GIS-generated land use maps, hazard maps, and the Community-Based Monitoring System (CBMS) data for households. The exposure database will enable the analysis on the potential number of households and individuals that will be severely affected by the hazards.

Population Exposure Indicators

Barangays Hoskyn and Rizal have high population densities, high percentage of informal settlers, high percentage of young and old dependents and high percentage of households living below the poverty threshold. Barangays San Miguel and Sta. Teresa have the highest populations in the municipality. These indicators are key considerations in terms of exposure of population in these areas (Fig. 8; Table 6).

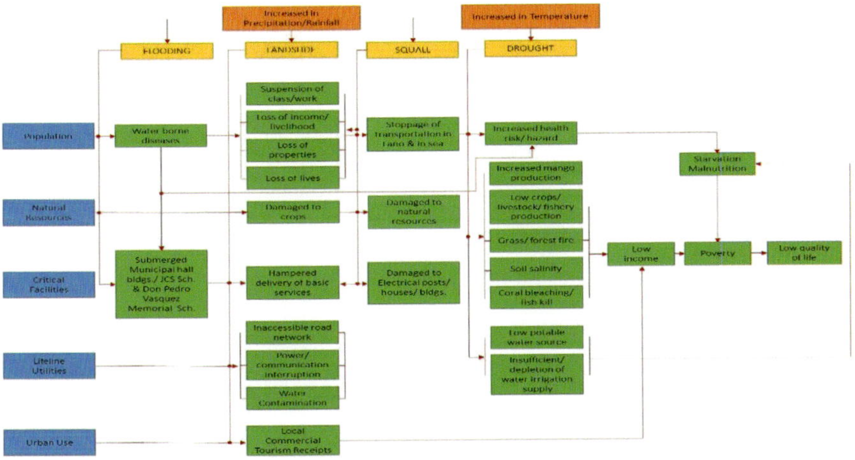

Fig. 7 Impact Chain Diagram of Jordan. *Source* Updating PDPFP and CLUPs: Workshop 2. October 2019

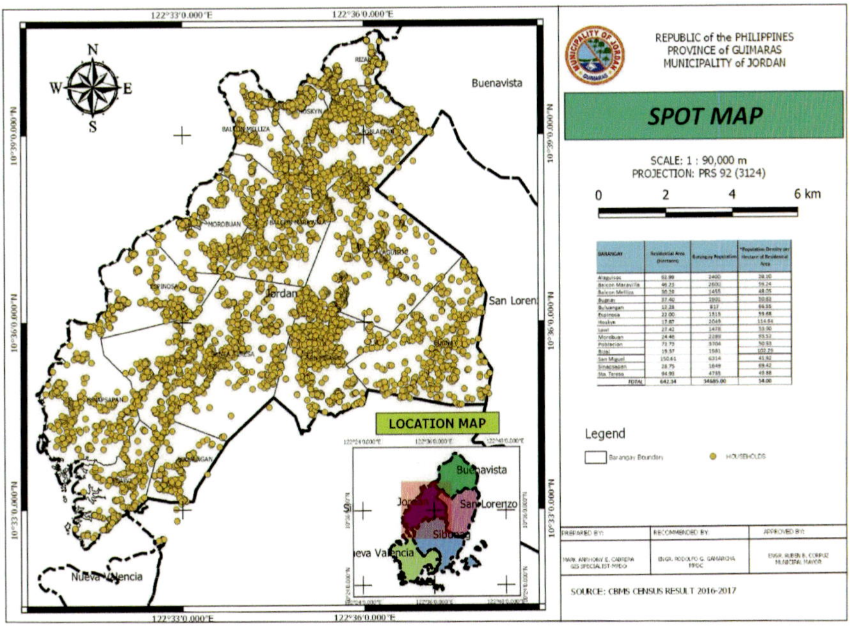

Fig. 8 Population exposure map of Jordan

Table 6 Population exposure and attributes database

Barangay	Exposure indicators			Sensitivity indicators					
A	B			C					
	Residential Area (Hectares)	Barangay Population	*Population Density per Hectare of Residential Area	Percentage of Informal Settlers	Percentage of population living in dwelling units with walls made of light to salvageable materials	Percentage of young (< 5 y.o.) and old (above 65) dependents	Percentage of people with disabilities	Percentage of Households Living below the poverty threshold	Percentage of malnourished individuals
Alaguisoc	62.99	2400	38.10	0.17	1.17	14.46	0.05	61.33	0.37
Balcon Maravilla	46.23	2600	56.24	0.96	0.96	15.99	0.07	75.08	1.24
Balcon Melliza	30.28	1455	48.05	8.72	2.45	13.88	0.09	57.77	2.03
Bugnay	37.40	1901	50.83	2.24	0.20	14.1	0.01	15.68	1.57
Buluangan	12.28	817	66.55	0.00	0.50	15.18	2.44	38.81	0.00
Espinosa	22.00	1313	59.68	0.66	0.00	14.93	0.06	52.46	1.30
Hoskyn	17.87	2049	114.64	3.09	1.23	14.01	1.02	47.74	1.29
Lawi	27.42	1478	53.90	0.00	0.57	16.71	0.05	49.72	0.56
Morobuan	24.48	2289	93.52	0.38	4.37	16.82	0.06	85.17	3.28
Poblacion	72.73	3704	50.93	10.41	2.41	12.88	0.02	39.83	0.29
Rizal	19.37	1981	102.29	54.02	0.96	15.5	1.06	77.78	0.99

(continued)

Table 6 (continued)

Barangay	Exposure indicators			Sensitivity indicators					
A	B			C					
	Residential Area (Hectares)	Barangay Population	*Population Density per Hectare of Residential Area	Percentage of Informal Settlers	Percentage of population living in dwelling units with walls made of light to salvageable materials	Percentage of young (< 5 y.o.) and old (above 65) dependents	Percentage of people with disabilities	Percentage of Households Living below the poverty threshold	Percentage of malnourished individuals
San Miguel	150.61	6314	41.92	5.28	1.76	13.7	0.28	43.57	1.49
Sinapsapan	23.75	1649	69.42	0.00	0.25	17.22	0.66	88.75	3.14
Sta. Teresa	94.93	4735	49.88	0.00	0.33	14.33	0.42	70.39	3.06

Barangay	Adaptive capacity				
A	D				
	Access to Post Disaster Financing	Philhealth Coverage	Household capacity to relocate or to retrofit	Government Capacity to Generate Job	Government Resources
Alaguisoc		414		Access to skills trainings from TESDA and livelihood program from the LGU	Budget available for skills training and livelihood program from the DSWD, Mayor's Office and Agriculture Office
Balcon Maravilla		402			
Balcon Melliza		207			

(continued)

Table 6 (continued)

Barangay	Adaptive capacity				
A	D				
	Access to Post Disaster Financing	Philhealth Coverage	Household capacity to relocate or to retrofit	Government Capacity to Generate Job	Government Resources
Bugnay		263			
Buluangan		131			
Espinosa		244			
Hoskyn		262			
Lawi		171			
Morobuan		401			
Poblacion		401			
Rizal		387			
San Miguel		705			
Sinapsapan		404			
Sta. Teresa		476			

Population Sensitivity

Barangay Hoskyn has a high percentage of households living below the poverty threshold and high population density which are sensitivity indicators that need to be considered. High population density exposes more people to hazards in a location. Owing to high poverty incidence, the residents also have limited adaptive capability and in relocating to safer locations.

Barangay Rizal has a high percentage of households living below the poverty threshold and has a high percentage of informal settlers. Morobuan has a remarkably high percentage of households living below the poverty threshold at 85.17%. Sinapsapan, Buluangan has the highest percentage of persons with disabilities and high percentage of young and old dependents. Sinapsapan, Morobuan and Lawi have the highest percentages of young and old dependents.

Population Adaptive Capacity

The percentage of population with Philhealth coverage is low for all barangays. The highest was only 24% (Sinapsapan) and the lowest at 10% (Sta.Teresa). All barangays have access to skills trainings from TESDA and livelihood program from the LGU. There is also budget available for skills training and livelihood program from the DSWD, Mayor's Office and Agriculture Office.

STEP 4: Climate Change Vulnerability Assessment

Conduct a CCVA—Identification of vulnerable areas and sectors by analyzing exposure, sensitivity and adaptive capacity to the various climate stimuli.

The climate change vulnerability assessment (CCVA) assesses the locality's vulnerabilities to various climate related stimuli. As a tool, it uses a qualitative approach in establishing the level of vulnerability of identified areas or sectors of interest. Vulnerability shall be based on the extent of exposure, and an analysis of the sensitivities and adaptive capacities. This will inform the identification of decision areas and be the subject of further detailing for the identification of area specific planning implications and policy interventions (HLURB Supplemental Guidelines on Mainstreaming Climate Change and Disaster Risks in the CLUP, 2015). In the guidelines, vulnerability of the system to the expected climate stimulus is the interplay of exposure, sensitivity, and adaptive capacity.

Vulnerability of Population to Increased Temperature

Based on the assessment, most barangays are categorized as low and only five barangays are under medium low. This indicates that in terms of increased precipitation, barangays are not highly vulnerable (Fig. 9).

STEP 5: Disaster Risk Assessment

Conduct a DRA—Identification of risk areas by analyzing hazard, exposure, and vulnerability.

The primary objectives of disaster risk assessment is to determine the risk areas and to analyze adaptive capacities of identified risk areas. This is done by undertaking

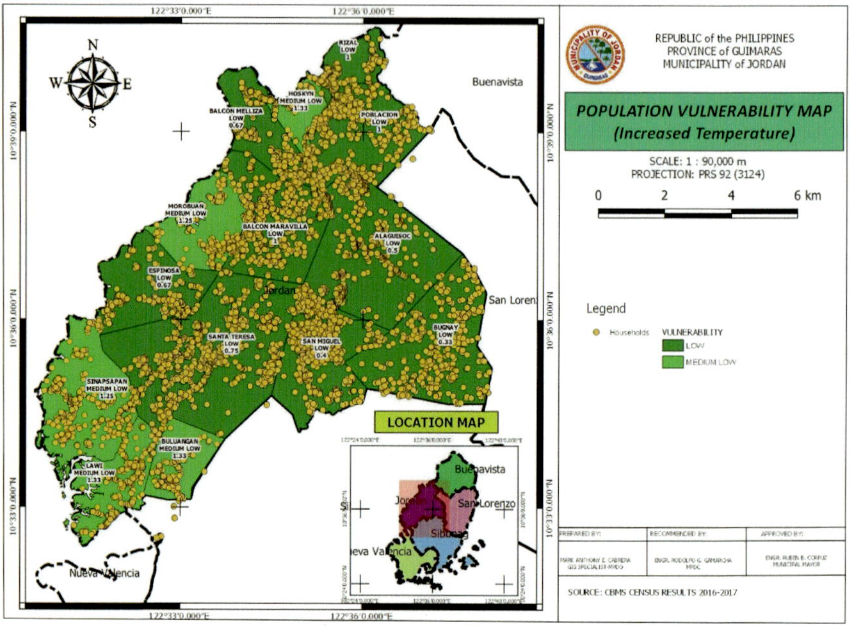

Fig. 9 Population vulnerability map (increased temperature)

an analysis of exposed elements, estimation of the severity of the consequence score and determination of risk scores. Based on the analysis, DRA summary decision areas and issues matrix risk maps will be prepared.

Flood Risk

There are six barangays where populations are subject to a high risk to flood hazards. These are Barangays Buluangan, Espinosa, Hoskyn, Morobuan, Poblacion, and Rizal. There are also population in areas identified as Moderate Risk in Barangays Balcon Melliza, Bugnay, Lawi, San Miguel, and Sinapsapan (Fig. 10).

STEP 6: Evaluating Vulnerability and Risk

Summarize findings—Identification of priority decision areas/sectors based on the combined level of risks and vulnerabilities, identification of risk management options, climate change adaptation and mitigation options.

Vulnerability of the Municipality to Climate Change

After completing the processes and steps, this section aims to identify the major decision areas based on the combined risks and vulnerabilities and to identify a menu of disaster risk reduction and climate change options within major decision areas. This was done by overlaying the risk and vulnerability maps (Fig. 11).

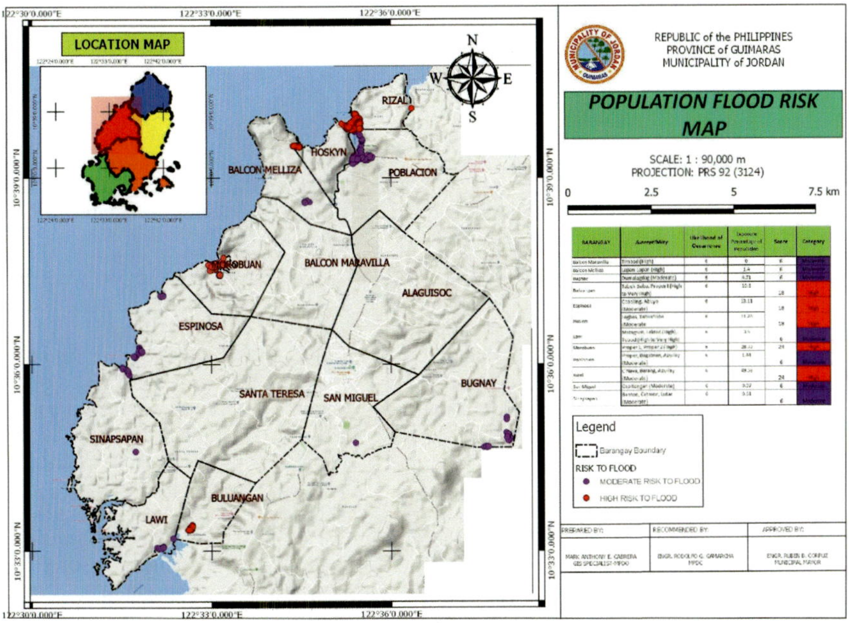

Fig. 10 Population flood risk map

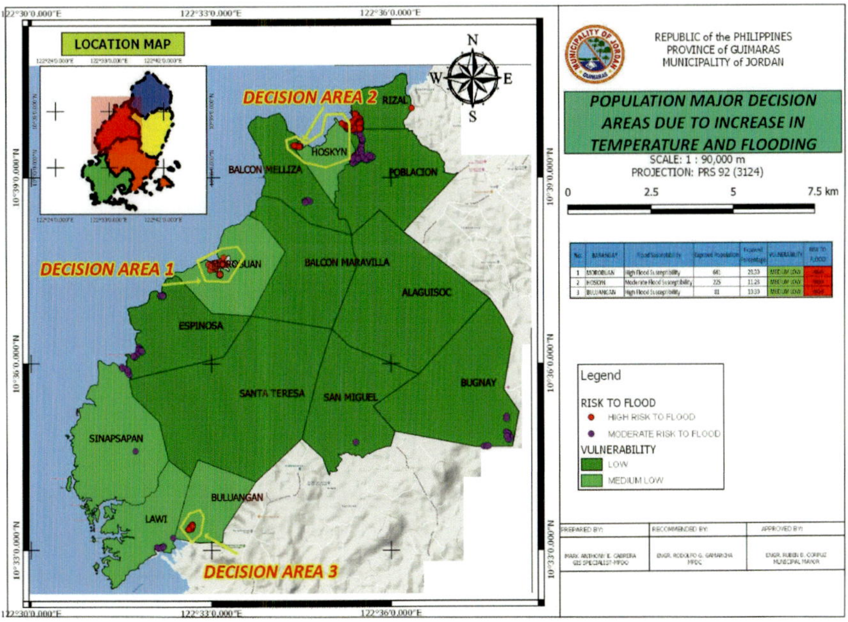

Fig. 11 Population major decision areas due to increase in temperature and flooding

7 Summary of Findings

See Table 7.

Table 7 Population decision area matrix (flood and increase in temperature)

Decision Area/s	Description	Problems/ Hazards	Impacts/ Implications	Policy interventions
1. Morobuan	• Coastal barangay where most of the residents are informal settlers living along the coastline • High risk to flooding as to vulnerability	• 146 households affected by flood and increase temperature • Presence of SC & children / PWD • Livelihood activity affected	• Can cause casualty/ injuries/ death • Presence of water-borne diseases • Increase malnutrition cases • Damaged on properties and income generating sources • Difficulty on mobilization/ evacuation	• Advocacy activities on disaster preparedness/ SIMEX/drills • Passage of resolution related thereto • Proposed relocation for affected families as per Local Shelter Plan • Implementation of Policy on preemptive/ Forced evacuation • Installation of early warning design • Protocol on camp/ evacuation management • Conduct of ocular inspection, assessment, and evaluation of buildings/ structure as to its structural stability and integrity

(continued)

Table 7 (continued)

Decision Area/s	Description	Problems/ Hazards	Impacts/ Implications	Policy interventions
2. Hoskyn	• Coastal barangay where most of the residents are informal settlers living along the coastline • Most areas are in 18% and above slopes	• 56 households affected by flood and increase temperature • Presence of SC and children / PWD • Livelihood activity affected • High population density • Problem on "dagsa" along coastline	• Difficulty on mobilization/ Evacuation • Damaged on properties • Increase exposure diseases/ malnutrition	• Advocacy activities on disaster preparedness/ SIMEX/drills • Passage of resolution related thereto • Provision of seawall • Repair and improvement of the damaged structures adopting disaster-resilient design • Enhance health programs and services • Enhanced programs and services for senior citizens/PWD/Youth/ children • Re-construction of dilapidated buildings/ structures adopting disaster-resilient design • Conduct annual maintenance activities • Construction of alternative road • Strict implementation of zoning ordinance • Implementation of RA 9003 (Ecological Solid Waste Management Act of 2000)
3. Buluangan	• Rural barangay with • Source of water supply for brgys. San Miguel, Sta. Teresa and Alaguisoc	• Contaminated potable water supply • 19 households affected of flood and increase in temperature	• Stoppage of water supply services • Increase in water-borne diseases • Hamper mobilization activities/ livelihood	• Advocacy activities on disaster preparedness/ SIMEX/drills • Passage of resolution related thereto • Provision of filtration facilities • Annual assessment/ maintenance of waterlines and other related facilities • Enhanced health programs and activities

8 Conclusion and Policy Recommendations

Water security enables economic growth and fosters a populace that is healthy and prosperous. In light of the likelihood that climate change will increase climate variability and the likelihood of floods and droughts, the following policy recommendations are provided to mitigate these risks. (ADB 2020):

- Make water the centerpiece of sustainable rural development through community protection and farmland flood retention. This will enable locally affordable investment, income generation, proper management and operation, and improved public welfare.
- Strengthen water systems to avoid water-related disasters and prepare for climate change. Rebuild better and use nature-based solutions to protect communities from the negative impact of future disasters.
- Pilot nature-based solutions for flood risk management. Flood risk mapping, land use planning, and early warning systems are cost-effective investments to limit water-related risk. Gray, green, and residual risk investments are complementary.

In addition, flood risk mapping, land use planning guidelines, including restrictions on land development in flood plains, and early warning systems are cost-effective investments to limit water-related disaster risks. Integrating structural approaches, nature-based solutions, early warning systems, and residual risk instruments are crucial efforts. Nature-based solutions can be a cost-efficient complement to gray infrastructure, building water disaster resilience and providing environmental, social, and economic co-benefits. Nature-based solutions may help countries "build back better" after the coronavirus disease 2019 (COVID-19) pandemic. Paying farmers for ecosystem services in exchange for protecting catchments and urban drainage systems are examples of nature-based solutions. (ADB 2020).

References

Asian Development Bank (2020) Asian water development outlook 2020: strengthening water security in asia and the pacific, Manila, Philippines

Babel MS, Shindea VR, Sharmab D, Dang NM (2020) Measuring water security: a vital step for climate change adaptation. Sci Total Environ 185:109400. https://doi.org/10.1016/j.envres.2020.109400

Jordan, Guimaras MDRRM Plan (2017–2019) Municipal Disaster Risk Reduction and Management Plan, Jordan, Guimaras, Philippines

Jordan, Guimaras CDRA (2020) Climate and disaster risk assessment, Jordan, Guimaras, Philippines

NEDA (2018) Philippine water supply and sanitation master plan 2018–2040. National Economic and Development Authority, Manila, Philippines

PEMSEA and Provincial Government of Guimaras, Philippines (2018) The second state of the coasts of Guimaras Province. Partnerships in Environmental Management for the Seas of East Asia (PEMSEA), Quezon City, Philippines

Flood Risk Assessment in Philippines: Iloilo City Flood Risk Management Plan

Chris Velmida Perfas Gianzon

Abstract Case study author brings focus to Iloilo City in the Philippines, telling the story of a highly urbanized area that is increasingly facing the negative impacts of flooding occurrence due to precipitation and river overflow. Case study author presents data on climate hazard and an analysis of vulnerability as the foundation for an evaluation of risk in the area and maps areas of flood risk in Iloilo City, Philippines. While the local Government of Iloilo City has taken major steps in addressing climate change vulnerability and impacts and flood hazards through tangible projects and the development of sustainability policy and institutional frameworks, the case study author identifies gaps that need to be addressed for flood management to be holistic and effective in addressing the impacts of climate change. Needed changes include local climate action and disaster management planning, a focus on flood control measures, and the development of the institutional capacity required for operation and maintenance over time.

Keywords Philippines and flood risk management · Climate and disaster risk assessment and Philippines · Iliolo city and disaster response · Climate risk assessment · Climate risk management · Disaster risk management · Riverine flooding · National disaster planning · Vulnerability · Adaptive capacity · Early warning systems

1 Introduction and Background

Flooding is one of the most destructive natural disasters exacerbated by the impacts of climate change in Iloilo City, Philippines. Flood risk occurrence is a combination of natural and anthropogenic factors, which calls for a better understanding of its spatial extent. The aim of this paper is to identify and map areas of flood risk in Iloilo City, Philippines.

C. V. P. Gianzon (✉)
Department of the Interior and Local Government Region VI, Iloilo City, Philippines
e-mail: cgianzon1025@gmail.com

K. Kim and C. Atkin (eds.), *Cross-Cultural Perspectives on Climate Change Adaptation*, Cities and Nature, https://doi.org/10.1007/978-3-031-50365-8_17

Natural disasters are regarded as the most serious challenge that must be addressed at the global, regional, and local levels. Climate change may increase the frequency, amplitude, and seasonality of extreme events such as floods, implying that concurrent flood hazards, which are essential for urban flood risk management, may occur more frequently in future (Duan et al. 2015a, b; Pedersen et al. 2012). Urbanization is also an essential factor in increasing flood risk in the cities through increasing runoffs affecting communities downstream (Cloke et al. 2013; Duan et al. 2016).

Floods are one of the world's most devastating natural disasters, claiming more lives and causing more property damage than any other natural phenomenon. As a result, floods are one of the greatest challenges to weather prediction.

In the case of developing countries, the Philippines is one of the countries that are often affected by floods brought by typhoons due to its location on the Pacific Rim. The Philippines was identified as the third most vulnerable country to climate change in the year 2017 by the World Risk Index (2017) for the highest disaster risk and exposure to natural calamities. Due to the geographical location of the Philippines, the country is visited by an average of 20 typhoons every year and five (5) or more of which are destructive.

As the Philippines is located in the tropics, many of the typhoons generated in the Pacific Ocean hit the Philippines every year, exposing the country to difficult geographical and metrological conditions. Many towns are located in the lowlands and are therefore more likely to be affected by floods, especially in the rainy seasons. Cities with low laying areas and within a coastal area are one of the most affected areas of the impacts of climate change. One of the cities with these conditions is the City of Iloilo in Western Visayas, Philippines.

Iloilo City, the capital of the Iloilo region in the Province of Iloilo, located in Panay (West Visayas) in the central Philippines, was expected to become the center for West Visayas's economic growth, having the seventh-largest population in the country at the time of the appraisal of the project. However, geographically, the city consists largely of plains with an average elevation of 3 m above sea level and is regularly affected by floods caused by typhoons and torrential rain. The city was also devastated by six flood cases between 1998 and 2001, affecting 119,028 households in total.

2 Study Area

Iloilo city is located on the central coastline of Iloilo province on the island of Panay in the Philippines. The city has a total landmass of 7834 hectares based on the approved Base Map as evaluated by the Department of the Environment and Natural Resources (DENR) Land Management Bureau with six districts composed of 180 barangays all classified as urban (Iloilo City Planning and Development Office 2018).

The average temperature in Iloilo city is around 32 degrees Celsius. The warmest month is May, with a mean temperature of 28.5 °C, while the coldest month is January, with a low of 26 °C. The city has two distinct seasons, akin to a modified Coronas type I climate: dry from November to April and rainy the rest of the year. During the rainy season, the average monthly rainfall is around 250 mm, with maximum rainfall of 300 mm around July and August. Tropical cyclones (TCs) are more common near the end of the year, around November, and less common around February.

Iloilo City is a highly urbanized local government unit with a total population of 484,290 as of 2021 (Iloilo City Population Office 2022). From the period 2010 to 2015, the urban population of Iloilo grew by 1.02% annually but between 2015 and 2021 the population was at 1.25% and is on the increasing trend. The Iloilo city urban population projections by 2030 will increase by 3.69%. Iloilo city is the seventh-largest population in the country and is expected to become the center of Western Visayas' economic growth.

Iloilo city, officially known as the City of Iloilo, is a highly urbanized and historic city on Panay Island in the Philippines. It was founded in the sixteenth century. It is the capital city of the same name province as well as the main city of the Iloilo–Guimaras Metropolitan Area (Fig. 17.1).

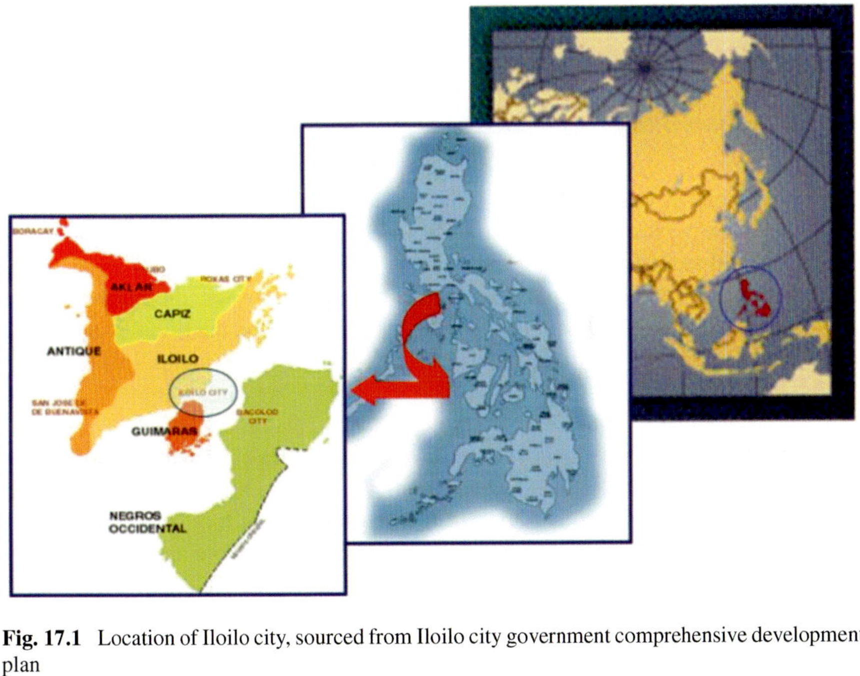

Fig. 17.1 Location of Iloilo city, sourced from Iloilo city government comprehensive development plan

Fig. 17.2 Location and boundary of Iloilo city, sourced from Iloilo city government comprehensive development plan

2.1 *Location and Geophysical Features*

The city of Iloilo is one of 16 cities found in the Western Visayas Region (Region VI). Located on the southwestern edge of Iloilo Province on the island of Pana; Iloilo City has a total area of 7834 ha.

The municipalities of Leganes and Pavia bound the city on the northeast, the municipalities of Pavia and San Miguel on the northwest, and the Iloilo Strait on the south. It is divided into six districts, where 180 barangays are situated–all of which are classified as urban.

The Latitude and longitude coordinates are 10.720321, and 122.562019, respectively. Iloilo City has more than 70% of its total land area classified as residential, commercial, and industrial. Iloilo's geographical feature and strategic location are significant factors in its eventual urbanization in the nineteenth century. Its strategic location is accorded by its river mouth facing a narrow strait protected by the islands of Guimaras from monsoon winds, typhoons, and tsunamis and as an inter-island port with a natural harbor gave impetus to the town's urbanization (Fig. 17.2).

2.2 *Topography and Slope*

About 90% of the city's landmass has an elevation of 2.64 m above sea level, while the remaining 10% has an elevation of 5.19 m. The slope category of the city falls between 0 to 3% and is nearly level with the land (Fig. 17.3; Table 17.1).

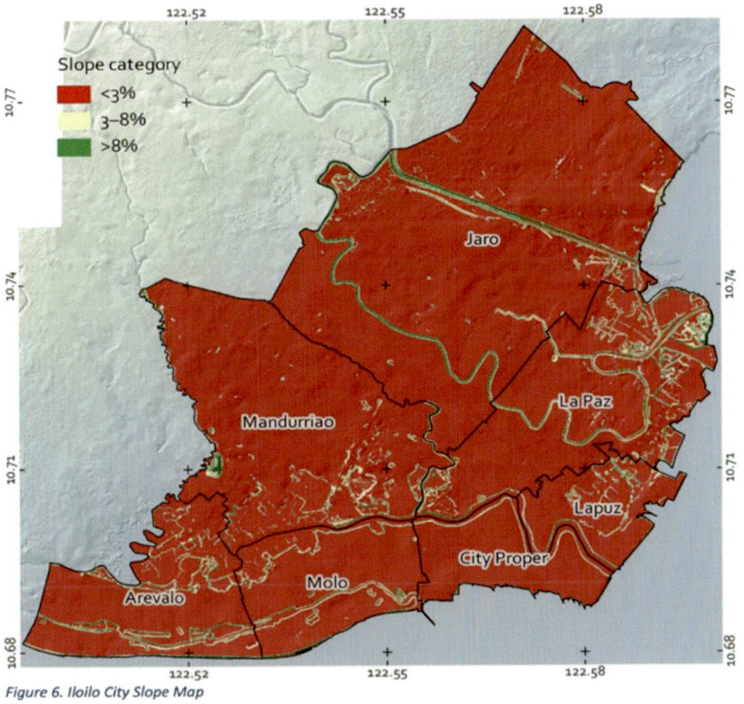

Slope category	Area, m²	Area, ha.	Percentage
<3%	65,701,510	6,570.1510	83.87%
3–8%	4,095,141	409.5141	5.23%
<8%	1,769,213	176.9213	2.26%
No data	6,774,136	677.4136	8.65%

Source: Ecological Profile 2018, CPDO

Fig. 17.3 Iloilo city slope map, sourced from Iloilo city government ecological profile, 2018

Table 17.1 Details of slope analysis map (Ecological Profile, 2018, CPDO)

Slope category	Area, m^2	Area, ha	Percentage
< 3%	65,701,510	6,570.1510	83.8%
3–8%	4,095,141	409.5141	5.23%
< 8%	1,769,213	176.9213	2.26%
No data	6,774,136	677.4136	8.65%

2.3 *Soil Type*

According to a 2010 study by the World Bank, Iloilo city was formed through alluvial deposits. A majority of the city's lower substratum soils are classified as clay and

Table 17.2 Details of slope analysis map (Government Ecological Profile, 2018, CPDO)

Soil type	Area (hectares)	Percentage (%)
Hydrosoil	241.92	3.5
Sand (beach)	1,347.90	19.7
Clay (Sta. Rita)	4,692.00	68.50
Sandy (loam)	570.3	8.30

loam, without stones and gravel, with other portions are made up of hydrosol and sand.

In profile, the surface soil (25–30 cm) is black to dark brown, moderately coarse, granular, and highly plastic clay. The lower substratum is brown silt loam friable and good fine granular without stones/gravels. The land as to formation and origin is of recent alluvial deposits. Table 17.2 shows the type of soil and the percentage per area in Iloilo city (Fig. 17.4).

2.4 River System

There are three major rivers in the City of Iloilo: the Jaro river, the Batiano river, and the Iloilo river. The upstream of the Jaro river serves as a potable water supply for the city[4]. The mouth of the Iloilo river has a natural river wharf, considered one of the country's best natural harbors, and is used for port and warehousing industries from the time of the Spanish occupation up to the present. In addition, Iloilo city is surrounded by beaches and coastal zones along the whole of the eastern and southern portions of its territory (Fig. 17.5).

3 Methodology for Hazard, Vulnerability, and Flood Risks Management

In this study, various basic thematic layers were created from different source including map, field study, satellite image and secondary data based on multicriteria analysis method.

This research features the employment of 6-Step CDRA methodology suggested by the Department of Human Settlements and Urban Development (DSHUD) (formerly HLURB) in its Supplemental Guideline for Mainstreaming DRR and CCA in CLUP (HLURB et al. 2015). This guideline proposes a systematic process of assessing risk that starts with data gathering on climate change and hazard information and analyzing their potential impacts to elements at risk.

The CDRA is a methodology to understand the potential impacts of hazards to people and their properties. It also allows assessment of people's vulnerabilities and

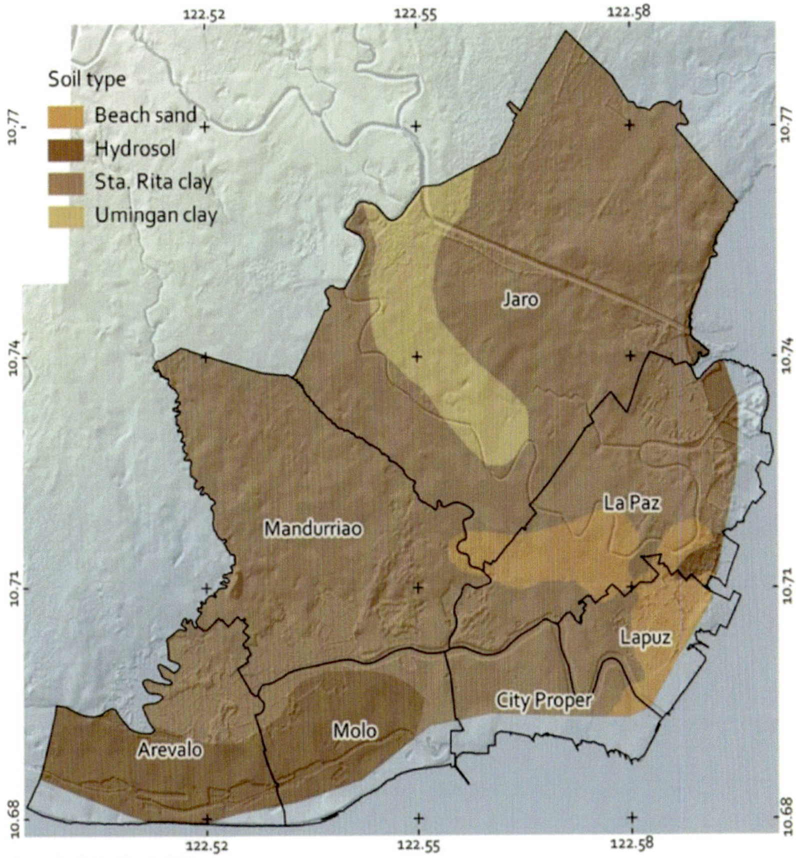

Fiaure 7. Iloilo Citv Soil Map

Soil Type	Area (Hectares)	Percentage (%)
Hydrosol	241.92	3.50%
Sand (Beach)	1,347.90	19.70%
Clay (Sta. Rita)	4,692.00	68.50%
Sandy (Loam)	570.3	8.30%

Source: Ecological Profile 2018, CPDO

Fig. 17.4 Iloilo city soil map, sourced from Iloilo city government ecological profile, 2018

capacities which largely determine the outcomes of disasters. It goes beyond the conventional "risk assessment" as it incorporates climate-related hazards and their prospective impacts on the behavior or characteristics of hazards.

Additionally, the 9-step flood risk assessment devised by Dr. Kim Kwi-Gon has been referenced to come up with a comprehensive Flood Risk Assessment study using maps and quantifiable factors with the following steps:

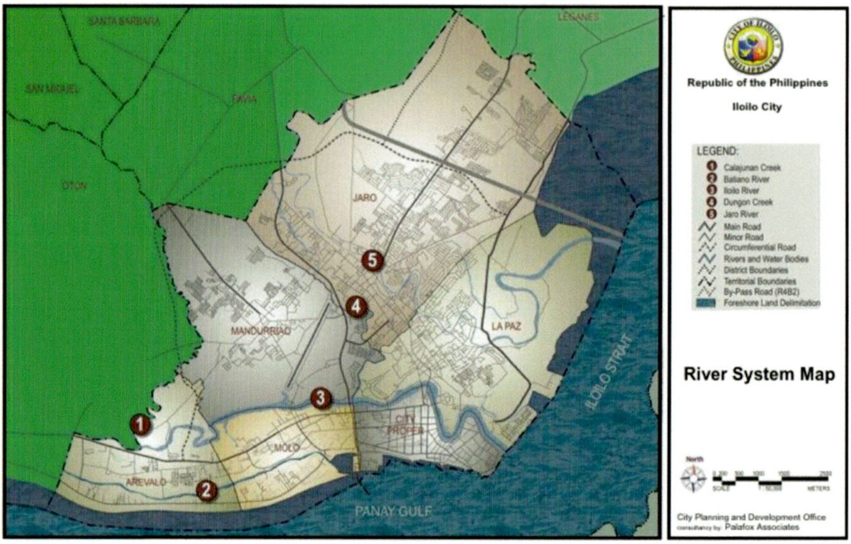

Fig. 17.5 River map system, sourced from Iloilo city government comprehensive development plan

Step 1: Hazard Identification
Step 2: Delineation of the boundary of landscape district and place unit (location map)
Step 3: Delineation of the boundary of the catchment area
Step 4: Preparation of an inventory map for each factor showing how it varies over the study area
Step 5: Creation of composite maps by overlay
Step 6: Analysis of composite maps to make inference relevant to flooding to risk assessment
Step 7: Evaluation of risk (probability of damage and losses) and risk mapping
Step 8: Planning for appropriate remediation and strategies and actions (measures)
Step 9: Monitoring with ICT application sand periodic evaluations.

The above methodologies for flood risk assessment will be used to develop analytical procedures which will be used to produce estimates of likely future damages and losses. With a combination of indicator data, use of Geographical Information Systems and statistical analysis of flood probabilities and populations at risk, a number of flood-related maps and figures will be generated.

4 Hazard Analysis

Iloilo city is surrounded by bodies of water. It is along the coastline of Panay Islands and has several rivers within the city's territory. Recent studies show that Iloilo City has a high risk of sea-level rise and coastal flooding due to climate change. It has the possibility of being wiped out from the map in 2050 (Climate Central, 2019). Iloilo city and its outlying areas, stretching over low and flat land, are vulnerable to large-scale flooding and mudslide disasters frequently during the rainy season (PIA, 2019).

Projected Climate Change in Iloilo City

The Philippine Atmospheric, Geophysical and Astronomical Services Administration -Department of Science and Technology projected climate change in the country using the mid-range emission scenario (A1B) developed by the Intergovernmental Panel on Climate Change. The A1 scenario encompasses rapid economic growth, a global population rise to nine billion then a gradual decline, the quick spread of new and efficient technologies, and extensive social and cultural interactions worldwide. The A1B scenario has a balanced emphasis on all energy sources, including both fossil and non-fossil fuels.

Based on projections, Iloilo city will be affected by three climate change drivers: increasing temperature, changes in rainfall in various seasons, and increasing occurrences of extreme events in 2020 and 2050.

Increasing Temperature

The temperature in Iloilo is projected to increase in future and by 2020, the temperature increase will range from 1.0 to 1.3 °C. This will likely affect crop yields, forestry, and coastal and marine life.

In Iloilo city, this may affect aquaculture activities and cause a heat island effect, which could affect public health and energy consumption. By 2050, the mean temperature is expected to rise between 2.0 °C up to 2.4 °C, worsening effects on natural resources including the human population (see Fig. 17.6). In Iloilo, stakeholders expect that the increase will directly affect watersheds and, consequently, the city's water supply. It is also expected to further run down the already declining agricultural production in city fringes, including fishery. More importantly, temperature increases could amplify the strength of storms.

The general trend of projections for all scenarios in both the mid and late century is an increase in temperature. The scenario under RCP 8.5 is hotter than that under RCP 4.5. And under RCP 8.5 all the months in both the mid and late century exceed the historical hottest of the year.

Changes in Precipitation

International Panel on Climate Change studies cite that increases in temperature catalyze evaporation, leading to heavier precipitation. Generally, Iloilo City is likely to experience wetter seasons in the coming years (Fig. 17.7).

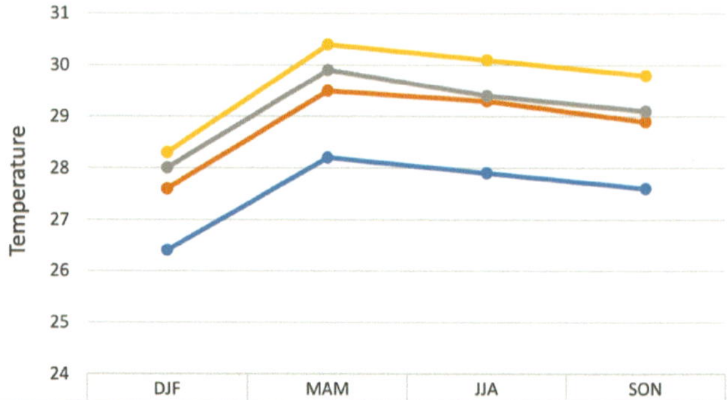

	DJF	MAM	JJA	SON
Baseline (1971-2000)	26.4	28.2	27.9	27.6
RCP 8.5 Lower Bound	27.6	29.5	29.3	28.9
RCP 8.5 Median	28	29.9	29.4	29.1
RCP 8.5 Upper Bound	28.3	30.4	30.1	29.8

Fig. 17.6 Iloilo temperature forecast. *Source* Iloilo City Government Comprehensive Development Plan?

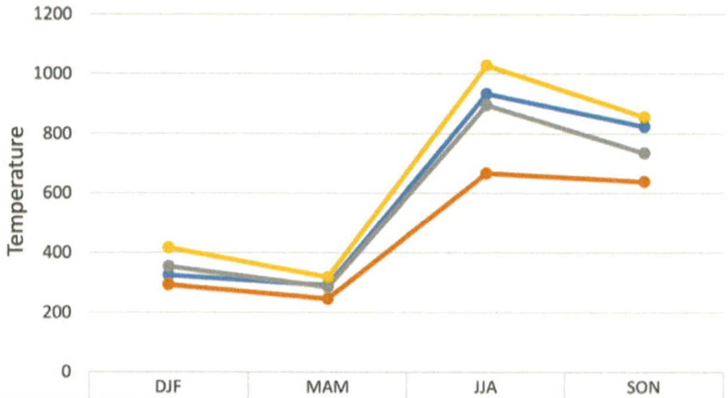

	DJF	MAM	JJA	SON
Baseline (1971-2000)	324	290.6	932.8	823.3
RCP 8.5 Lower Bound	293	246	667.9	639.9
RCP 8.5 Median	353.2	284.6	895	734.6
RCP 8.5 Upper Bound	416.8	318.2	1028.6	856.2

Fig. 17.7 Rainfall changes in 2050 (2036–2065) based on RCP 8.5 for Iloilo city, sourced from Iloilo city government comprehensive development plan

Increase in Extreme Events

The temperature increase by 2020 is also seen to contribute to the high increase in the number of days with a temperature higher than 35 °C at 211 and 569%. By 2050, an extreme temperature increase is expected during the months of March, April, and May, and June, July, and August. The days with extreme rainfall of more than 200 mm per day will remain almost constant with a 1-day increase from 4 days in 2020 and back to 4 days by 2050. By 2020, there will be days with extreme rainfall of > 200 and > 300 mm.

In terms of frequency of extreme daily rainfall, Iloilo can expect at least 10 days with a high daily rainfall of 101–150 mm per day in the next decade. While the frequency declines by 2050, the city cannot afford to neglect flood mitigation since its history shows a high susceptibility of many areas to flooding, even with a much lower rainfall volume (Table 17.3).

Mapping of the Study Site

Hazard Maps (Flood Focused)

Iloilo City is exposed to several hydrometeorological and geological hazards particularly flooding, storm surge, and drought. Current scientific findings are also in agreement that climate change further exacerbates the extremities of hydrometeorological hazards. Typhoons, for instance, become stronger, more frequent, and larger in terms of the geographical scope of impact. Hydrometeorological hazards and drought that may affect the cities can become more unpredictable which makes risk assessment an even more challenging task. It is, therefore, necessary to map the spatial extent of hazards that affect the city to identify the exposed populations and other elements at risk.

According to the hazard maps from MGB, DOST-NOAH, and PHIVOLCS, the hydrometeorological hazards that affect Iloilo city include flood and storm surges.

Table 17.3 Rainfall and temperature in Iloilo city, Philippines

	Observed baseline (1971–2000)	Change in 2020 (2006–2035)	Change in 2050 (2036–2065)	Findings
No of days with Temp > 35 °C	460	1431	3076	Increasing number of hotter days
No. of dry days (rainfall < 2.5 mm)	7839	5227	3075	Lesser but still significant number of drier days
No of days with Extreme rainfall > 200 mm	4	5	4	Higher possibility of flooding due to excessive rainfall
Sea level rise	Project change by 2100 relative to 1986–2005 Global mean sea level: 20 cm for RCP8.5			Greater exposure of coastal areas to sea-level rise and other hazards such as storm surge and tsunamis

Exposure to Flood

Based on the hazard maps provided in the CDRRMP of Iloilo city, a huge portion of the city, both urban and rural barangays, is exposed to different levels of the flood (Fig. 17.8).

Based on Fig. 17.9, there are 41 barangays that have very high and high susceptibility to flooding, namely: Montinola, Calubihan, TayTay Zone 2, Desamparados, Cuartero, Ma. Cristina, San Roque, Simon Ledesma, Democracia, Tabuc Suba, Cubay, Balabago, M.V. Hechanova, Buhang, Quintin Salas, Tacas, Dungon A, San Isidro, Sambag, Ungka, Dungon B, Lanit, Buntatala, Tagbak, Camalig, Arguelles, EL 98, Muntinola, Tabucan, San Rafael, Bolilao, Bakhaw, Buhang Taft North, Santa Rosa, PHH Blk 17, PHH Blk 22, San Isidro, Banuyao, Hinactacan, Magsaysay Village, So-oc (CDRRMP of Iloilo City, 2018).

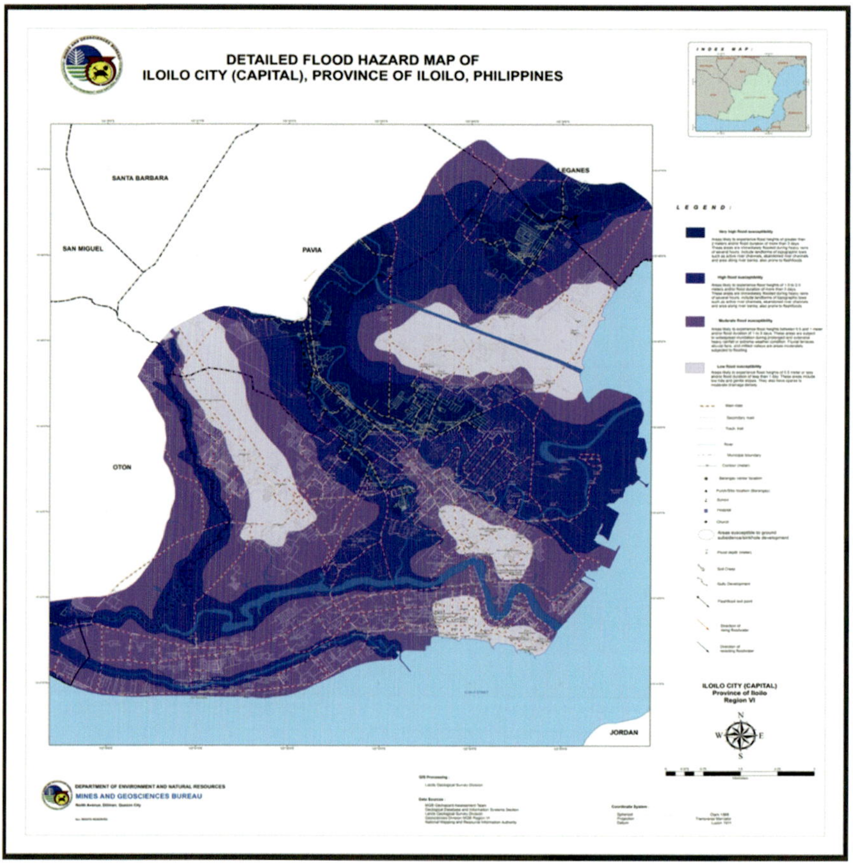

Fig. 17.8 Flood hazard map of Iloilo City

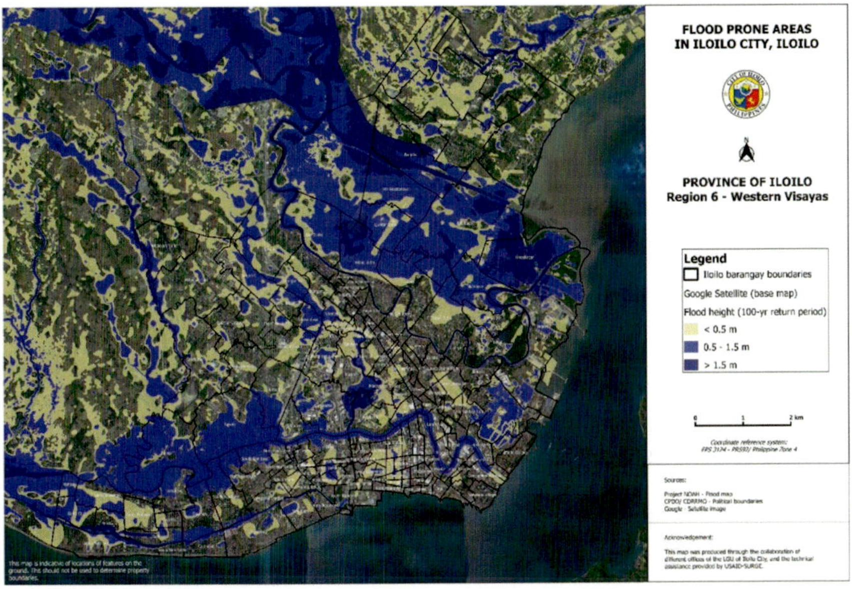

Fig. 17.9 Exposure to 100-year flooding, sourced from Iloilo city government comprehensive development plan

4.1 Exposure to More Frequent and Extreme Flooding

The flood hazard map produced by DOST-NOAH for Iloilo city is providing another different scenario of exposure to flood hazards. The map above shows a 100-year flood hazard scenario for the city. The "100-year" means that there is a single chance (1%) of the occurrence of a major flood in a 100-year period. Major river tributaries' exposure to hazards ranges from < 0.5 m to > 1.5 m flood heights. This 100-year flood does not mean that it will only happen once in a century. As the climate becomes more unpredictable and extreme, a 100-year flood might happen more frequently.

5 Vulnerability Assessment and Mapping

This Vulnerability and Adaption Assessment report contains a multi-sector view of the hazards and impacts of climate change on Iloilo City, and the perceived level of threat to the development potential of the city. Cross-referencing this threat with existing initiatives to avert adverse impacts of these hazards, this report also describes the adaptation and disaster risk-reduction measures that the city has been pursuing.

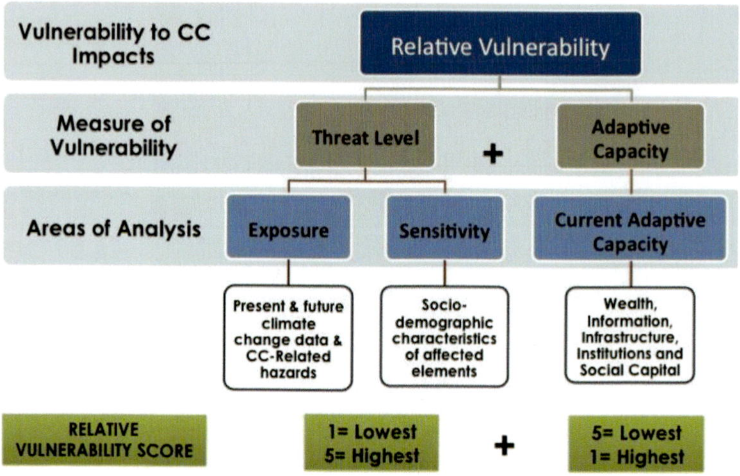

Fig. 17.10 UN habitat vulnerability and adaptation assessment. *Source* UN Habitat Vulnerability and Adaptation Assessment

Vulnerability Framework

The key components of United Nations Habitat's Vulnerability and Adaptation Assessment are the evaluation of the city's relative vulnerability to various climate-related hazards, focusing on the three determinants that show the specific hazards and systems affected (Fig. 17.10).

Exposure is defined as the degree of climate stress upon a particular unit of analysis (i.e., the city or parts thereof), and may be characterized as a long-term change in climate conditions, or changes in climate variability, including the magnitude and frequency of extreme events focusing on the city. These determinants are indicated by empirical data on regional climate variability.

The second component is sensitivity, which shows the level of impacts of climate-induced, natural, and human- made hazards affecting various systems in a city. It is the degree to which systems are affected and refers to characteristics inherent to the system, or those factors and conditions that predispose an element or the system to risks. In this report, these refer largely to the physical and socio-economic attributes of exposed elements like people, places, and institutions.

The third component is adaptive capacity, demonstrating the city's autonomous and planned actions to climate change. This determinant zeroes in on mostly planned measures that directly or indirectly mitigate or temper the impacts of climate change—which is an integral to long-term system adjustments or adaptation. This assessment examined the present measures and how these would likely fare with increased risks due to changes in exposure and/or sensitivity factors (Table 17.4).

Combining the scores from the threat level and adaptive capacity, the relative vulnerability of the city of Iloilo by hazard is 6.42 out of 10 (Fig. 17.11).

Table 17.4 Relative vulnerability by hazard

Climate change hazard	Ave. threat level (5-highest)	Ave. adaptive capacity (1-highest)	Relative vulnerability rating (10-most vulnerable)
Flooding	3.83	2.69	6.52
Storm surge	2.56	3.15	5.71
Typhoon	3.65	2.64	6.29
Sea level rise	3.2	3.48	6.68
Drought	3.7	3.22	6.92
Ave. score			6.42

Source Climate and disaster risk assessment tool 6 step method (scoring)

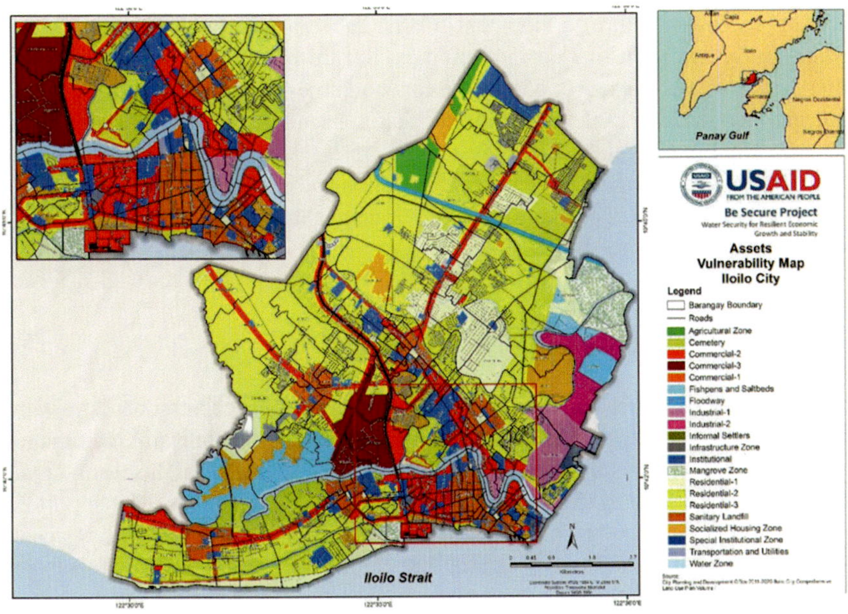

Fig. 17.11 Exposed urban uses based on the distribution of land uses. *Source* USAID

Iloilo City is considered a highly urbanized city characterized by an increasing number of populations, economic activities, and building footprints. The current land use map shows the variety of land uses in the city. The following generalizations can be made:

- The City Proper District, the eastern section of Mandurriao, and the western sections of La Paz and Lapuz are dominated by commercial land uses (i.e., trading and shopping areas.) with medium-density residential areas.

- Lapuz and La Paz District house most of the industrial zones in the areas, i.e., ports and energy power plants.
- Other districts in the north and west remain dominantly residentials with a long stretch of commercial uses along major roads.
- A significant number of critical institutions that provide social services are found in and near the City Proper District.

6 Flood Risk Management

The results from steps 1–9 of Dr. Kim's framework are critical in mainstreaming DRR including CCA in land use and development planning. Hazard and risk maps allow the identification of areas that are relatively safe from hazards and thus can be the focus of present and future development efforts. The same maps identify areas that are hazardous for human habitation and activities. Hazard-prone areas particularly in the cities, however, also present opportunities for economic development because of their strategic location, favorable topographical features, and historical significance. It is thus important to formulate and apply spatial strategies and development options to ensure that those areas remain viable for habitation and economic development. This step identifies major decision areas (MDAs) or management (Fig. 17.12; Table 17.5).

6.1 Adaptive Capacity Analysis

Urban resilience describes the ability of a city to withstand or accommodate stress and shocks such as climate change impacts and natural hazards, while still maintaining its function. At the urban scale, resilience will depend on the ability to maintain essential assets as well as to ensure access to services and functions that support the well-being of citizens. This is particularly prevalent for populations lacking access to financial, material, and social capital that can be used to buffer the stresses and enable long-term adjustments (Table 17.6).

Urban populations depend on interrelated and interdependent urban systems (e.g., infrastructure, ecosystems, institutions, and knowledge networks) that support and are supported by social agents—both private and public.

The resilience of a city depends on the fragility of the urban system and the capacity of social agents to anticipate and take action to adjust to changes and stresses considering constraints on resource access and supporting systems.

The city's adaptive capacity relies on regular services, programs, national policies, and civil society organization networks for support. Current measures addressing climate risks are mandate-driven and concentrate on social services disaster responses directed toward addressing the impacts of flooding. This is evident in the structural measures pursued by the city (e.g., flood control).

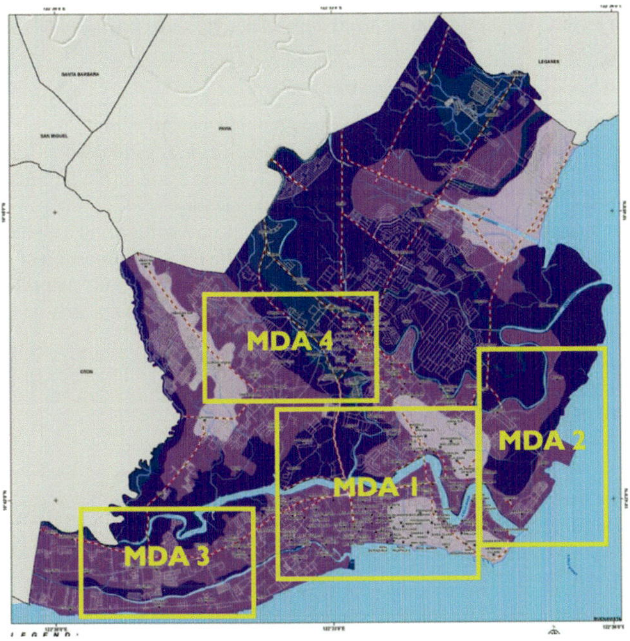

<table>
<tr><th>MDAs</th><th>Situational analysis</th><th>Risk management options</th></tr>
<tr><td>MDA-1</td><td><ul><li>Commercial development is unprecedented despite exposure to flood</li><li>Major roads and transportation facilities are high risk to flood</li><li>Huge concentration of (fragile) settlements is exposed to flood</li><li>Critical point facilities (public and private institutions providing social services are present in the area</li></ul></td><td rowspan="3">Risk reduction through implementation of preparedness measures such as evacuation drill, establishment of Early Warning System (EWS), and putting in place structural measures.

Risk insurance for those who are willing and capable
Risk avoidance or elimination through relocation of households living in fragile shelters without support mitigating measures, and critical public and private facilities and infrastructures</td></tr>
<tr><td>MDA-2</td><td>Critical industrial uses (i.e. Panay Energy Development Corporation and ports) are highly susceptible to flood</td></tr>
<tr><td>MDA-3 and 4</td><td>Huge concentration of settlements is exposed to flood.</td></tr>
</table>

Fig. 17.12 Map of major decision areas (MDAs) for rain-induced flood and possible risk management options

Private sector participation is also minimal, except for the infrastructure sector, implying heavy dependence on local drawn resources and aid from international agencies. Noticeable in these measures is the absence of a focused target based on earlier assessments of threat levels. As impacts are context-specific, adaptive measures should likewise be so.

Table 17.5 Major decision areas (MDAs) for rain-induced flood including situational analysis and risk management options

MDAs	Situational analysis	Risk management options
MDA-1	• Commercial development is unprecedented despite exposure to flood • Major roads and transportation facilities are high risk to flood • Huge concentration of (fragile) settlements is exposed to flood • Critical point facilities (public and private institutions providing social services are present in the area	Risk reduction through implementation of preparedness measures such as evacuation drill, establishment of Early Warning System (EWS), and putting in place structural measures Risk insurance for those who are willing and capable Risk avoidance or elimination through relocation of households living in fragile shelters without support mitigating measures, and critical public and private facilities and infrastructures
MDA-2	Critical industrial uses (i.e., Panay Energy Development Corporation and ports) are highly susceptible to flood	
MDA-3 and 4	Huge concentration of settlements are exposed to flood	

Table 17.6 Adaptive capacity at the local level

Adaptive capacity at the local level	
Characteristic	Features that reflect a high adaptive capacity
Asset base	Availability of key assets that allow the system to respond to evolving circumstances
Institutions and entitlements	Existence of an appropriate and evolving institutional environment that allows fair access and entitlement to key assets and capitals
Knowledge and information	The system creates an enabling environment to foster innovation, experimentation and the ability to explore niche solutions in order to take advantage of new opportunities
Flexible forward-looking decision-making and governance	The system is able to anticipate, incorporate and respond to changes with regards to its governance structures and future planning

Source Iloilo City Government Local Climate Change Action Plan

Building knowledge to inform the public and private sector decisions is also limited to information and education campaign activities. Recently, there has been a deliberate effort to improve decision parameters through the use of risk data generated by national agencies and some development organizations. As for governance, typical processes seem to apply like usual planning tasks and regulations. Pro-active measures for all hazards are not readily discernible.

7 Conclusion and Recommendations

The local government of Iloilo city has taken major steps in addressing climate change vulnerability and impacts and flood hazards through tangible projects and policy and institutional framework that focuses on sustainability. However, there are still, several gaps that need to be addressed for flood management to be holistic and dynamic to the ever-changing demand of impacts of climate change.

Recommendations:

1. Ensure that the crafted local climate change action plans of the local government units move beyond mere compliance with the existing policies. With the understanding of risk and integration of the risk assessment in their respective local development planning process, the locality can strengthen the local risk management, governance, and investments in climate change adaptation and mitigation.
2. Ensure the sustainability of existing flood control measures and employ monitoring and evaluation.
3. Pay attention to the institutional aspects of the operation and its maintenance.
4. Ensure the active participation of the different group sectors in the planning and implementation stage, that target affected groups and sectors are involved.
5. Strengthen the information dissemination system relative to hazards of the locality.
6. Support the development of proactive and flexible innovations to address demands of climate change impact

References

Cloke HL, Wetterhall F, He Y, Freer JE, Pappenberger F (2013) Modelling climate impact on floods with ensemble climate projections. Q J R Meteorol Soc 139(671):282–297

Duan W, He B, Takara K, Luo P, Hu M, Alias NE, Nover D (2015a) Changes of precipitation amounts and extremes over Japan between 1901 and 2012 and their connection to climate indices. Clim Dynam 45:2273–2292

Duan W, He B, Nover D, Fan J, Yang G, Chen W, Meng H, Liu C (2016) Floods and associated socioeconomic damages in China over the last century. Nat Hazards 1–13

Duan W, He B, Takara K, Luo P, Hu M, Alias NE, Nover D (2015b) Changes of precipitation amounts and extremes over Japan between 1901 and 2012 and their connection to climate indices. Clim Dynam 45:2273–2292

IPCC (2007) Climate Change 2007: synthesis report. Contribution of Working Groups I, II and III to the fourth assessment report of the Intergovernmental Panel on Climate Change. Core Writing Team, Pachauri, R. K., and Reisinger, A, 104. IPCC Geneva, Switzerland

Iloilo City Local Government Unit. Philippines. Climate and Disaster Risk Assessment: Iloilo City. Iloilo City, Western Visayas, Philippines

Iloilo City Local Government Unit. Philippines. Comprehensive Development Plan 2020–2025. Iloilo City, Western Visayas, Philippines

Iloilo City Local Government Unit. Philippines. Comprehensive Land Use 2021–2029. Iloilo City, Western Visayas, Philippines

Iloilo City Local Government Unit. Philippines. Disaster Risk Reduction and Management Plan. Iloilo City, Western Visayas, Philippines

Iloilo City Local Government Unit. Philippines. Local Climate Change Action Plan 2014-208. Iloilo City, Western Visayas, Philippines

Iloilo City Local Government Unit. Philippines. Socio-Economic Profile 2018–2028. Iloilo City, Western Visayas, Philippines

Pedersen AN, Mikkelsen PS, Arnbjerg-Nielsen K (2012) Climate change-induced impacts on urban flood risk influenced by concurrent hazards. J Flood Risk Manage 5:203–214

Flood Risk Assessment and Green Recovery Planning Case Studies—Middle East and South America

Flood Control and Its Management in the Badakhshan Province of Afghanistan

Kambiz Karimi

Abstract Case study author, Kambiz Karimi takes a look at the broader country context and then brings focus to the Badakhshan province, an area that is one of the most climate-related disasters prone zones in Afghanistan. In Badakhshan province, landslides, avalanches, droughts, floods, and earthquake are recurring hazards, causing immense loss of live and property every year. According to National Vulnerability and Resource Assessment (NVRA) 2007–2008, Badakhshan province has the highest level of inequality in Afghanistan, and is also one of the most food insecure provinces of Afghanistan. Flood reduction strategies must include strengthening poverty-alleviation programs—and targeting them to those residents most in need—to break the vicious cycle of poverty and disasters. They should aim to both maximize benefit coverage to the poor and minimize leakage of the benefits to the non-poor. Policy reforms proposed in the case study include: development of disaster management strategy at the national and local levels, improved water management and use efficiency, land and water management at the watershed level, and strengthened local planning and regulations among others.

Keywords Climate risk assessment and Afghanistan · Climate risk management and Badakhshan province · Disaster risk management · Adaptation · Resilience · Vulnerability · Adaptive capacity · Indicator analysis · Structural and non-structural measures · Hazard reduction measures · Nature-based solutions · Comprehensive urban planning · Watershed planning · Water management · Early warning systems

---

K. Karimi (✉)
National Environmental Protection Bureau, Kabul, Afghanistan
e-mail: kambizkarimi1994@gmail.com

K. Kim and C. Atkin (eds.), *Cross-Cultural Perspectives on Climate Change Adaptation*, Cities and Nature, https://doi.org/10.1007/978-3-031-50365-8_18

1 National Level Analysis

1.1 Geography

Afghanistan is located in Southern-Central Asia. At over 652,864 km^2 (World Bank 2011), it is the world's 41st largest country (CIA 1991). There is no coastline, as it is landlocked. Afghanistan shares its longest land border (the Durand Line) with Pakistan to the east and south, followed by borders with Tajikistan to the north-east, Iran to the west, Turkmenistan to the north-west, Uzbekistan to the north, and China to the north-east; India recognizes a border with Afghanistan through Pakistani-administered Kashmir (Fig. 1).

The geography in Afghanistan is varied, but is mostly mountainous and rugged, with some unusual mountain ridges accompanied by plateaus and river basins (Fisher 2002). It is dominated by the Hindu Kush range, the western extension of the Himalayas that stretches to eastern Tibet via the Pamir Mountains and Karakoram Mountains in Afghanistan's far north-east. The Hindu Kush ends at the west-central highlands, creating plains in the north and south-west, namely the Turkestan Plains

Fig. 1 Afghanistan exact location

and the Sistan Basin; these two regions consist of rolling grasslands and semi-deserts and hot windy deserts, respectively (The Christian Science Monitor 2010). Forests exist in the corridor between Nuristan and Paktika provinces and tundra in the north-east. The country's highest point is Noshaq, at 7492 m above sea level. The lowest point lies in Jawzjan province along the Amu River bank, at 258 m above sea level.

Despite having numerous rivers and reservoirs, large parts of the country are dry. The endorheic Sistan Basin is one of the driest regions in the world (Fisher 2002). The Amu Darya rises at the north of the Hindu Kush, while the nearby Hari Rud flows west toward Herat, and the Arghandab River from the central region southward. To the south and west of the Hindu Kush flow a number of streams that are tributaries of the Indus River (Fisher 2002), such as the Helmand River. Afghanistan receives heavy snow during the winter in the Hindu Kush and Pamir Mountains, and the melting snow in the spring season enters the rivers, lakes, and streams (History of Environmental Change … 2007). However, two-thirds of the country's water flows into the neighboring countries of Iran, Pakistan, and Turkmenistan. As reported in 2010, the state needs more than US $2 billion to rehabilitate its irrigation systems so that the water is properly managed (Reuters 2012).

1.2 Demographics

The population of Afghanistan was estimated at 32.9 million as of 2019 by the Afghanistan Statistics and Information Authority. About 23.9% of them are urbanite, 71.4% live in rural areas, and the remaining 4.7% are nomadic (Worldometers 2020) (Fig. 2).

The current population growth rate is 2.37%, one of the highest in the world outside of Africa. This population is expected to reach 82 million by 2050 if current

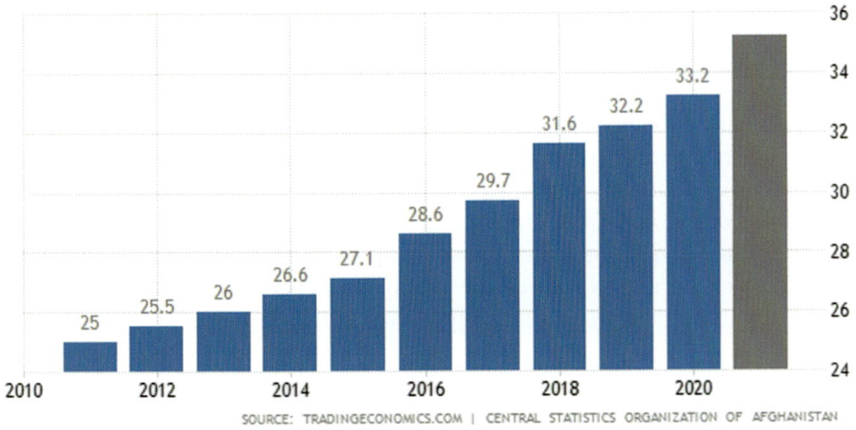

Fig. 2 Afghanistan population growth 2010–2020

population trends continue (Population Reference Bureau 2013). The population of Afghanistan increased steadily until the 1980s. Afghanistan's healthcare has recovered since the turn of the century, causing falls in infant mortality and increases in life expectancy. This, along with other factors such as returning refugees, caused rapid population growth in the 2000s that has only recently started to slow down.

1.3 Economic Profile

Afghanistan's economic growth is primarily dependent on development assistance from donors, while the protracted conflict and political instability undermine private sector investment and a range of economic activities (World Bank n.d.). In 2019, the country experienced 4% growth in GDP, up from less than 1% in 2015. While the overall GDP grew, Afghanistan's per capita GDP has contracted from USD 755 in 2014/15 down to USD 560 in 2019/2018. The service sector contributed 57.4% of the total GDP, followed by the agricultural sector at 25.5% and the industrial sector at 12.1%. However, the agricultural sector employs a disproportionately large number of people, providing jobs for 44.3% of the overall population and 52.6% of the rural population. Without changes in the ongoing conflict and stabilization of the country's politics, it is expected that Afghanistan will continue to experience slow economic growth (Olson et al. 2001). The country's economy is also expected to suffer from losses and damages caused by the impacts of climate change owing to the vulnerabilities of critical economic sectors such as agriculture (Fig. 3).

1.4 Climate of Afghanistan

Afghanistan has a continental climate with harsh winters in the central highlands, the glaciated north-east (around Nuristan), and the Wakhan corridor, where the average temperature in January is below − 15 °C (5 °F) and can reach − 26 °C (−15 °F) and hot summers in the low-lying areas of the Sistan Basin of the south-west, the Jalalabad basin in the east, and the Turkestan plains along the Amu River in the north, where temperatures average over 35 °C (95 °F) in July and can go over 43 °C (109 °F) (Fisher 2002). The country is generally arid in the summers, with most rainfall falling between December and April. The lower areas of northern and western Afghanistan are the driest, with precipitation more common in the east. Although proximate to India, Afghanistan is mostly outside the monsoon zone, except the Nuristan Province which occasionally receives summer monsoon rain (Kladnik 2017).

Changes in Temperature

Climate reanalysis from model outputs was combined with observations from weather stations in Afghanistan for air temperature and precipitation. From 1950

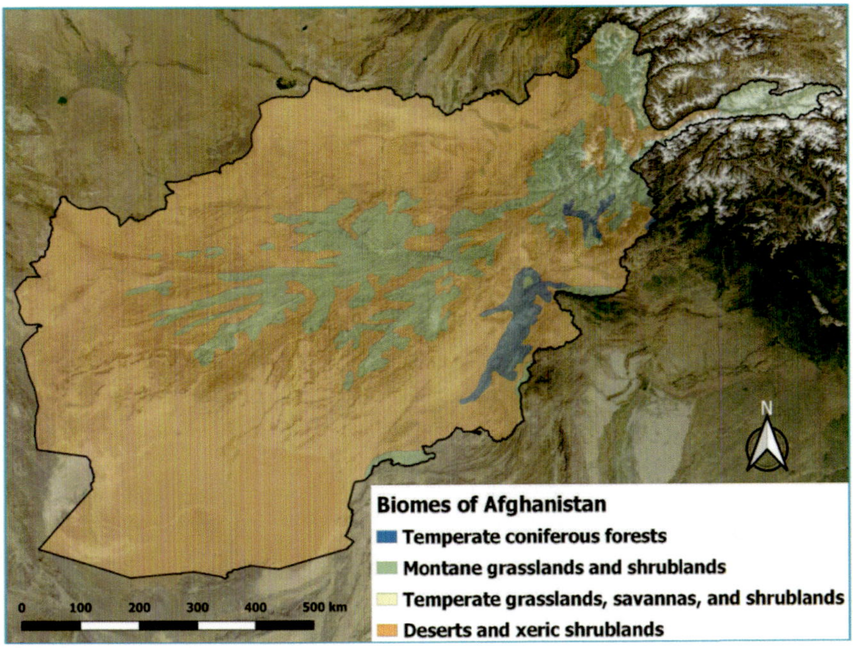

Fig. 3 Biomes of Afghanistan

to 2010, Afghanistan's mean annual temperature increased significantly by 1.8 °C—approximately double the global mean annual temperature increase for the same period, with much higher warming of 2.4 °C in the south-west of the country. Over the same period, the overall quantity of mean annual precipitation did not change significantly. However, detailed analysis of the seasonality of precipitation reveals that spring precipitation decreased significantly by − 27% across the country as a whole, with a decrease of − 30% or more in the north, central highlands and east of Afghanistan. This decrease in spring precipitation is particularly relevant for rain-fed agriculture and dependent on sufficient rainfall during this period. Indeed, the regions most affected by the decrease of spring precipitation are also those that are the major contributors to the country's agricultural production. Somehow winter precipitation and heavy precipitation events showed some significant changes in trend during the period 1950–2010. Extreme precipitation is reported as the main reason for recent increases in flooding, landslides, and related hazards, and in some cases the sole cause of such hazards. In addition, higher temperatures lead to rapid melting of snow in spring and also cause droughts that harden soils and thereby reduce the permeation of water during heavy precipitation events. Both of these phenomena increase the risk of flash flooding (Figs. 4 and 5).

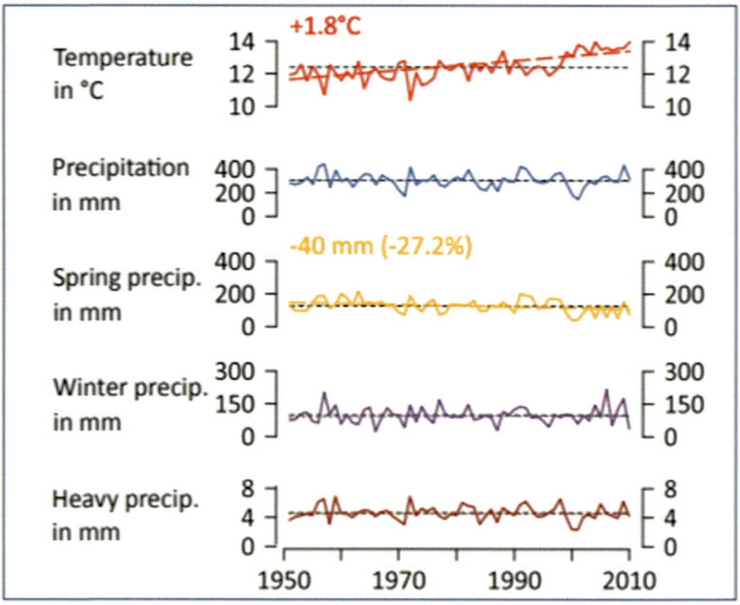

Fig. 4 Trends of temperature and precipitation for Afghanistan from 1951 to 2010. Derived from Reanalysis Data (GSWP3)

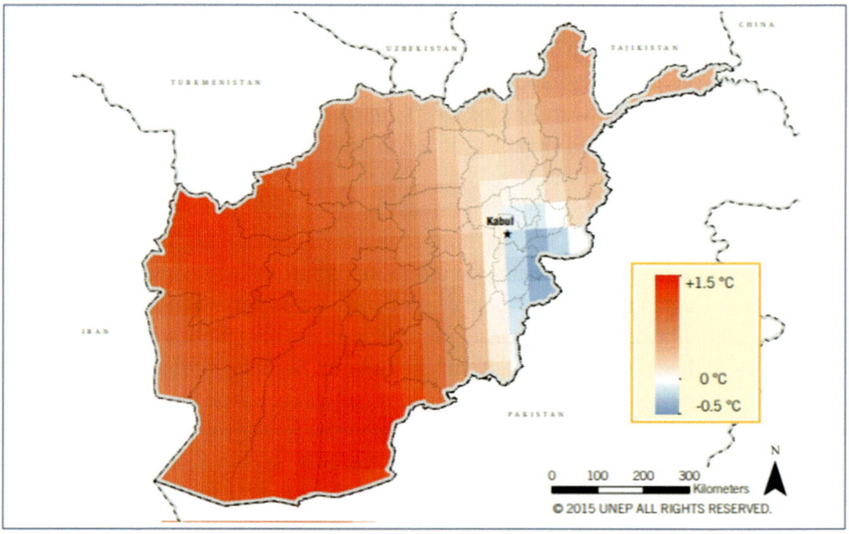

Fig. 5 Change in temperature between 1981–2010 and 1951–1980 in Afghanistan. Derived from Reanalysis Data (GSWP3)

Precipitation

Average precipitation has not changed significantly when considered over the whole of Afghanistan for all months. However, a deeper analysis of spring and winter precipitation reveals that the changes are simply leveled out since spring precipitation has significantly decreased by almost a third and winter precipitation has slightly increased, though not by a statistically significant amount. The decrease in spring (March–May) is particularly relevant for agriculture, since spring crops are typically rain-fed and dependent on sufficient rainfall during this period. The regions with most significant agriculture (east, central highlands, and north) are affected by the decrease in spring precipitation. The trends show the strongest decrease in the central highlands of almost 40% from 1950 until 2010. In the Hindu Kush and the South, the decrease is less distinct during these months, and there is even a slight increase in the Wakhan and the far South. Winter precipitation, contrastingly, has increased in most parts of the countries, or only slightly decreased, and none of the regional trends is significant. Heavy precipitation has often been reported to have increased during the last decades, leading to more frequent and intense flooding, landslides, and other related hazards. In contrast, this study could not find an increase in heavy precipitation (defined as 95th percentile of annual rainfall) between 1950 and 2010. However, an increase in heavy precipitation is just one reason among many that can lead to these aforementioned catastrophes. For example, rising temperatures might lead to more rapid and earlier spring snow melt, causing an increased risk of flash flooding. Also, droughts can exacerbate the impact of heavy precipitation by hardening soils and reducing their permeability, which can cause flash floods or landslides. In addition, increasing exposure and/or vulnerability against certain hazards might have increased the risk (Fig. 6).

1.5 Climate Change-Induced Hazards

One of the primary impacts of climate change is the potential increase in frequency and/or severity of natural hazards. Afghanistan is highly prone to such hazards that result in considerable losses and damages to lives, livelihoods, and productive assets. During the formulation of Afghanistan's National Adaptation Program of Action (NAPA) (NEPA and UNEP 2009), widespread consultation identified the impacts that natural hazards have on lives, human health, food security, livelihoods, water availability, and the environment. Through this process, expert participants stated that all hazards are occurring with increasing frequency and that floods, droughts, earthquake, avalanche, landslide, and rising temperatures are affecting thousands of Afghans and causing millions of dollars of damages. For example, since the 1950s, floods and droughts combined have caused over 5000 deaths, affected over 21 million people, and caused more than USD 540/year million in losses and damages. (Figs. 7, 8 and 9).

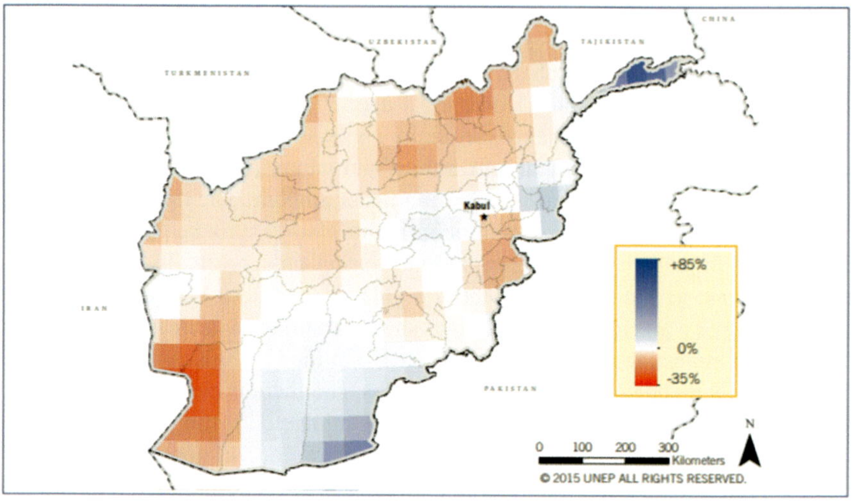

Fig. 6 Change in annual precipitation between 1981–2010 and 1951–1980 in Afghanistan. Derived from Reanalysis Data (GSWP3)

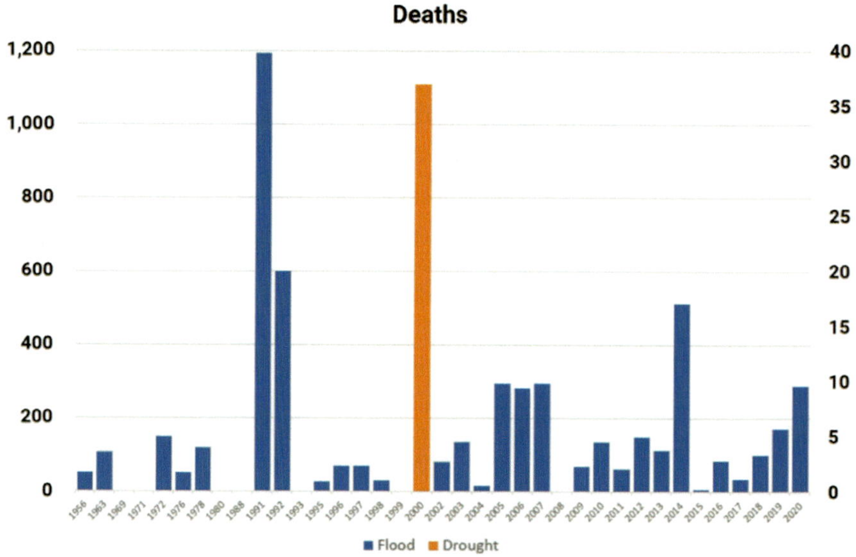

Fig. 7 Number of deaths due to climate disasters

Frequency of Disasters in Afghanistan

Globally, the total number of natural disasters around the world significantly increased from a total of 903 occurrences in the 1970s to 4476 during 2000–2009;

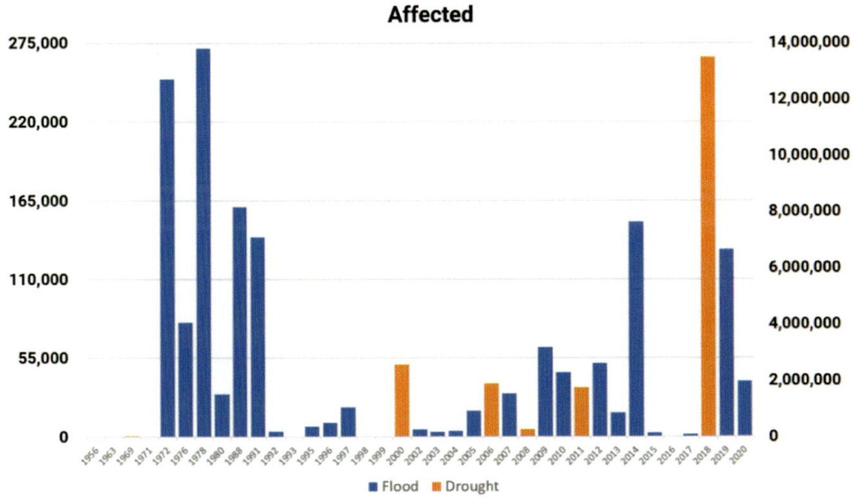

Fig. 8 Number of affected people due to climate hazards. *Source* EM-DAT. CRED/UC Louvain

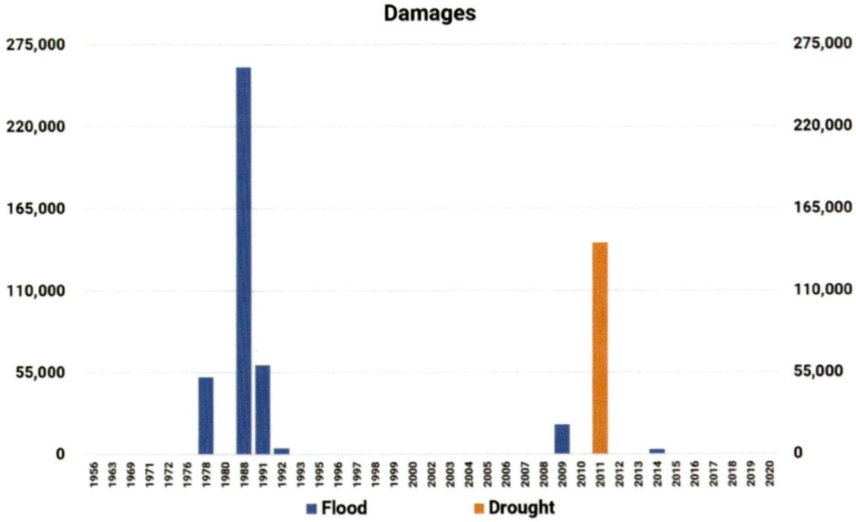

Fig. 9 Number of damaged people by flood and drought. *Source* EM-DAT. CRED/UC Louvain

meanwhile a total number of 2553 occurrences are recorded over the last seven years. Afghanistan is one of 15 most vulnerable countries to the flood risk worldwide (Fig. 10).

The increasing trend in the number of natural disasters was mostly driven by the increase in incidences of flood, storm, and epidemic, possibly in direct relation to

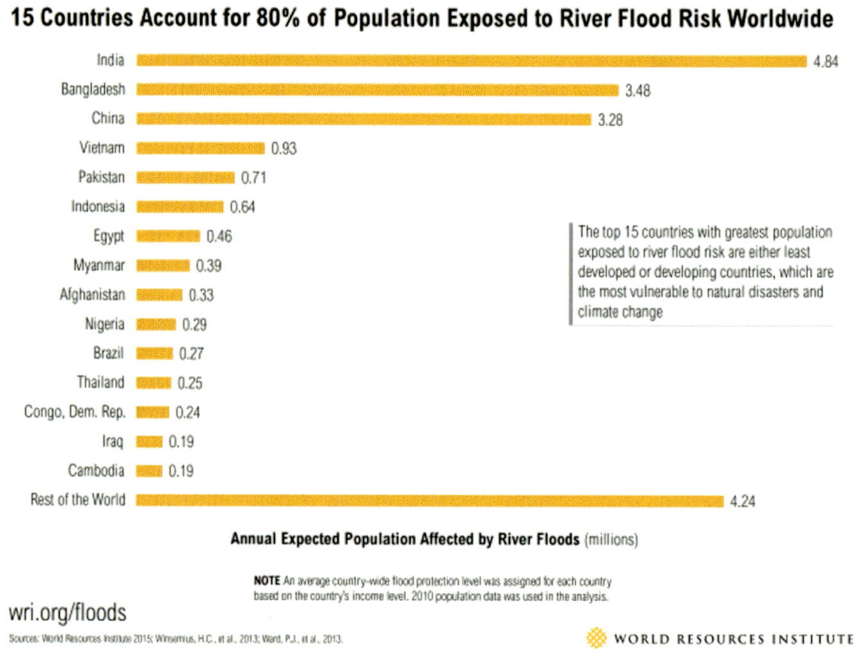

Fig. 10 Top 15 countries with the highest population exposed to river flood risk worldwide

the impacts of global warming. In OIC countries flooding was again by far the most threatening type of disaster, followed by epidemic, earthquake, storm, landslide, and drought (SESRIC, 2014) (Fig. 11).

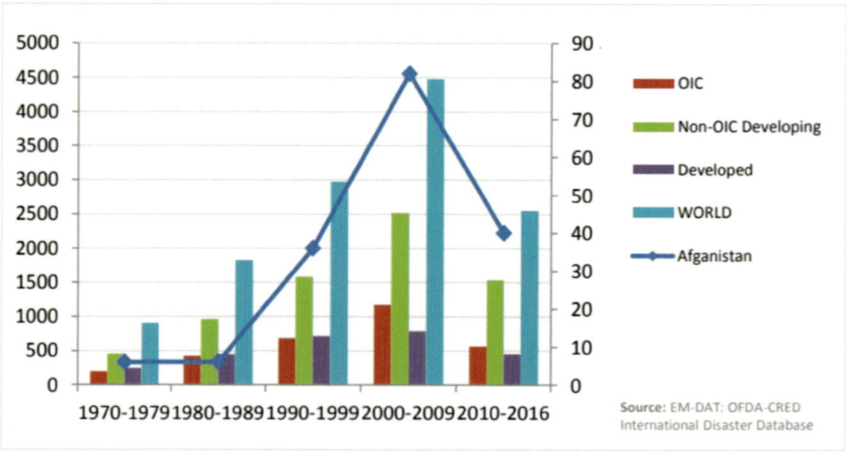

Fig. 11 Trends in natural disaster (1970–2016)

Similarly, the total number of disasters recorded in Afghanistan shows an increasing trend over the period under consideration. On aggregate, 1.3% of total disasters in the world and 5.3% of all disasters in the OIC countries were recorded in Afghanistan during 1970–2016. While Afghanistan had a share of 0.7% in total number of natural disaster incidents in the world all throughout the 1970s, their share increased to 1.6% during the last six years. While some of the increase can be attributed to improved availability of data on natural disasters, it is also evident that the country experiences increasingly more natural disasters. The major drivers of such a fast increase in the number of natural disaster incidents in Afghanistan were flood, earthquake, landslide, and epidemic, respectively, in order of frequency. In aggregate terms, 79 flood incidents, 30 earthquakes, 20 landslides, and 20 epidemics are recorded as the most frequently observed natural disasters during the period under consideration. Out of 170 disasters recorded, flood and earthquake correspond to 46% and 18% of all disasters in Afghanistan (Fig. 12).

Fatal and Non-fatal Impacts

A number of natural disasters have caused severe fatal impacts in Afghanistan. Since 1970, more than 21,000 people were killed by different types of natural disasters in Afghanistan, corresponding to 0.6% in the world. The share of Afghanistan in the world fluctuated over the decades, hitting a record high of 2% in 1990s but

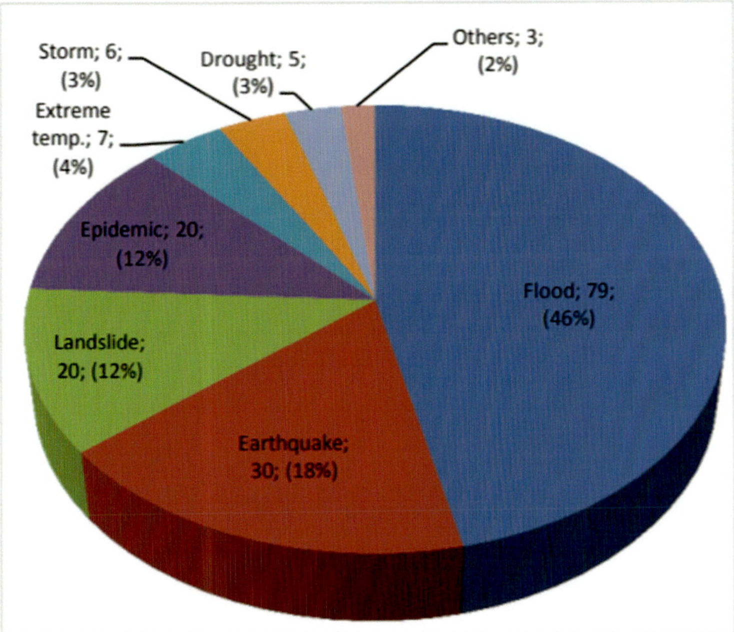

Source: EM-DAT: The OFDA/CRED International Disaster Database.

Fig. 12 Rate and number of disasters in Afghanistan

decreasing to less than 1% after 2000. With respect to the fatal impacts of different types of disasters in Afghanistan, earthquake was the deadliest type of natural disaster during 1970–2016. It killed more than 9400 people, followed by flood with more than 4500 lives lost and epidemics with more than 3800 people dying. Fatal impacts of landslide, extreme temperature, storm, drought, and mass movement remained rather limited during 1970–2016 in Afghanistan (Figs. 13, 14 and 15).

Non-fatal Impacts

The non-fatal impacts of natural disasters in Afghanistan, in comparison with the other country groups, are reported in Figs. 16 and 17. These figures refer to the number of people who have been injured and/or left homeless by a disaster. The figure shows the number of injured people due to natural disasters, and there is an upward trend in Afghanistan during the last decade, by increasing from less than 2000 people in the 2000s to nearly 60,000 people during 2010–2016. This dramatic increase was mostly driven by the devastating earthquake that took place in 2015. Nonetheless, this increase proves that Afghanistan remains highly vulnerable to natural disasters, and there is a need to develop strategies to reduce such vulnerabilities.

Fig. 13 Total population, at flood risk population, and buildings

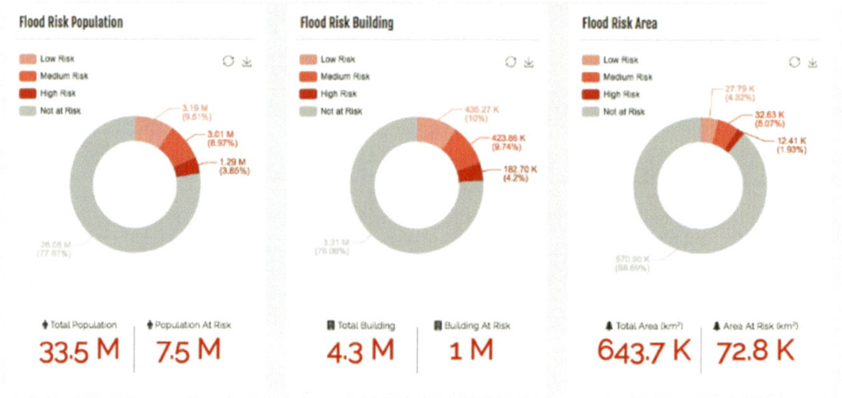

Fig. 14 Rate of vulnerability of Afghan people against flood

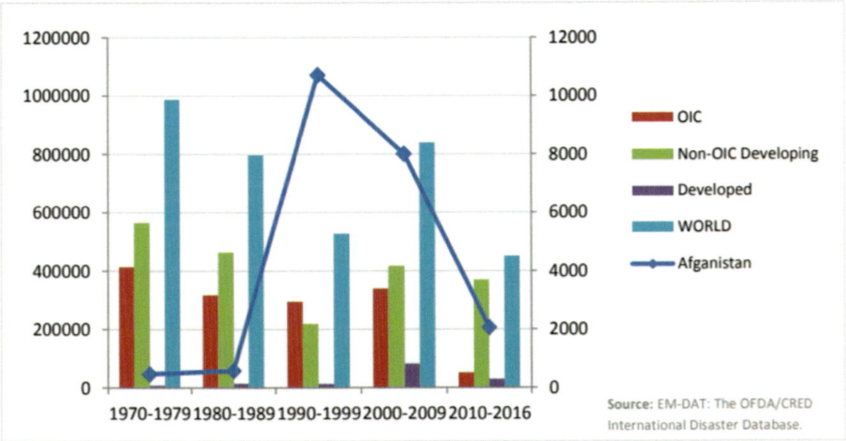

Fig. 15 Fatal impacts of natural disaster (1970–2016)

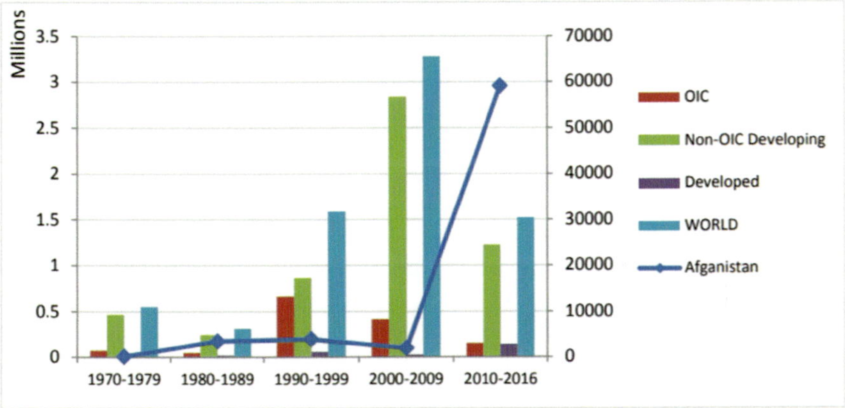

Fig. 16 Total number of injured people in Afghanistan due to natural disasters (1970–2016)

Economic Impacts

Economic costs of natural disasters in Afghanistan and other country groups during 1970–2016 are reported in Fig. 8. It shows that the cost of damages substantially increased in OIC countries in the 1980s, recorded as US $269,000. Although there was a downward trend in 1990s and 2000s, during the last seven years economic cost of natural disasters increased again in Afghanistan and estimated to reach US $148,000 during 2010–2016.

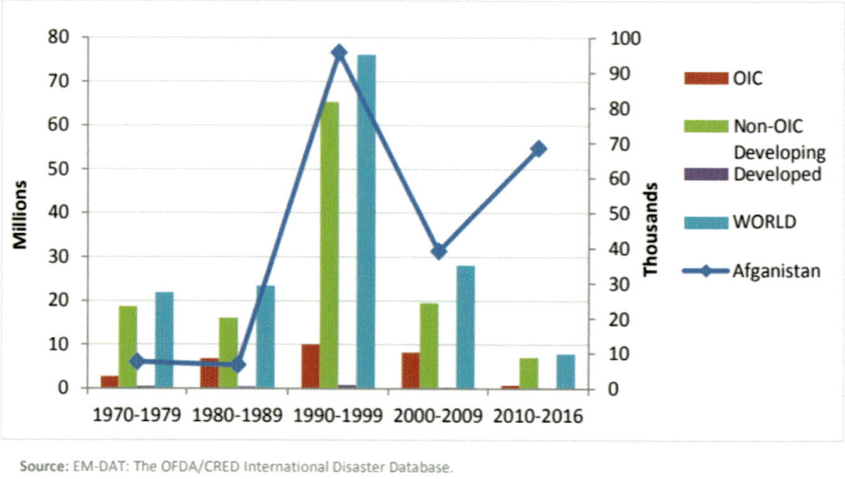

Fig. 17 Total number of homeless people in Afghanistan due to natural disasters (1970–2016)

2 Badakhshan Province Analysis

2.1 Geography

Badakhshan is located in the extreme north-east of Afghanistan between coordinates (N 37.91, E 71.58 and N 37.93, E70.17) in the north and (N 35.52, E70.88 and N 36.66, E 70.25) in south. The province covers an area of 44,059 km^2. It has external borders to Tajikistan in the north, Pakistan and Kashmir in the south-east, and China in the far end of Wakhan corridor. Pyanzh tributary of River Amu demarcates the border with Tajikistan. Internally the province borders Takhar, Nuristan, and Panjshir (Fig. 18).

High mountains and steep river valleys are the main geographical features of the province. Nearly nine-tenths of the province (89.9%) is mountainous or semi-mountainous terrain, while one-tenth of the area (9.7%) is made up of flat or semi-flat land, as shown in Table 1.

Badakhshan is one of the most remote, least developed, and disaster-prone regions owing to its geography. The eastern Pamir mountains lie between 3000 and 7000 m above the sea level and are covered with snow and glaciers throughout the year. The Hindu Kush range, which forms the border with Kashmir and Pakistan, has 38 summits higher than 7000 m, including Afghanistan's highest peak Noshaq (7492 m). In the south, Hindu Kush mountains effectively cut off Badakhshan from the rest of Afghanistan, the only road being the steep Anjuman Pass which leads over into the Panjshir Valley. Communications and road connections inside the province are difficult due to the mountainous nature of the terrain, and some roads remain closed for up to six months of the year. Rough topography and poor-quality roads make the emergency operation extremely difficult during winters.

Fig. 18 Area of the study. *Source* Mapsofworld.com

Table 1 Topography of study area

Topography					
Flat	Mountainous	Semi-mountainous	Semi-flat	Not reported	Total
6.9%	75.7%	14.2%	2.8%	0.3%	99.9%

Climate, Fauna, and Flora

The climate of the Badakhshan is typical of a continental interior mountainous region. The winters are characterized by cold temperatures (minus 0–40 °C) heavy snow and substantial amounts of precipitation (81–255 mm). The summers are warm (28–37.5 °C) and dry (precipitation 24–100 mm). Spring is a transition period, with increasing temperatures and some rain. Badakhshan has "Potential Natural Vegetation" which means that vegetation cover should remain as if it is not modified by human activity. However, because of continued exploitation such as animal grazing, farming, deforestation, and uprooting shrubs for fuel, much of the original natural vegetation is found only in a few remote areas of Badakhshan now. The destruction of the natural vegetation has resulted in the degradation and erosion of the soil cover in some areas. Many areas exhibit signs of long-lasting desertification caused by human activity. Common vegetation includes meadow and birch forests. Juniper woodlands in lower slopes are found in north-western and central areas. Montane grasslands and

Table 2 General features of six climatic zones in Afghanistan including Badakhshan

Zone	Name	Precip. (mm)	Dry (months)	Frost (months)
1	Badakhshan (without Wakhan)	300–800	2–6	1–9
2	Central and northern mountains	200–600	2–9	0–8
3	Eastern and southern mountains	100–700	2–9	0–10
4	Wakhan corridor and Pamir	< 100–500	2–5	5–12
5	Turkestan plains	< 100–400	5–8	0–2
6	Western + south-western lowlands	< 100–300	6–12	0–3

shrublands and meadows are found in the high mountains in the northern and south-western regions. The Toga Forest, an important and distinctive wetland ecosystem of the dry lands of central Asia, is found in Koran Wa Munjan district and stretches along the Kokcha River. This ecosystem consists of reeds interspersed with the Tamarix and Salix trees. The Toga Forest also has potential for eco-tourism. Wildlife such as snow leopard, brown bear, wolf, Marco Polo sheep, and Siberian ibex are found in the province and are included in the IUCN's Red List of Endangered Species. Furthermore, the region hosts several indigenous plant species, birds, insects, and other animals that are of high scientific interest, be it for their local specificity or their coping capacity with the harsh living conditions (Table 2).

2.2 *Hazards in Badakhshan Province*

Badakhshan is one of the most climate-related disasters-prone zones in Afghanistan. Landslides, avalanches, droughts, floods, and earthquake are principle recurring hazards, causing immense loss of live and property every year. Contributing factors to vulnerability include lack of awareness and skills, inadequate preparedness and early warning, remoteness and limited connectivity, absence of mitigation practices, and public service in hazardous areas. Every year these disasters result in loss of lives, population displacement, and destruction of houses, agricultural lands, community assets, and critical infrastructure (Fig. 19).

Flood, Flood Management, and Response to Flood in the Area of the Study

Badakhshan is geologically one of the most vulnerable regions to natural disasters. It is laced with large faults such as Chaman, Moqur, Harirod, Central Badakhshan, and Darwaz which meet near the province and make it the highest seismic risk region in the country. The heavy atmospheric precipitation, unconsolidated soft deposits rolling stones, and steep slopes in mountainous terrain have made the province susceptible to multiple disasters such as landslides and flashfloods. The ongoing deforestation and overgrazing of pastures have further led to the soil erosion and ecosystem degradation (Fig. 20).

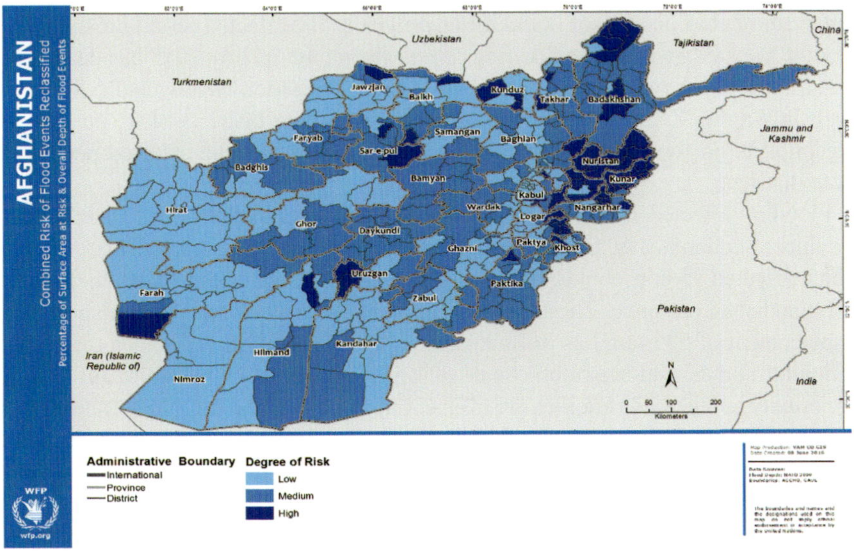

Fig. 19 Flood map of Afghanistan. *Source* USAID

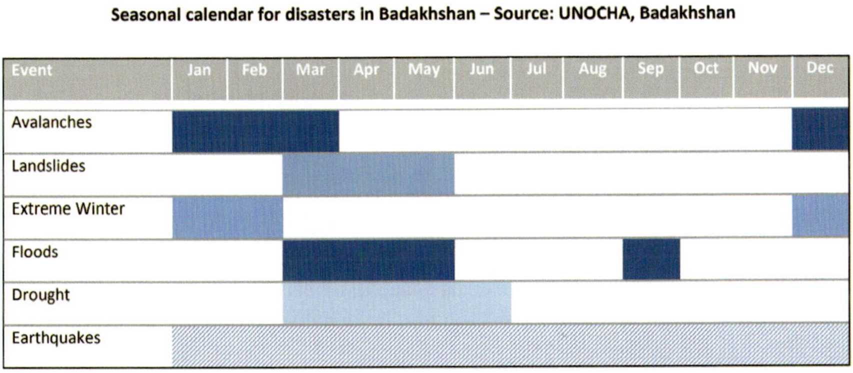

Fig. 20 Seasonal calendar for disasters in Badakhshan. *Source* UNOCHA, Badakhshan

Based on UNOCHA records in Badakhshan, every year floods kill and injure people and destroy and damage houses, household items, livestock, community assets, agricultural lands, gardens, and critical infrastructures such as roads, bridges, and canals et al. further blocking the means of communication. Floods result in internal displacements and create unhygienic conditions causing waterborne and communicable diseases outbreaks.

Flooding have affected many households in remote districts of Badakhshan province, but due to accessibility issues, timely and appropriate food responses could not be provided to the affected people in Shukai, Maimai, Nusai, Wakhan, Shighnan,

Kufab, and Zebak districts of Badakhshan province. The affected people are highly in need of food assistance. The flooding during the month of July 2015 has destroyed about 1433 jeribs of agriculture lands, 57 km of irrigation canals, and 6.5 km of roads in Warduj, Zebak, Shighnan, Shuhada, and Taulqan districts which puts negative impacts on food availability, access, and utilization of the affected people in these districts.

Flooding affected 167 families in Shighnan and Warduj districts of Badakhshan province. Sudden/heavy snow melting caused river water flow (river bank erosion) which affected 75 households in Zebak district and 283 families in Shuhda districts of Badakhshan province. These massive flooding and river water flows also caused displacement of 525 families in Badakhshan province and also damaged the roads, agriculture lands, and irrigation canals of 1297 families in the month of July 2015. Reportedly another 129 families are affected by floods in Maimai, Nusai, and Kufab districts of Badakhshan province. While the RAFs have not been received this data was confirmed through phone by IOM.

2.3 *Climate Disaster-Related Vulnerability in the Area of the Study*

Poverty and inequality: According to National Vulnerability and Resource Assessment (NVRA) 2007–2008, Badakhshan has the highest level of inequality in Afghanistan. Therefore, it is particularly important to incorporate poverty-alleviation programs and DRM to break the vicious cycle of poverty and disasters. The design of these policies should be attentive to the differences in the level of inequality. For Badakhshan, the high poverty rate means that many people in the province are poor, but there are also many people with consumption levels significantly above the poverty line, and the program design needs to be attentive to the concern of benefit leakage. They should aim to both maximize benefit coverage to the poor and minimize leakage of the benefits to the non-poor.

Food insecurity of those in poverty: Badakhshan is one of the most food insecure provinces of Afghanistan. Agriculture is restricted by the relatively cold climate which allows for only one crop in most areas. The mountainous terrain does not allow for sufficient arable land, and the region is prone to flash floods, avalanches, and landslides. The province was also affected by drought followed by unusually high snowfall in 2011 adding to the food insecurity. It is highly correlated with low levels of food consumption. In the case of Badakhshan, this is certainly true and Badakhshan has the second highest rate of calorie deficiency, which is the proportion of people consuming less than 2100 cal per day. One part of explanation for this difference can be seen from the inequality indicator. Poor in Badakhshan are worse off and also means that calorie deprivation is much more severe in Badakhshan (Photo 1).

Photo of the damages caused by the recent flooding in Badakhshan province
Photo credit: IOM Badakhshan/Mashal Darwish

Photo of the damages caused by the recent flooding in Shuhada district of Badakhshan province
Photo credit: IOM Badakhshan/Mashal Darwish

Photo 1 Photo of the damages caused by the recent flooding in Badakhshan province and in Shuhada district of Badakhshan province. *Photo Credit* IOM Badakhshan/Mashal Darwish

2.4 *Study Area Adaptation Priorities and Recommendations*

The following are high-priority policy reforms and other flood reduction measures that should be considered in the study area to lessen the risk of adverse consequences.

Adaptation priorities:

1. Improve water management and use efficiency
2. Land and water management at the watershed level
3. Developmentt of horticulture
4. Development of disaster management strategies
5. Climate-related research and early warning systems
6. Strengthen local planning and regulations
7. Storm water management planing

Flood reduction measures:

8. Flood control measures
9. Restoring and preserving natural drainage systems
10. Building houses in flood safe areas
11. Planting trees
12. Construction of roads
13. Provide primary transportation
14. Increase the number of health care centers.

References

CIA (1991) CIA Factbook—area: 41, 26 Nov 1991. Archived from the original on 31 Jan 2014. Accessed 4 Feb 2012

Fisher WB (2002) Afghanistan: physical and social geography. In: The Far East and Australasia 2003. Psychology Press, pp 59–60. ISBN 9781857431339

History of environmental change in the Sistan Basin 1976–2005 (PDF). Archived (PDF) from the original on 7 Aug 2007. Accessed 20 July 2007

Kladnik D (2017) Terraced landscapes. Založba ZRC. ISBN 9789610500193

NEPA and UNEP (2009) National capacity needs self-assessment for global environmental management and national adaptation program of action for climate change

Olson DM et al (2001) Terrestrial ecoregions of the world: a new map of life on earth. Bioscience 51:933–938

Population Reference Bureau (2013) Afghanistan. Population Reference Bureau. Archived from the original on 2 Dec 2013. Accessed 29 Dec 2009

Reuters (2012) Snow may end Afghan drought, but bitter winter looms, 18 Jan 2012. Archived from the original on 30 Dec 2013

The Christian Science Monitor (2010) Afghanistan's woeful water management delights neighbors, 15 June 2010. Archived from the original on 14 Nov 2010. Accessed 14 Nov 2010

World Bank (2011) Land area (sq. km). World Development Indicators. Archived from the original on 29 Oct 2013. Accessed 13 Oct 2011

World Bank (n.d.) Afghanistan country overview
Worldometers (2020) Afghan population estimates 2020. Archived from the original on 26 Nov 2020. Accessed 27 Nov 2020

Assessment of Early Warning Systems in Afghanistan: Avalanches and Flash Floods in Shighnan District of Badakhshan Province, Afghanistan

Mohammad Nazir Safi

Abstract The residents of Shighnan district of Afghanistan live on the ridges of high mountains, putting them at greater danger of climate-related natural disasters such as avalanches, landslides, and floods. Case study author, Nazir Safi, focuses on the power of community developed and implemented early warning systems as a critical element of climate risk management. Current efforts to implement a community-based early warning system while not without challenges have had important successes in Shighnan district. After evaluating the current state and efficacy of the community based early warning systems he identifies key challenges to their impact and efficacy including: Limitations in the mobile network coverage and, the need to establish trust in the accuracy of the information as well as confidence that the evacuation centers would be able to meet their needs.

Keywords Afghanistan and climate disaster · Community-based early warning systems · Climate risk management and Shighnan district · Disaster risk management · Adaptation · Vulnerability · Adaptive capacity · Hazard reduction measures · Comprehensive urban planning

1 Introduction

The linkages between climate change adaptation, disaster risk reduction, and sustainable development became clearer and internationally recognized with the passage of the Paris Agreement on climate change, the sustainable development goals (SDGs), and the Sendai Framework for DRR (NEPA 2016). All three global frameworks mention EWS specifically, emphasizing the need of monitoring and early warning in lowering climate and risk vulnerabilities and increasing resilience.

M. N. Safi (✉)
National Environmental Protection Agency, Kabul, Afghanistan
e-mail: mohammadnazirsafi2019@gmail.com

Hallym University, Chuncheon-si, Gangwon-do, Republic of Korea

K. Kim and C. Atkin (eds.), *Cross-Cultural Perspectives on Climate Change Adaptation*, Cities and Nature, https://doi.org/10.1007/978-3-031-50365-8_19

The Community-based early warning system (CBEWS) is a community-based early warning system that is built utilizing a "people-centered" approach. CBEWS equips individuals and communities with the tools and plans they need to use and administer an integrated system of tools and plans that provide early warnings on hazards and help them avoid danger. The system is critical in helping hazard-prone communities become more resilient (UNEP et al. 2020).

All four aspects described for an early warning system must be considered for an effective CBEWS. The key aspect of a CBEWS is its "people-centered" approach. The planning, implementation, monitoring, and dissemination of early warning information should include all members of vulnerable populations and key players. Furthermore, for improved functioning and sustainability, the community should take ownership of the system (UNEP et al. 2020).

The UNFCCC has designated Afghanistan, a member of the World Meteorological Organization (WMO), as the least developed country. Natural catastrophes strike Afghanistan frequently, claiming lives, livelihoods, and property (AMD 2020). This has resulted in widespread food insecurity and population flight from the worst-affected areas in recent decades. Floods and droughts have killed over 5000 individuals and affected over 21 million people since the 1950s. Floods and flash floods, on the other hand, caused an estimated $600 million in economic losses during this time period, much exceeding the losses caused by the drought (EM-DAT 2020).

2 Background of Study and Objectives

Hydro meteorological dangers have made Afghanistan susceptible (including floods, landslides, droughts and extreme weather). According to a UNDP survey of 200 nations' vulnerability to various natural catastrophes, Afghanistan was ranked 17th most vulnerable to floods (ANDMA 2015). According to AMD data, about 15,000 people died in Afghanistan as a result of hydro meteorological disasters between 1980 and 2015, with floods and flash floods accounting for approximately 31% of all deaths (AMD 2020) (Fig. 1).

According to the Afghanistan Natural Disasters Atlas, approximately 7.5 million people in the country are at risk of flooding. Out of the 7.5 million people, 1.29 million are at high risk of flooding, while 3.1 million are in average danger and 3.19 million are in low danger (ANDMA 2018).

Flooding in 14 northern regions of the country in May 2014 harmed 90,000 people, displaced 20,000 people, and caused economic damages of over $100 million USD (ANDMA 2018) (Fig. 2).

Objective of Study

- To analyze the CBEWS for Avalanche and flash flood in Shighnan district.
- To assess the effectiveness and efficiency of CBEWS on Shighnan district.
- To identify the Avalanche and flash flood prone area in Shighnan district.

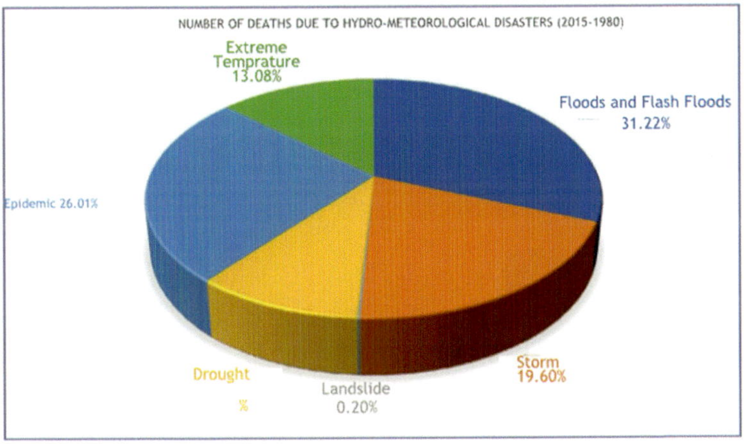

Fig. 1 Deaths statistics due to climate-induced natural disasters in Afghanistan (Afghanistan Meteorological Department [AMD] 2020)

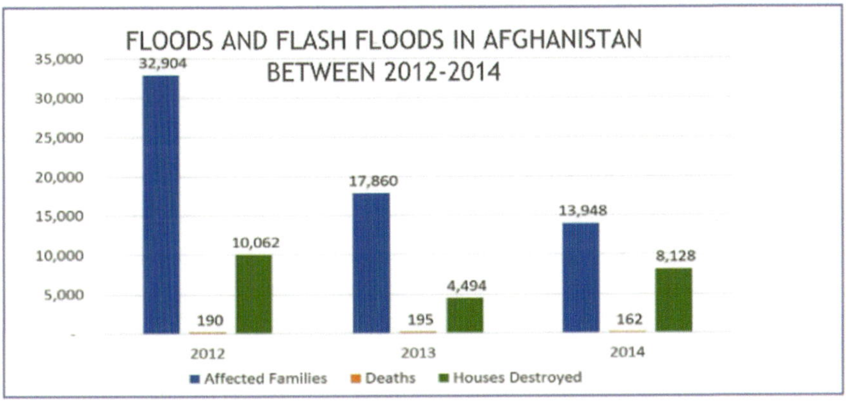

Fig. 2 Floods and flash floods impacts in Afghanistan (ANDMA 2018)

3 Conceptual Framework and Methodology

In this research, I have applied the The Standard Methodology for Climate-Related Disaster and Green Recovery which has been prepared by Professor Dr. Kwi-Gon Kim. The methodology has eight standard steps which described in chart (Fig. 3).

In addition, effective early warning systems, according to the 3rd International Conference on Early Warning, must be people-centered and include the following four interrelated core elements:

- Risk awareness
- Services for monitoring and warning

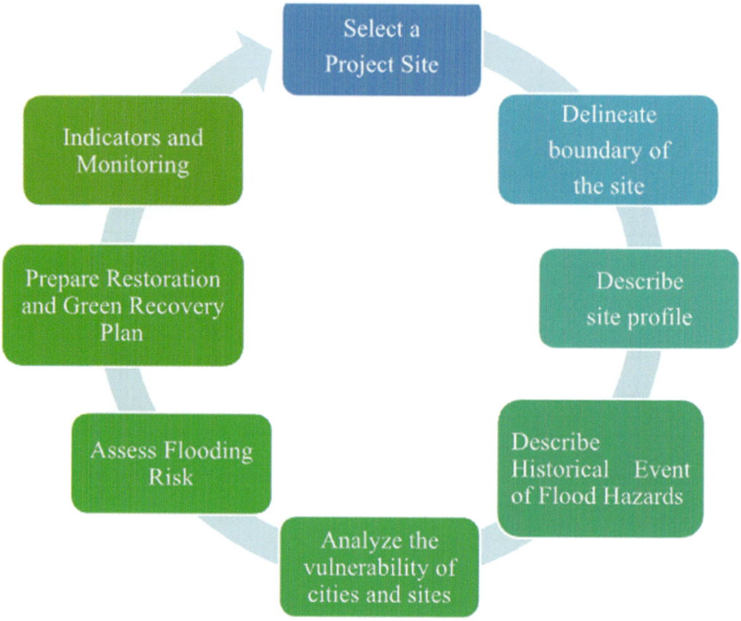

Fig. 3 Disaster Risk Assessment and Green Recovery Framework (Dr. Kwi-Gon Kim)

- Dissemination and communication
- Capability to respond.

The incorporation and interplay of four major elements: risk knowledge, monitoring and warning services, dissemination and communication, and reaction capabilities are critical for the efficiency of the people-centered EWS (United Nations Environment Programme [UNEP] et al. 2020) (Fig. 4).

4 Shighnan District Description

Shighnan is one of the 28 districts of Badakhshan province's north-eastern Badakhshan province, located 150 km east of the province's capital, Faizabad. Shighnan district covers 3545.266 km^2 and has a population of 3,184,767 persons. The district of Shighnan comprises 59 settlements, with Bahshar serving as its administrative center. A total of 3200 homes dwell in this district, with 2080 of them being vulnerable to avalanches (Agha Khan Agency for Habitat [AKAH] 2020a).

The climate of Shighnan is humid continental. There is more rain in the winter than in the summer. Shighnan has an average yearly temperature of 8.0 °C (46.4 °F). Almost all transportation routes in the area are closed due to its high mountains and inhospitable geography, which causes many challenges for its residents, especially

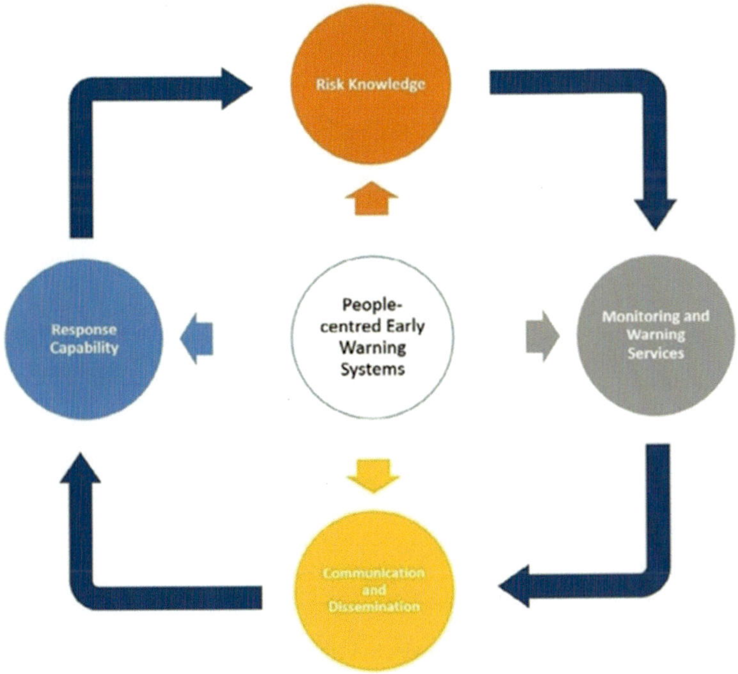

Fig. 4 People-centered early warning system (UNEP et al. 2020)

during the winter when it snows. Summers are short, hot, and dry in the district, and winters are harsh and long. This district receives about 480 mm of precipitation per year on average, mostly in the form of snow. This district is particularly affected by the chilly Siberian cold waves (UNEP et al. 2020).

Since 2017, AKAH has connected the Shighnan district of Badakhshan province to regular EWS. To date, 56 villages out of total 59 villages of Shighnan districts are covered under the CBEWS (Fig. 5).

5 Shighnan District Historical

Background and Vulnerability Assessment

Most of the population in the Shighnan district live on the ridges of high mountains, putting them at greater danger of climate-related natural disasters such as avalanches, landslides, and floods.

The two most common natural disasters are avalanche and landslide, followed by floods and flash floods. Avalanches occur more than four times each year on average in each village, resulting in the loss of life and property. At the same time, floods

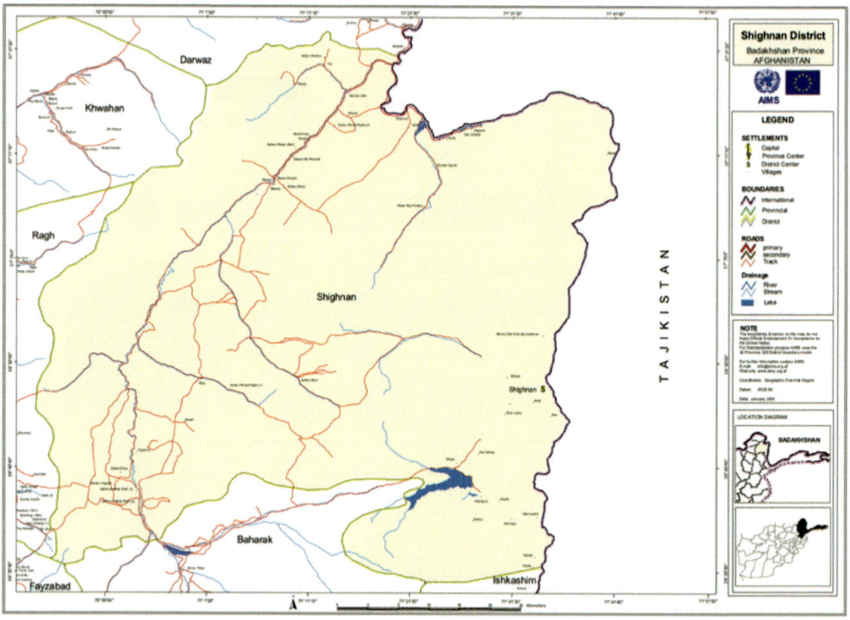

Fig. 5 Shighnan district of Badakhshan province, Afghanistan (Afghanistan Information Management Service [AIMS] 2005)

occur on average every two years. At the district level, however, the ratio of flood and avalanche frequency is 1–3/5–10 each year.

A total of 9.76 thousand individuals and 736 km^2 in the Shighnan district have been recognized as avalanche prone (Fig. 6).

Flood being one of the moderate risks to the residents of Shighnan district, threaten 5 k people and 113 km^2 area of Shighnan district (Fig. 7).

The statistics of avalanche and flood are provided in Fig. 8.

A summary of the historical avalanche and flood incidents in Shighnan district is provided in Table 1.

6 Assessment of Community-Based Early Warning System in Shighnan District

The AKAH office in Badakhshan connected the Shighnan district to regular EWS in 2018. As previously stated, the Shighnan district of Badakhshan province is a mountainous region, making the majority of its residents exposed to climate-related natural calamities.

As previously stated, an EWS is made up of four main components. The following are the primary parts of the CBEWS installed in the Shighnan district.

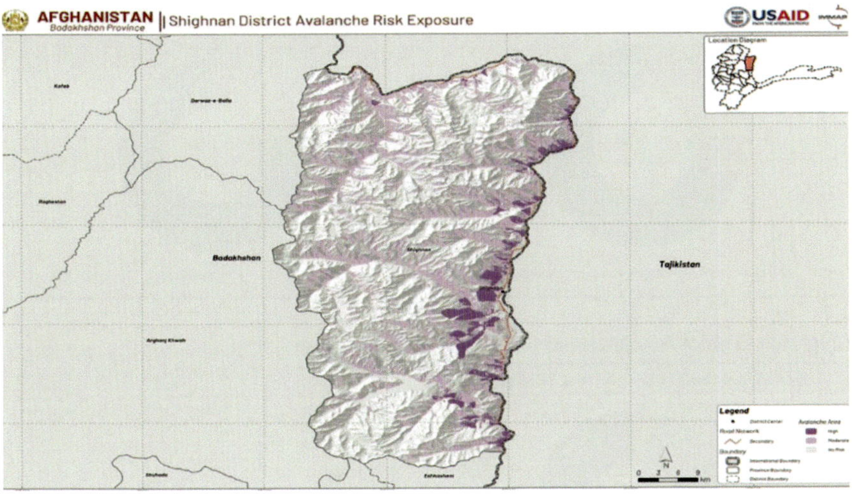

Fig. 6 Avalanche-prone area map of Shighnan district (iMMAP 2020a)

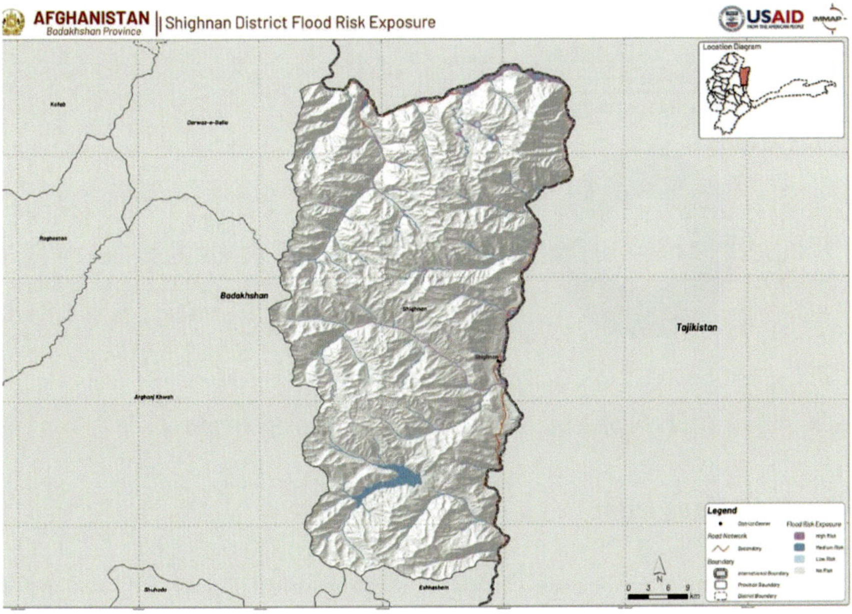

Fig. 7 Flood prone area map of Shighnan district (iMMAP 2020b)

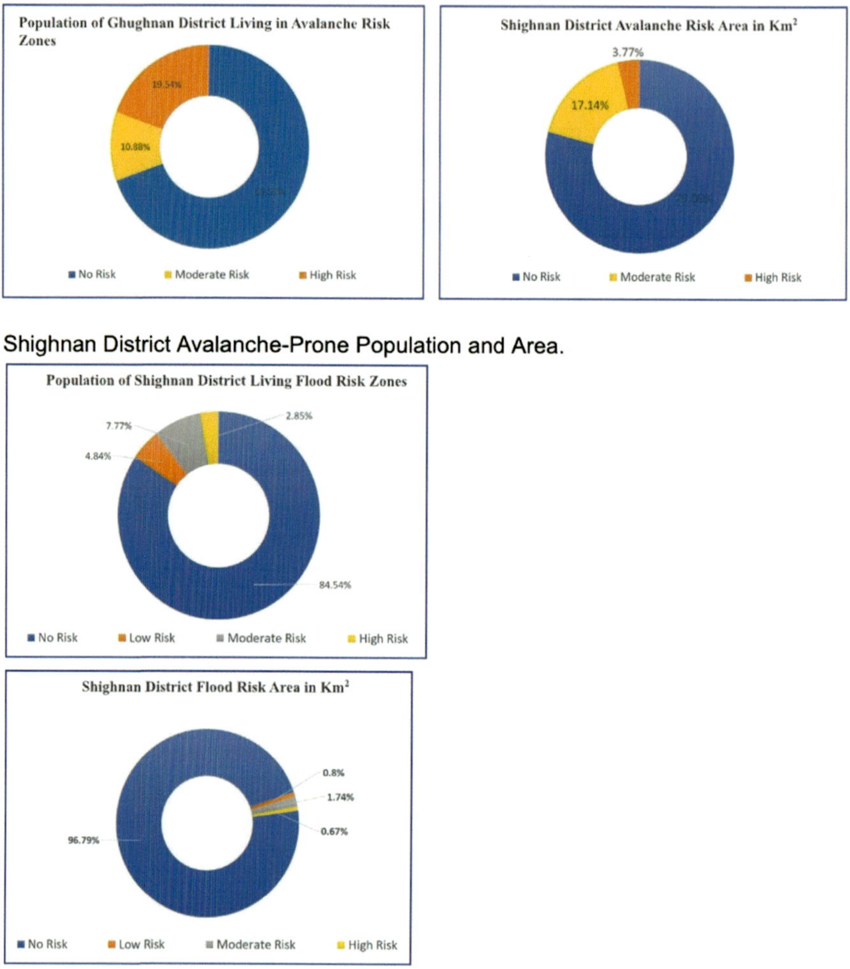

Fig. 8 Shighnan district avalanche-prone population and area (iMMAP 2020c)

6.1 *Risk Knowledge*

The risk mapping procedure was carried out using historical data as well as extensive discussions and interviews with local residents. A team of AKAH professionals was deployed to the area for a transaction walk exercise to determine the communities' vulnerability and risk exposure objectively after studying historical data and working closely with the locals. National and international specialists construct a risk and vulnerability profile of the community based on the conclusions of this team, which is referred to as the village disaster management plan (VDMP). This VDMP contains detailed socioeconomic and demographic information for each community, as well as elements at danger and a safe haven in the event of a natural disaster. Each VDMP

Table 1 Summary of the historical climate-induced natural disaster events in Shighnan district (UNEP et al. 2020)

#	Type of disaster	Year	Losses and damages			
			Loss of lives	Loss of agricultural land (ha)	Loss of livestock	Loss and damages of properties
1	Avalanche in Badam area	1945	1	12	No record available	1 ha pasture 300 m irrigation canal
2	Avalanche in Badam area	1985	0	7	100 sheep and goats	9 houses 250 m canals 1 ha pasture
3	Flood in Amu river	2010	0	3	11 sheep	1500 trees 1 km road
4	Flood from Koh-e-Safid area	2012	0	5	20 sheep and goats	Main irrigation canal of Badam, Tandir and Nimdah areas 500 trees
5	Flood from Badam area	2012	0	4	0	4 houses 15 gardens 30 m irrigation canal 40 m main road
6	Flood from Bawar Dara	2014	0	677	0	120 m main irrigation canal 4 drinking water reservoirs 7 small hydropower plants 17 watermills

classifies the area in three colors (red, yellow, and green), with each color highlighting the area's vulnerability to numerous natural disasters, mainly avalanche and flood (UNEP et al. 2020) (Fig. 9).

Following the preparation of the VDMP, a village seminar is held in which all people of the community are invited to attend. During the seminar, a community emergency response team (CERT) is formed, which is made up of 25 regular individuals, both men and women, who are trained and outfitted with basic emergency response equipment. This squad is also trained in first aid, search, and rescue, and has the necessary equipment.

In the Shighnan district, AKAH has formed a new team named the search and rescue team (SART). This group is in charge of raising resident awareness and training the CERT teams.

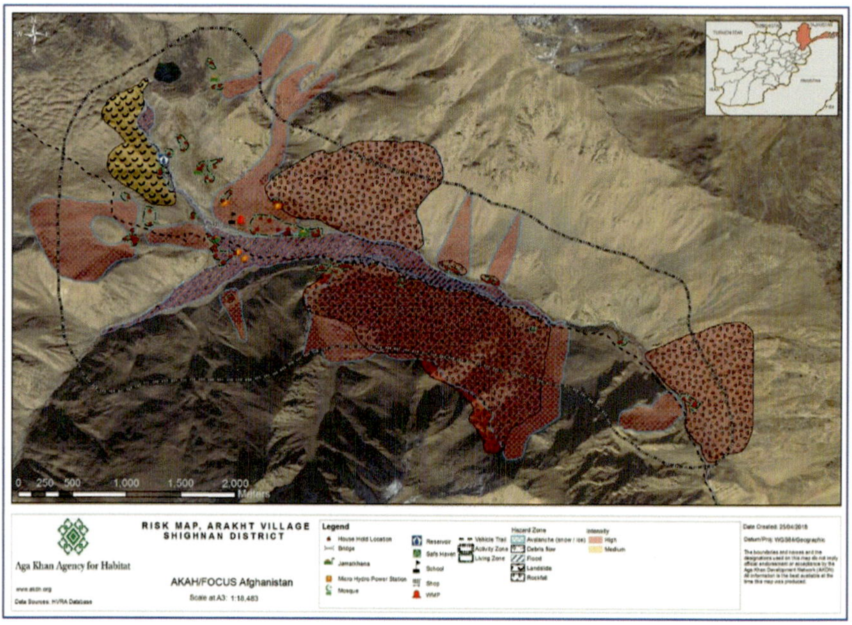

Fig. 9 An example of risk map of a village of Shighnan (AKAH 2020a)

6.2 Monitoring and Warning Services

Once the VDMP is developed and the vulnerable areas are identified, the feasibility studies were done to establish the EWS. As a finding of the feasibility study, two areas of Shighnan district Arakht and Werezn were selected for the installation of the WMPs. Arakht is the upper stream area of the Shighnan district and the Werezn in the lower stream area (UNEP et al. 2020) (Figs. 10 and 11).

WMPs provide update data of some of the key meteorological parameters on a daily basis, such as:

- 24-h rainfall amount,
- 24-h snowfall amount,
- total snow depth on the ground,
- wind speed,
- wind direction,
- daily maximum/minimum temperature.

The above-mentioned parameters' data is compiled from WMPs by local volunteers and relayed to the AKAH HQ in Kabul via phone. There are five persons in the field who are in charge of monitoring the WMPs. These individuals were given instruction on how to gather data and how to use the equipment installed in each WMP. In addition, the caretakers (WMPs observers) are outfitted with personal protective equipment (PPE) and technical instruments.

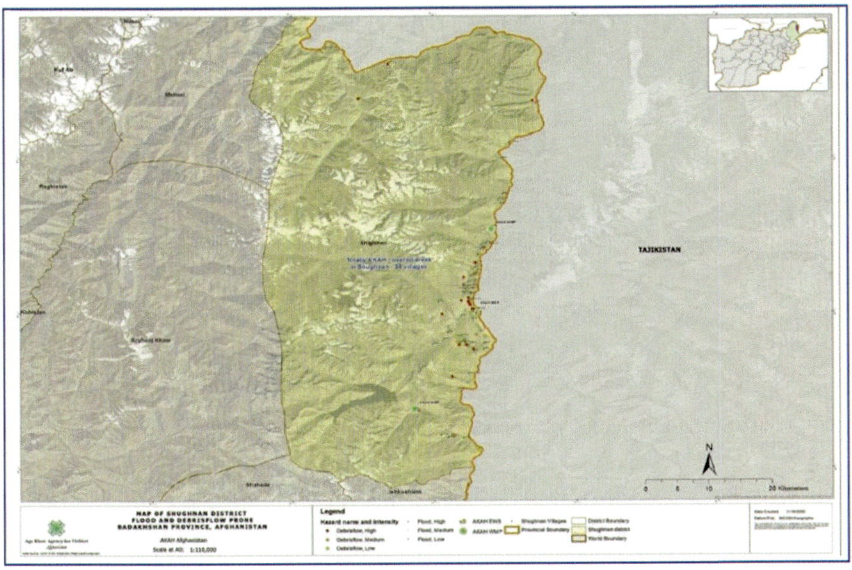

Fig. 10 Map of Shighnan district area covered by AKAH for FEWS (AKAH 2020c)

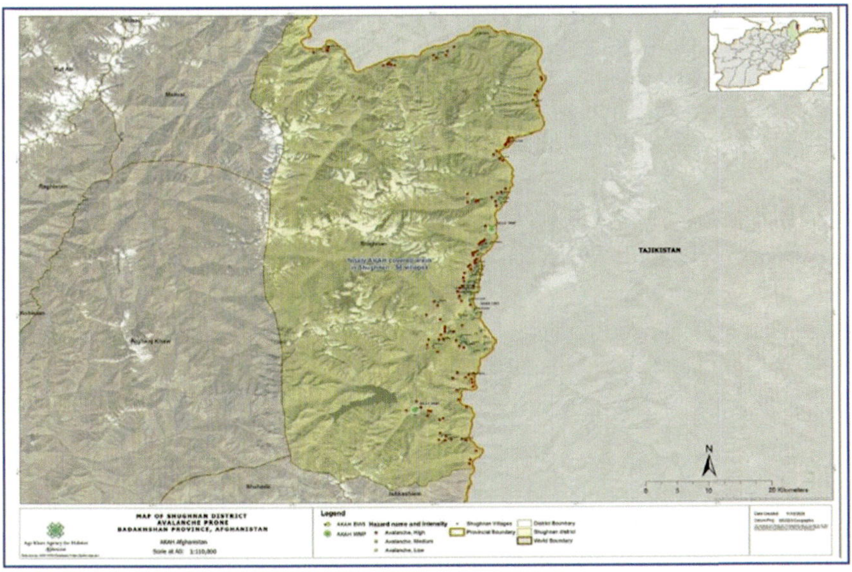

Fig. 11 Map of Shighnan district avalanche-prone area and the locations of AKAH WMPs (AKAH 2020b)

The data is entered into the system and the cloud-connected server once it arrives at HQ. After the data is entered into the system, a team of national and international consultants reviews it and compares it to other international weather forecasting sources. The warning threshold is determined by this group of experts (UNEP et al. 2020).

If the HQ is unable to send alerts for whatever reason, the local observers have been taught and are given instructions on how to do so. For the avalanche threshold, for example, if the snowfall exceeds 30 cm in 6 h and is projected to continue at a rate of more than 5 cm/h for the next 6 h, the temperature climbs above zero, or wind is forecasted, they issue avalanche warnings (UNEP et al. 2020).

6.3 Communication and Dissemination

Once the team at AKAH HQ has decided on the warning threshold, they inform the heads of communities and the various teams that have been established in the field in case action is required.

Mobile phones are the primary mode of communication, which is not always the ideal option for a rural place like the Shighnan district. The use of mobile phones in Shighnan district is modest, and different people utilize different mobile networks, including Tajik networks.

The AKAH uses additional communication methods as a backup, especially in crises, such as radio, Thuraya, Codan, VCN, and so on. In Fig. 12, you'll find a list of EWS communication networks in the Shighnan district (AKAH 2020d).

6.4 Response Capability

It is critical for individuals to understand the warnings in order to respond appropriately. In addition, practice drills for evacuation drills and other preventative measures are required. This will assist citizens in better understanding their risks and how to respond to specific warnings. In this context, AKAH has developed numerous organizations for emergency reaction to warnings in the Shighnan district, including (UNEP et al. 2020):

- Community Emergency Response Team (CERT)

The teams were educated for two days on how to coordinate and implement disaster-related evacuation plans, as well as how to organize rescue operations such as caring for casualties, finding victims, administering first aid to injured people, and searching for trapped victims. During the training, there were numerous practical exercises to allow members to master the skills and use them in the event of an emergency in the territory. To respond to any crisis, the teams are prepared with basic first aid, search

#	Village	Communication Network
1	Andeez	Roshan, MTN, MegaFon, Tcell
2	Astana-Meyanshar	VCN
3	Bawar-Rejistek	Roshan, MTN, MegaFon, Tcell
4	Dehmurghan	Roshan, MTN, MegaFon, Tcell
5	Dehshar	Roshan, MTN, MegaFon, Tcell
6	Gharjween-Newad	VCN
7	Khawar-Redaj	Roshan, MTN, MegaFon, Tcell
8	Lochid-Parveez-Bezokh	Roshan, MTN, MegaFon, Tcell
9	Meyanshar	VCN
10	Nimda	Roshan, MTN, MegaFon, Tcell
11	Pastive	Roshan, MTN, Tcell, VCN (2G/3G)
12	Sarchashma	Roshan, MTN, Tcell, (2G/3G)
13	Sawan	Roshan, MTN, Tcell, (2G/3G)
14	Sebdeej	Roshan, MTN, Tcell, (2G/3G)
15	Senorg	Roshan, MTN, Tcell, (2G/3G)
16	Shive	Roshan, MTN, Tcell, (2G/3G)
17	Vonshar-Tirew	VCN
18	Werizn	MegaFon,Tcell Roshan, MTN, Afghantelecom, /(2G)
19	Yarkh	MegaFon, Tcell
20	Yazgam	VCN
21	Chasnood payan	MegaFon, Tcell
22	Darmarakht	Megafon
23	Arakht	VCN
24	Shoran	MegaFon,Tcell
25	Sheduge	MegaFon, Tcell
26	Sheduge Dara	VCN
27	Chasnood bala, Newadek	MegaFon, Tcell
28	Robat	MegaFon, Tcell
29	Pojwar	MegaFon, Tcell
30	Chaweed	MegaFon, Tcell, Salam
31	Shaikha (Khezri Roshan)	VCN

Fig. 12 Emergency communication network for EWS in Shighnan district (AKAH 2020d)

Number	Item	Description
1	Shovel	A -14gauge iron sheet with wooden handle
2	Crowbars	Five feet long and Wt. 8kg
3	Axe	2.5 kg with wooden handle
4	Stretcher	Folding made of tarpaulin
5	Hummer	12 pound with wooden handle
6	Saw	Medium size
7	First aid box	
8	Picks	Super China, 2.3 kg with wooden handle
9	Telephone	Megaphone with battery
10	Flashlights	Medium size with wider beam and batteries
11	Waistcoat	
12	Rope	1 meter long
13	Bag	
14	Compass	With protective cap KD 351
15	Avalanche pole	For search and rescue works
16	Avalanche shovel	For search and rescue works
17	Whistle with lanyard	
18	Backpack	Waterproof -30litre volume
19	Stationary	Notebook, pens and pencils
20	Protective Goggles	Protective glasses to be used by the search and rescue team
21	Light Waist	With reflecting stripes
22	Headlight	LED with rechargeable batteries

Fig. 13 CERT toolkit for avalanche and flood preparedness and avalanche safety equipment (AKAH n.d.)

and rescue instruments, and personal protective equipment (PPE). The following equipment is available to each CERT team (UNEP et al. 2020) (Fig. 13).

Evacuation centers are in place. Some people do not accept the warnings, but they are evacuated by the heads of the CDCs and the CERT teams.

- Avalanche Preparedness Team (AVPT)

Ten community members from each community who currently hold CERT memberships are chosen to become AVPT members. The goal of the one-day training this team receives is to increase the AVPT members' ability to plan for evacuation before, during, and after the disaster.

To assist the teams in comprehending the avalanche level and the likelihood of avalanche occurrences, basic avalanche instruments are provided (UNEP et al. 2020).

- School Emergency Response Team (SERT)

Another volunteer group in the Shighnan district is the SERT team, which helps schools respond to climate-related natural disasters. This group will receive two days of First Aid and Search and Rescue training. The SERTs are also outfitted with basic first aid and search-and-rescue equipment, as well as personal protective equipment (PPE). To date (November 2020), 120 people have been trained as SERT members, 56 of whom are male and 64 of them are female (UNEP et al. 2020).

6.5 *Results and Impacts*

Since February 2017, when AKAH launched EWS in Shighnan, until the winter of 2019, roughly 902 families have been evacuated, primarily owing to avalanche hazards in the area. In total, approximately 6745 people were safely evacuated throughout that time period. Male residents of Shighnan district numbered 3277, while female residents numbered 3462 (UNEP et al. 2020) (Fig. 14).

Based on this data, Shedoj Dara and Werezn are two of the more frequently evacuated villages of Shighnan district. The data provided in figure is only for 2017 and 2019. Since Afghanistan was facing severe drought in 2018, there is no evacuation recorded for that particular year (UNEP et al. 2020).

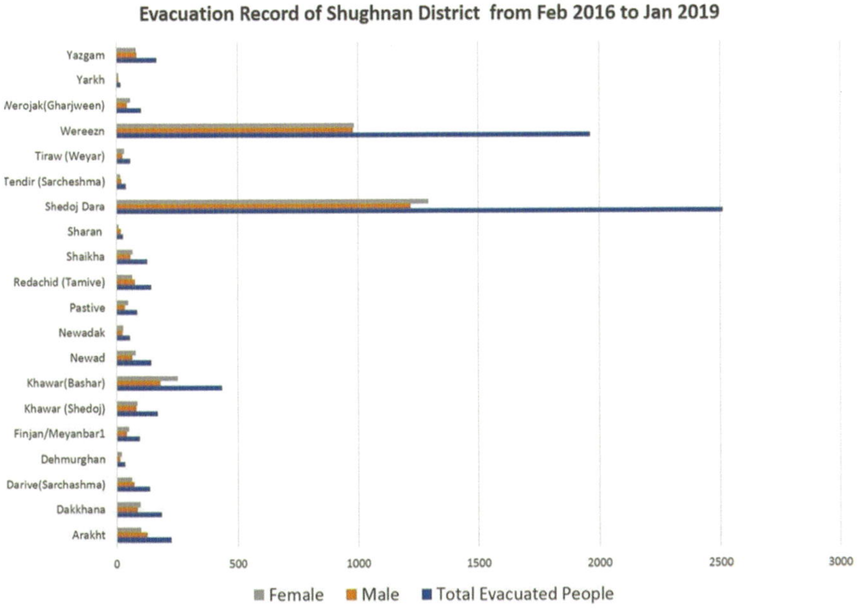

Fig. 14 Shighnan district evacuation report per villages (UNEP et al. 2020)

7 Challenges and Recommendation

The early warning system (EWS) is a unique example of its form. However, there are some challenges still existing. One of the challenges at the beginning of the establishment of the system was the lower level of the trust of people on the warnings issued. Even after gaining the trust of the residents, they did not accept the evacuation since they were not sure of the condition of the evacuation centers. However, there are still some challenges to overcome. One of the first challenges in establishing the system was people's lack of trust in the warnings given. Residents refused to accept the evacuation even after winning their trust because they were unsure of the status of the evacuation centers.

Another main challenge against this system is the limitation of the communication network. Most of the areas of the Shighnan district are partially covered by mobile networks. This fact sometimes causes challenges against the dissemination of the warnings to the public.

Furthermore, one of the obstacles to the system's long-term viability is the lack of an annual budget for equipment maintenance and repair. To yet, AKAH has used its own limited resources to keep the system running well.

As described under the impacts section, this is one of the most successful examples of the community-based early warning system (CBEWS) which could be replicated in other natural disasters prone areas of the country. However, AKAH has already expanded to include regions like Bamyan, Baghlan, Panjshir, and Takhar. This system is a great example of a community based EWS that is driven and directed by local residents. The cost of operation is substantially lower, and it does not require employees; anyone may operate it.

References

Afghanistan Information Management Service [AIMS] (2005) Shighnan district. https://www.ecoi.net/en/document/1256343.html

Afghanistan Meteorological Department [AMD] (2020) Accessed Afghanistan Meteorological Department: http://www.amd.gov.af/disaster-effects-in-afghanistan/

Afghanistan National Disaster Management Authority [ANDMA] (2015) ANDMA strategy for 2015–2019. Afghanistan National Disaster Management Authority, Kabul. https://www.refworld.org/cgi-bin/texis/vtx/rwmain/opendocpdf.pdf?reldoc=y&docid=5b28f2214

Afghanistan National Disaster Management Authority [ANDMA] (2018) Afghanistan natural disasters risk reduction strategy in line with Sendai framework. Afghanistan National Disaster Management Authority, Kabul. https://andma.gov.af/sites/default/files/2020-03/Persian_ASDRR.pdf

Agha Khan Agency for Habitat [AKAH] (2020a) A questionnaire on CBEWS in Shughnan, 17 Oct 2020 (Noori MM, Interviewer)

Agha Khan Agency for Habitat [AKAH] (2020b) Map of AKAH WMPs in Shughnan district, 20 Nov 2020 (Habitat AK, Compiler). Agha Khan Agency for Habitat, Kabul

Agha Khan Agency for Habitat [AKAH] (2020c) Map of Shughnan district area covered by AKAH for FEWS, 20 Nov 2020. Kabul

Agha Khan Agency for Habitat [AKAH] (2020d) Emergency communication network for EWS in Shughnan district, 20 Nov 2020. Kabul

Agha Khan Agency for Habitat [AKAH] (n.d.) CERT toolkit for avalanche and flood preparedness and avalanche safety equipment. Kabul

EM-DAT (2020) Centre for research and the epidemiology of disasters. Accessed The International Disaster Database: https://www.emdat.be/database

iMMAP (2020a) Shughnan district avalanche risk exposure map, 18 Nov 2020. Kabul

iMMAP (2020b) Shughnan district flood risk exposure map, 18 Nov 2020. Kabul

iMMAP (2020c) Shughnan district flood prone population statistics, 19 Nov 2020. Kabul

National Environmental Protection Agency [NEPA] (2016) Afghanistan: climate change science perspectives. National Environmental Protection Agency & UN Environment, Kabul. Accessed 15 May 2022. https://postconflict.unep.ch/publications/Afghanistan/UNEP_AFG_CC_Science_perspect

United Nations Environment Programme, Afghanistan National Environmental Protection Agency, Afghanistan National Disaster Management Authority (2020) Current state of early warning system in Afghanistan—an outlook of the flood early warning system. https://wedocs.unep.org/20.500.11822/42292

Proposed Study Design for Spatio-Temporal Drought Assessment Using Standardized Precipitation Evapotranspiration Index (SPEI) and Adaptation Over Mantaro Valley, Peru

Del Piero Raphael Arana Ruedas

Abstract Case study author Del Piero R. Arana Ruedas. This paper presents a proposed study design for a Spatio-temporal drought assessment using Standardized Precipitation Evapotranspiration Index (SPEI) and Adaptation over Mantaro Valley, Peru. An in-depth literature review is undertaken followed by a description of the types of data that will be used to undertake a risk assessment analysis with proposed objective to create a risk assessment at multiple time intervals for Mantaro Valley. Arana proposes that the results be analyzed and used as a source of insights and needed changes to the current Peruvian National Adaption Plan (NAP). In addition, the Regional Adaptation Plan, where the Mantaro Valley is located will be contrasted with the drought risk assessment result to understand its effectiveness to finally provide adaptation strategies evaluated throughout social, economic, and environmental criteria.

Keywords Peru · Mantaro Valley and climate change · Standardized Precipitation Evapotranspiration Index (SPEI) · Disaster risk management · Adaptation · Vulnerability · Adaptive capacity · Hazard reduction measures

1 Introduction

This paper presents a proposed study design for a Spatio-temporal drought assessment using Standardized Precipitation Evapotranspiration Index (SPEI) and Adaptation in the Mantaro Valley of Peru.

D. P. R. Arana Ruedas (✉)
Universidad Continental, Huancayo, Junin, Peru
e-mail: darana@continental.edu.pe

K. Kim and C. Atkin (eds.), *Cross-Cultural Perspectives on Climate Change Adaptation*, Cities and Nature, https://doi.org/10.1007/978-3-031-50365-8_20

Atmospheric Carbon Dioxide (CO_2) concentration started to be measured in 1958 at the Mauna Loa, Hawaii; resulting in the Keeling Curve, which shows the accumulation of CO_2 in the atmosphere, based on part per million (ppm) as united of measure (Howe 2015). The first atmospheric CO_2 measure at Mauna Loa Observatory revealed a concentration close to 315 ppm in 1958, drastically increasing to 417.43 ppm by April 2022 in a global monthly mean CO_2 (NOAA 2022).

Furthermore, CO_2 is considering the main greenhouse gas (GHG) as well as Methane (CH_4), Nitrous Oxide (N_2O), and Sulfur Hexafluoride (SF_6), which concentrations have continued to increase in the atmosphere. These gases are commonly known as greenhouse gases (GHG) because of their capability to keep the temperature on Earth under conditions where life can over exist, mostly known as the greenhouse effect (Thøgersen et al. 2021).

Unfortunately, the GHG emissions have been rising the global temperature, known as global warming, over the years (Shi et al. 2021). It is also essential to consider the Global Warming Potential (GWP), which measures the radiative force (heat absorbed) over a period, and the Global Temperature Potential (GTP) (IPCC 2014).

In effect, the Intergovernmental Panel on Climate Change (IPCC) refers that it is unequivocal that human influence has warmed the atmosphere, ocean, and land. A rapid change in the atmosphere, hydrosphere, cryosphere, and biosphere has occurred (IPCC 2021).

Several effects due to climate change have been demonstrated. They are disrupting ecological interactions (Fontúrbel et al. 2021), melting glaciers at an unprecedented speed (Magnason and Smith 2020), which is recognized as one of the leading indicators of climate change (Gan et al. 2015). Sea level rise affects many people in coastal areas (Robert and Schleyer-Lindenmann 2021), and ocean acidification affects its resources and ecosystem (Greenhill et al. 2020). Severe extreme weather events like heatwaves, wildfires, storms (Howe 2021), or precipitation patterns changes with variable frequency and severity increase the risk of floods and droughts (Abdulla 2020; Ezaz et al. 2022; Sarkar and Maity 2020). This last one strongly affects agriculture activities and fresh water supply (Gumbo et al. 2021; Kalair et al. 2019; Lychuk et al. 2021; Mengistu et al. 2021; Qi et al. 2021; Shagega et al. 2020; Zhuang et al. 2018).

One of the most affected countries are those where vulnerable characteristics given by the United Nations Framework Convention on Climate Change (UNFCCC) are observed (UNTC 1992). For instance, Peru has reported seven of nine vulnerable characteristics on its Nationally Determined Contributions (NDC) and National Adaptation Plan (NAP) (UNFCCC 2021).

Within the country, there are several studies based on the Peruvian Tropical Andes (PTA) and its effects on climate change. The PTA shows that geographic changes cause various climate conditions (Zárate Malpica and Miranda Zambrano 2017). In that sense, the PTA has a very complex climate system, which causes a significant Spatio-temporal variability (Callañaupa Gutierrez et al. 2021).

Vulnerable farmers in the PTA, already affected by poverty with a small land, mainly develop subsistence agriculture based on rain-fed cropping systems (Sanabria et al. 2014). In that sense, the most relevant impact on the PTA is due to mean temperature and precipitation changes, which increase agricultural vulnerability (Lozano-Povis et al. 2021). According to Jurt et al. (2015), future water availability for human use and agriculture irrigation represents the most relevant concern in PTA communities.

This water availability is threatened due to frequent and severe droughts. The estimated yearly economic loss due to climate change in Peru goes around 7.3 and 8.6% of the gross domestic product (GDP) by 2050 under the worse climate scenario where the agricultural sector could lose up to 33% of income every year (Gubler et al. 2020).

A significant part of this agricultural loss comes from areas in the PTA like the Mantaro Valley, which is composed of four provinces (Jauja, Concepcion, Chupaca, and Huancayo) due to the drought and frost risk events (Callañaupa Gutierrez et al. 2021).

In order to face such climate change issues, there are two main approaches, Climate Change Mitigation (CCM) and Climate Change Adaptation (CCA) (Hritonenko and Yatsenko 2022; Jiang et al. 2020). Adaptation has been defined by IPCC (2014) as "the adjustment of current or projected climate and its effects in human/natural systems" to enhance coping capacity and reduce vulnerability (Sharifi 2020).

In that sense, this study aims to elaborate on spatio-temporal drought risk assessment over Mantaro Valley to provide regional/local climate change information using the Standardized Precipitation Evapotranspiration Index (SPEI) given by the World Meteorological Organization (WMO 2016). Then, explain the link of drought assessment information with the climate change adaptation strategies that could be applied to increase resilience to climate change, promoting sustainable development in the Peruvian Tropical Andes, Mantaro Valley.

2 Literature Review

2.1 Climate Change in the Peruvian Tropical Andes

Agriculture

Regional-scale analysis reveals tendencies of more frequent and severe droughts (Le et al. 2020). For instance, Peru considered a megadiverse country is at the same time one of the most vulnerable ones in South America regarding climate variability. According to Gubler et al. (2020), smallholder farmers in the Peruvian Tropical Andes (PTA) are suffering from droughts. These droughts are enormously threatening 81% of the eight primary food producers in the country (Wongchuig et al. 2018).

The Peruvian Tropical Andes (PTA) is located in South America, hosting 25% of global biodiversity. The PTA creates Inter-Andean valleys among their mountain. One of these valleys, Mantaro Valley, is located in the Mantaro River Basin

(MRB), where hydroelectric stations produce 35% of the Peruvian energy, and several families practice agricultural activity as their livelihood (Correa et al. 2016).

There are solid reasons to provide farmers with timely and effective climate information under climate-changing conditions to ensure their adaptative response (Ponce 2020) in the PTA, Mantaro Valley.

Glaciers

The Peruvian Tropical Andes (PTA) is the host of 70% of tropical glaciers worldwide (Wood et al. 2021). Which ones are located in elevated regions where the air warming is higher than in lower places (Correa et al. 2016; Vuille 2013). Such an increase in mean temperature is triggering glacier retreat (Ponce 2020) like those in the Mantaro River Basin (MRB) (Correa et al. 2016), increasing the Earth's energy absorption due to glacier melting and snow cover loss (Correa et al. 2016). In effect, socio-economical systems are strongly affected in high mountain regions like the PTA (Emmer et al. 2021).

Furthermore, glaciers in the PTA are relevant because of their contribution to freshwater, ecosystem development, truism, and self and community cultural identity, where political and poverty issues remain (Chevallier et al. 2011; Jurt et al. 2015). Huaytapallana Glacier, is located in Mantaro Valley.

2.2 *Culture*

The cultural and natural history of the Peruvian Tropical Andes (PTA) is interconnected. Human cultures, knowledge systems, religions, and social interactions of the population reflect an Andean cosmology (Anderson et al. 2012).

People are firmly concerned about the climate change issue; however, from awareness to actions still existing gaps. Behavioral and brain science reveals that climate change is related to moral judgments, provoking a lack of action. For instance, people mainly divide climate actors into those who suffer climate harm and those who cause it. Fortunately, PTA communities started to correlate scientific evidence and natural and cultural spheres on different social scales (Jurt et al. 2015).

2.3 *El Niño–Southern Oscillation (ENSO)*

One of the most relevant natural phenomena in Peru is the El Niño–Southern Oscillation (ENSO) which forces changes in precipitations and temperature due to variations in sea surface temperature (SST) along the Pacific Ocean, influencing somewhat the South America Summer Monsoon (SASM) (Viveen et al. 2019). ENSO causes a strong westerly current over the Peruvian Tropical Andes (PTA), significantly inhibiting moisture transport from the Amazon basin and bringing about drought conditions (Vuille 2013). The climate variability in the PTA has been reported to be dependent on altitude, topography, ENSO, and the Pacific Decadal Oscillation (PDO) (Veettil and Kamp 2021).

2.4 *Importance of Climate Change Variability*

Temperature and Precipitation

Several changes are noticed in the PTA, the mean air temperature value is currently exceeding historical meteorological data (Michelutti et al. 2015), and precipitation patterns can be related to topographic forcing, which considerably varies between mountains and valleys (Saavedra et al. 2020). In that sense, the PTA has a very complex climate system, which causes a significant spatio-temporal variability of climatic conditions (Callañaupa Gutierrez et al. 2021).

According to Anderson et al. (2012), the temporal variability of climatic conditions such as temperature and precipitation are evident in the PTA under interannual and decennial time changes impulse by the interactions between tropical pacific and amazon influences. Recently, the range of natural climate variability in the PTA has begun to exceed historically documented thresholds. For instance, in the central inter-Andean valleys of Peru above 3500 m, the number of heavy rain events and early freezes has recently increased.

Spatio-Temporal Consideration

The Climate Impact Drivers (CID) are not homogenous on Earth, varying considerably from region to region (Sharma and Goyal 2020). Climate variability analysis and future climate projections are complicated in highland regions due to their diverse orography and complex climate system (Eccel and Tomozeiu 2015). It is essential to understand that future climate projections remain uncertain (Drobinski et al. 2018), mainly in terms of precipitation patterns change (Sharma and Goyal 2020). Furthermore, climate impacts might vary according to altitude conditions in places like the Mantaro Valley (Correa et al. 2016).

Droughts

Definition

An international definition of drought varies worldwide; however, based on Javed et al. (2021), Karatayev et al. (2022), Leng et al. (2015), Wang et al. (2022) and Zhao et al. (2021), it can be defined as a complex slowly developing a climatological event that frequently occurs in almost all climate regimes with observed trends of warming. It is characterized by below-normal precipitation patterns, reducing water supply and increasing evapotranspiration over an extended period. In addition, drought can remain for a long time impacting on environment, economic, and social spheres.

Drought Types and Characteristics

As well as drought definition, its types of presence have been widely studied, mainly dived in (i) meteorological, (ii) agricultural, (iii) hydrological, and (iv) socioeconomic drought (Nikraftar et al. 2021; Wang et al. 2022; Zhao et al. 2021). It is based on below-normal precipitation patterns over a medium-long term, lack of soil moisture to meet crops cycle, reduced surface–subsurface water streamflow and groundwater storage, and affected livelihood due to supply–demand water disequilibrium, respectively (Karatayev et al. 2022; Vorobevskii et al. 2022; Wang et al. 2022).

2.5 *National Adaptation Plan*

In order to successfully achieve the planning phase evidenced in the Peruvian National Adaptation Plan and Nationally Determined Contributions; identifying concerning signs in climate change variability and future projections is crucial to provide forward appropriate adaptation strategies (Jahangir et al. 2022). Policymakers, climate advocates, scientific, and general stakeholders are interested in regional/local climate change information (Wang et al. 2017), which is a valuable tool for them to develop intelligent adaptation strategies (Kim et al. 2020; Sharma and Goyal 2020) under a bottom-up approach.

3 Prospective Methodology

3.1 *Selection of the Study Area*

Junin region is located in the central area of the Peruvian Tropical Andes (PTA), with 44,329 km^2. It represents 3.4% of the Peruvian territory involving two of the three geographic regions, Andes and Forest, with 20,821 km^2 (47%) and 23,508 km^2 (53%). The Andes area is where the Mantaro Valley is mainly divided into four provinces (Huancayo, Chupaca, Concepción, and Jauja), and the Huaytapalla Glacier is located. Table 1 contains surface and population data of the Mantaro Valley.

Table 1 Mantaro Valley surface and population characteristics

Province	Districts	Surface (km^2)	Population
Jauja	34	3749	88,405
Huancayo	28	3561	595,183
Concepción	15	3075	59,138
Chupaca	9	1144	57,604

Source Banco de Reserva del Peru (BCR), 2020

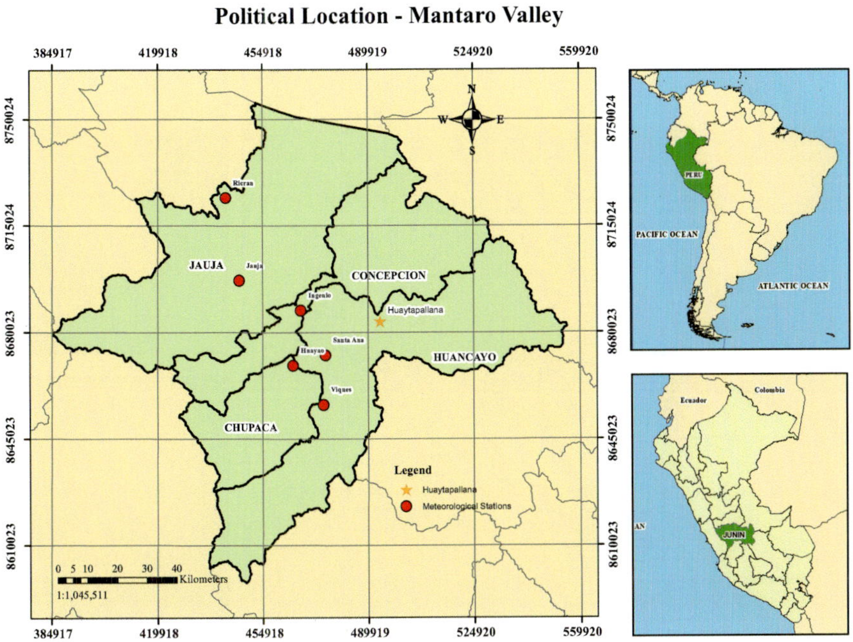

Fig. 1 Mantaro Valley location and monitoring stations map

In the PTA, the climate is also changing along with the altitude. In that sense, Peru is divided into eight natural regions based on altitudinal characteristics. The Peruvian Natural Regions based on altitude levels *msnm*-masl and whose names have origin in the Inca's language Quechua: (i) *Janca* (white due to ice in the mountains), (ii) *Puna* (highland evil), (iii) *Suni* (broad land), (iv) *Quechua* (mild weather or valley), (v) *Yunga* (hot valley), and (vi–vii) *Selva Alta-Baja* (high-low forest) (Correa et al. 2016). According to it, the Mantaro Valley is primarily located in the *Quechua* region, while the Huaytapallana is between *Quechua and Janca.*

Nevertheless, as detailed previously in this study, several changes have been occurring due to climate change, exacerbating the analysis challenge for the orography and complex climate system. In that sense, Fig. 1 evidence the political location where the study took place, Fig. 2 show a hydrographical map, Fig. 3 show a Orography map, Fig. 4 show a Land Use map.

3.2 *Data Collection Method and Sources*

The data will be collected from six different meteorological stations regarding precipitation and temperature data (Fig. 1). The data comes from (1990–2021) and will

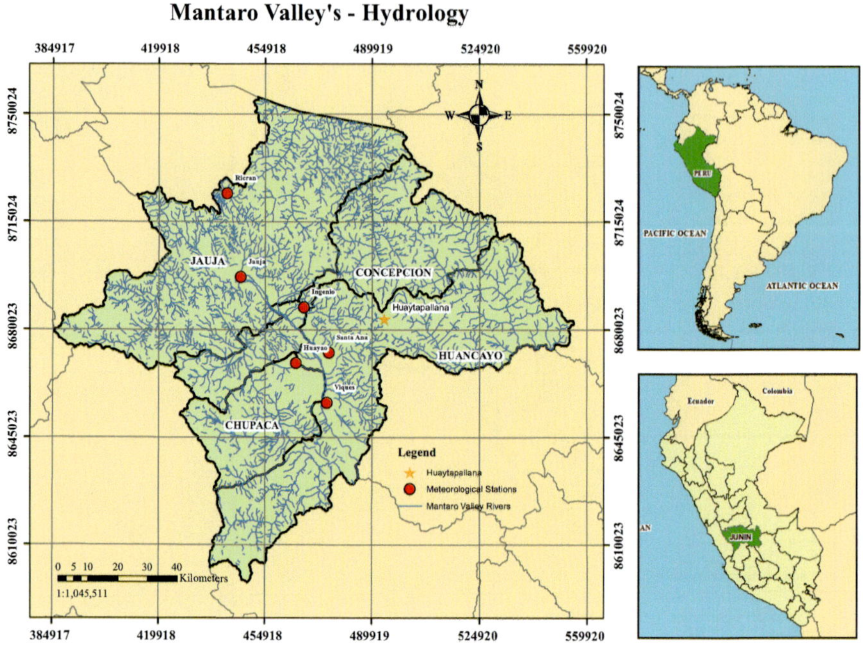

Fig. 2 Mantaro Valley watershed and rivers map

be provided for the Peruvian National Meteorological and Hydrological Service (SENAMHI) (1990–2021).

3.3 *Data Analysis*

First of all, data will be reviewed. The missing one will be addressed using the interpolation technique and fulfill a complete data set. Furthermore, the study aims to analyze droughts. In order to do so, it is essential to consider temperature, precipitation, and evapotranspiration, especially in places with a very complex climate system and changing orography like the Mantaro Valley.

In that sense, based on the handbook of drought indicators and Indices of the World Meteorological Organization (WMO). The Standardized Precipitation Evapotranspiration Index (SPEI) was chosen. In order to explain the reasons why this methodology is suitable with the study, see Table 2.

The time scale analysis is important because the useable water resource vary considerable from the arrival time until its availability. In that sense, accumulative water deficit for a given time scale is substantially important. Which, also provide better information to define if the drought event is meteorological, agricultural, or hydrological.

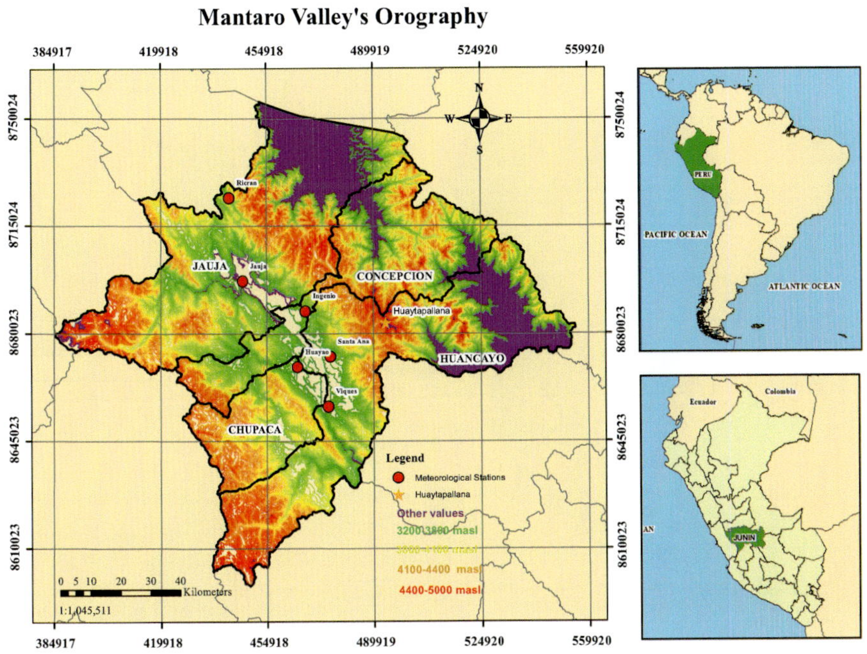

Fig. 3 Mantaro Valley orography map

Furthermore, SPEI is specifically suitable for the present study due to its accuracy, global warming consideration, and no need for wind speed, soil water, sun radiation data, which are not available in the study area. The SPEI calculation can be done at Rstudio software due to its pacakage availability which provides (i) Potential Evapotranspiration (PET), (ii) Water Balance (WB), and (iii) SPEI values for 1-, 3-, 6-, 12-, 24-, and 48-months' time scale for each station. Nevertheless, it is important to understand the calculation of the Potential of Evapotranspiration which is explained as follows:

Potential Evapotranspiration (PET) Calculation

See Fig. 5.

Monthly Water Surplus or Deficit

Di = Pi − PETi; It is a parameter for obtaining a drought index however, if PET = 0 the parameter is not defined letting $T°$ information useless.

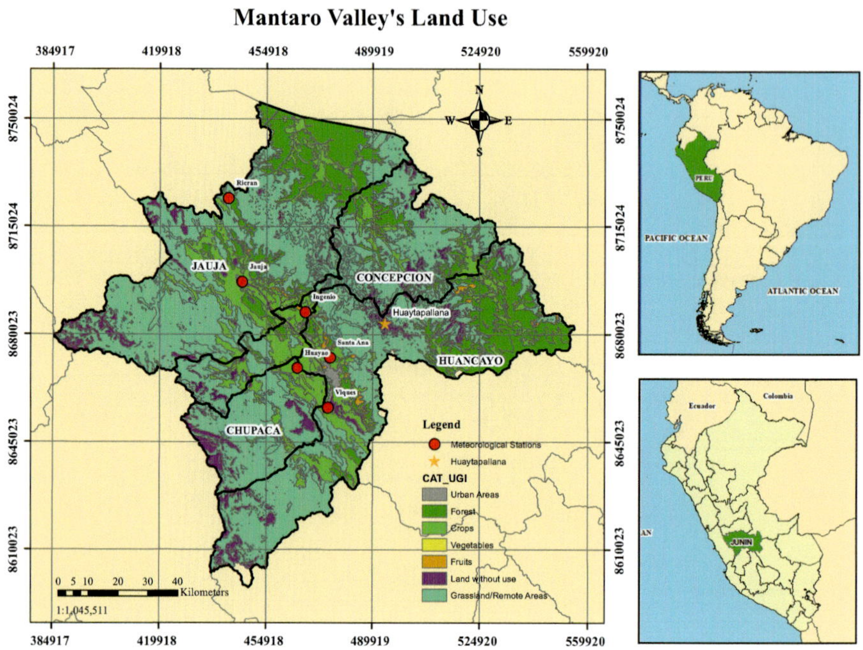

Fig. 4 Mantaro Valley land use map

Table 2 Drought risk assessment characteristics

Palmer Drought Severity Index (PDSI)	The Standardized Precipitation Index (SPI)	The Standardized Precipitation Evapotranspiration Index (SPEI)
Nature		
Based on soil water balance analysis	Based on precipitation probabilistic approach	Based on water balance and temperature
Strength		
Enables wetness and dryness values-based on water balance taking into account evaporation	It is easy to calculate and can be comparable in different time scale	Combines the accuracy of PSDI regarding evaporation under changing $T°$ with the easy calculation and the different time scale analysis of SPI
Weakness		
It has problem in spatial comparability and short time scale analysis	It does not take warming considerations ($T°$)	Deemphasizes the role of $T°$ in droughts when potential evapotranspiration (PET) = 0

a. Solar declination

$\delta = 0.4093 sen\left(\frac{2\pi J}{365} - 1.405\right)$; J = Average Julian day of the montl

b. Hourly angle of sun rising

$\omega = arccos(- tan\varphi * tan\delta)$; φ = Latitude in radians

c. Maximum number of sun hours

$$N = \left(\frac{24}{\pi}\right) * \omega$$

d. Number of days of the month

NDM taking into account leap-years

e. Correction coefficient computed

$$K = \left(\frac{N}{12}\right) * \left(\frac{NDM}{30}\right)$$

f. Monthly-mean temperature

$$T = \left(\frac{\Sigma\ daily\ T^{o}\ values}{days\ of\ the\ month}\right)$$

g. Heat index

$$i = \left(\frac{T}{5}\right)^{1.514} \quad ; I = \Sigma\, 12\ montyly\ (i) values$$

h. Coefficient

$$m = \left(6.75 * 10^{-7} * I^{3}\right) - \left(7.71 * 10^{-5} * I^{2}\right) + \left(1.79 * 10^{-2} * I\right) + 0.492$$

i. Monthly Potential Evapotranspiration

$$PET = 16K\left(\frac{10T}{I}\right)^{m}$$

Fig. 5 Potential evapotranspiration (PET) calculation *Source* (Vicente-Serrano et al. 2010)

4 Adaptation Actions Analysis

Finally, the result will be analyzed and explained along with the current Peruvian National Adaption Plan (NAP) and how it could get involved in the actions proposed. In addition, the Regional Adaptation Plan, where the Mantaro Valley is located, will be contrasted with the drought risk assessment result to understand its effectiveness to finally provide adaptation strategies evaluated throughout social, economic, and environmental criteria.

5 Expected Results

Drought risk events for 1-, 3-, 6-, 9-, 12-, 24-, 48-months' time scale (maps and graphs) for each station between 1990–2021 over Mantaro Valley and wet events, related to floods; which is particulary less applied for.

Worldwide adaptation actions under similar drought conditions.

Practical climate adaptation action to tackle drought in Mantaro Valley.

References

Abdulla F (2020) 21st century climate change projections of precipitation and temperature in Jordan. Procedia Manuf 44(2019):197–204. https://doi.org/10.1016/j.promfg.2020.02.222

Anderson E, Marengo J, Villalba R, Halloy S, Young B, Cordero D, Gast F, Jaimes E, Ruiz D (2012) Consecuencias del cambio climático en los ecosistemas y servicios ecosistémicos de los Andes Tropicales. In: Climate change and biodiversity in the tropical Andes, p 410

Callañaupa Gutierrez S, Segura Cajachagua H, Saavedra Huanca M, Flores Rojas J, Silva Vidal Y, Cuxart J (2021) Seasonal variability of daily evapotranspiration and energy fluxes in the Central Andes of Peru using eddy covariance techniques and empirical methods. Atmos Res 261. https://doi.org/10.1016/j.atmosres.2021.105760

Chevallier P, Pouyaud B, Suarez W, Condom T (2011) Climate change threats to environment in the tropical Andes: glaciers and water resources. Reg Environ Change 11(SUPPL. 1):179–187. https://doi.org/10.1007/s10113-010-0177-6

Correa SW, Mello CR, Chou SC, Curi N, Norton LD (2016) Soil erosion risk associated with climate change at Mantaro River basin, Peruvian Andes. CATENA 147:110–124. https://doi.org/10.1016/j.catena.2016.07.003

Drobinski P, Silva ND, Panthou G, Bastin S, Muller C, Ahrens B, Borga M, Conte D, Fosser G, Giorgi F, Güttler I, Kotroni V, Li L, Morin E, Önol B, Quintana-Segui P, Romera R, Torma CZ (2018) Scaling precipitation extremes with temperature in the Mediterranean: past climate assessment and projection in anthropogenic scenarios. Clim Dyn 51(3):1237–1257. https://doi.org/10.1007/s00382-016-3083-x

Eccel E, Tomozeiu R (2015) Increasing resolution of climate assessment and projection of temperature and precipitation in an alpine area. Theoret Appl Climatol 120(3–4):479–493. https://doi.org/10.1007/s00704-014-1181-4

Emmer A, Le Roy M, Sattar A, Veettil BK, Alcalá-Reygosa J, Campos N, Malecki J, Cochachin A (2021) Glacier retreat and associated processes since the Last Glacial Maximum in the Lejiamayu Valley, Peruvian Andes. J S Am Earth Sci 109:1–12. https://doi.org/10.1016/j.jsames.2021.103254

Ezaz GT, Zhang K, Li X, Shalehy MH, Hossain MA, Liu L (2022) Spatiotemporal changes of precipitation extremes in Bangladesh during 1987–2017 and their connections with climate changes, climate oscillations, and monsoon dynamics. Glob Planet Change 208:103712. https://doi.org/10.1016/j.gloplacha.2021.103712

Fontúrbel FE, Nespolo RF, Amico GC, Watson DM (2021) Climate change can disrupt ecological interactions in mysterious ways: using ecological generalists to forecast community-wide effects. Clim Change Ecol 2:100044. https://doi.org/10.1016/j.ecochg.2021.100044

Gan R, Luo Y, Zuo Q, Sun L (2015) Effects of projected climate change on the glacier and runoff generation in the Naryn River Basin, Central Asia. J Hydrol 523:240–251. https://doi.org/10.1016/j.jhydrol.2015.01.057

Greenhill L, Kenter JO, Dannevig H (2020) Adaptation to climate change–related ocean acidification: an adaptive governance approach. Ocean Coast Manag 191:105176. https://doi.org/10.1016/j.ocecoaman.2020.105176

Gubler S, Rossa A, Avalos G, Brönnimann S, Cristobal K, Croci-Maspoli M, Dapozzo M, van der Elst A, Escajadillo Y, Flubacher M, Garcia T, Imfeld N, Konzelmann T, Lechthaler F, Liniger M, Quevedo K, Ramos H, Rohrer M, Schwierz C, Wüthrich B (2020) Twinning SENAMHI and MeteoSwiss to co-develop climate services for the agricultural sector in Peru. Clim Serv 20. https://doi.org/10.1016/j.cliser.2020.100195

Gumbo AD, Kapangaziwiri E, Chikoore H, Pienaar H (2021) Assessing water resources availability in headwater sub-catchments of Pungwe River Basin in a changing climate. J Hydrol Reg Stud 35:100827. https://doi.org/10.1016/j.ejrh.2021.100827

Howe JP (2015) This is nature; this is un-nature: reading the keeling curve. Environ Hist 20(2):286–293. https://doi.org/10.1093/envhis/emv005

Howe PD (2021) Extreme weather experience and climate change opinion. Curr Opin Behav Sci 42:127–131. https://doi.org/10.1016/j.cobeha.2021.05.005

Hritonenko V, Yatsenko Y (2022) Sustainable adaptation and mitigation in regions and cities: review of decision-support methods. Resour Conserv Recycl Adv 13:200066. https://doi.org/10.1016/j.rcradv.2022.200066

IPCC (2014) AR5 synthesis report: climate change. https://www.ipcc.ch/report/ar5/syr/

IPCC (2021) AR6 climate change: the physical science basis. https://www.ipcc.ch/report/sixth-assessment-reportworking-group-i/

Jahangir MH, Haghighi P, Danehkar S (2022) Downscaling climate parameters in Fars province, using models of the fifth report and RCP scenarios. Eco Inform 68:101558. https://doi.org/10.1016/j.ecoinf.2022.101558

Javed T, Zhang J, Bhattarai N, Sha Z, Rashid S, Yun B, Ahmad S, Henchiri M, Kamran M (2021) Drought characterization across agricultural regions of China using standardized precipitation and vegetation water supply indices. J Clean Prod 313:127866. https://doi.org/10.1016/j.jclepro.2021.127866

Jiang C, Zheng S, Ng AKY, Ge YE, Fu X (2020) The climate change strategies of seaports: mitigation vs. adaptation. Transp Res Part D Transp Environ 89:102603. https://doi.org/10.1016/j.trd.2020.102603

Jurt C, Burga MD, Vicuña L, Huggel C, Orlove B (2015) Local perceptions in climate change debates: insights from case studies in the Alps and the Andes. Clim Change 133(3):511–523. https://doi.org/10.1007/s10584-015-1529-5

Kalair AR, Abas N, Ul Hasan Q, Kalair E, Kalair A, Khan N (2019) Water, energy and food nexus of Indus Water Treaty: water governance. Water-Energy Nexus 2(1):10–24. https://doi.org/10.1016/j.wen.2019.04.001

Karatayev M, Clarke M, Salnikov V, Bekseitova R, Nizamova M (2022) Monitoring climate change, drought conditions and wheat production in Eurasia: the case study of Kazakhstan. Heliyon 8(1). https://doi.org/10.1016/j.heliyon.2021.e08660

Kim G, Cha DH, Lee G, Park C, Jin CS, Lee DK, Suh MS, Ahn JB, Min SK, Kim J (2020) Projection of future precipitation change over South Korea by regional climate models and bias correction methods. Theoret Appl Climatol 141(3–4):1415–1429. https://doi.org/10.1007/s00704-020-03282-5

Le MH, Kim H, Moon H, Zhang R, Lakshmi V, Nguyen LB (2020) Assessment of drought conditions over Vietnam using standardized precipitation evapotranspiration index, MERRA-2 re-analysis, and dynamic land cover. J Hydrol Reg Stud 32:100767. https://doi.org/10.1016/j.ejrh.2020.100767

Leng G, Tang Q, Rayburg S (2015) Climate change impacts on meteorological, agricultural and hydrological droughts in China. Glob Planet Change 126:23–34. https://doi.org/10.1016/j.gloplacha.2015.01.003

Lozano-Povis A, Alvarez-Montalván CE, Moggiano N (2021) Climate change in the Andes and its impact on agriculture: a systematic review. Sci Agropecu 12(1):101–108. https://doi.org/10.17268/SCI.AGROPECU.2021.012

Lychuk TE, Moulin AP, Lemke RL, Izaurralde RC, Johnson EN, Olfert OO, Brandt SA (2021) Modelling the effects of climate change, agricultural inputs, cropping diversity, and environment on soil nitrogen and phosphorus: a case study in Saskatchewan, Canada. Agric Water Manag 252(3):106850. https://doi.org/10.1016/j.agwat.2021.106850

Magnason AS, Smith L (2020) On time and water. https://books.google.com.pe/books/about/On_Time_and_Water.html?id=3EjQDwAAQBAJ&redir_esc=y

Mengistu D, Bewket W, Dosio A, Panitz HJ (2021) Climate change impacts on water resources in the Upper Blue Nile (Abay) River Basin, Ethiopia. J Hydrol 592:125614. https://doi.org/10.1016/j.jhydrol.2020.125614

Michelutti N, Cooke CA, Hobbs WO, Smol JP (2015) Climate-driven changes in lakes from the Peruvian Andes. J Paleolimnol 54(1):153–160. https://doi.org/10.1007/s10933-015-9843-5

Nikraftar Z, Mostafaie A, Sadegh M, Afkueieh JH, Pradhan B (2021) Multi-type assessment of global droughts and teleconnections. Weather Clim Extrem 34:100402. https://doi.org/10.1016/j.wace.2021.100402

NOAA (2022) Global monitoring laboratory—carbon cycle greenhouse gases. https://gml.noaa.gov/ccgg/trends/

Ponce C (2020) Intra-seasonal climate variability and crop diversification strategies in the Peruvian Andes: a word of caution on the sustainability of adaptation to climate change. World Dev 127:104740. https://doi.org/10.1016/j.worlddev.2019.104740

Qi P, Xia Z, Zhang G, Zhang W, Chang Z (2021) Effects of climate change on agricultural water resource carrying capacity in a high-latitude basin. J Hydrol 597:126328. https://doi.org/10.1016/j.jhydrol.2021.126328

Robert S, Schleyer-Lindenmann A (2021) How ready are we to cope with climate change? Extent of adaptation to sea level rise and coastal risks in local planning documents of southern France. Land Use Policy 104:105354. https://doi.org/10.1016/J.LANDUSEPOL.2021.105354

Saavedra M, Junquas C, Espinoza JC, Silva Y (2020) Impacts of topography and land use changes on the air surface temperature and precipitation over the central Peruvian Andes. Atmos Res 234:104711. https://doi.org/10.1016/j.atmosres.2019.104711

Sanabria J, Calanca P, Alarcón C, Canchari G (2014) Potential impacts of early twenty-first century changes in temperature and precipitation on rainfed annual crops in the Central Andes of Peru. Reg Environ Change 14(4):1533–1548. https://doi.org/10.1007/s10113-014-0595-y

Sarkar S, Maity R (2020) Increase in probable maximum precipitation in a changing climate over India. J Hydrol 585:124806. https://doi.org/10.1016/j.jhydrol.2020.124806

Shagega FP, Munishi SE, Kongo VM (2020) Assessment of potential impacts of climate change on water resources in Ngerengere catchment, Tanzania. Phys Chem Earth 116:102804. https://doi.org/10.1016/j.pce.2019.11.001

Sharifi A (2020) Trade-offs and conflicts between urban climate change mitigation and adaptation measures: a literature review. J Clean Prod 276. https://doi.org/10.1016/j.jclepro.2020.122813

Sharma A, Goyal MK (2020) Assessment of the changes in precipitation and temperature in Teesta River basin in Indian Himalayan Region under climate change. Atmos Res 231:104670. https://doi.org/10.1016/j.atmosres.2019.104670

Shi X, Chen J, Gu L, Xu CY, Chen H, Zhang L (2021) Impacts and socioeconomic exposures of global extreme precipitation events in 1.5 and 2.0 °C warmer climates. Sci Total Environ 766:142665. https://doi.org/10.1016/j.scitotenv.2020.142665

Thøgersen J, Nørnberg P, Finster K, Knak Jensen SJ (2021) Greenhouse gas capture by triboelectric charging. Chem Phys Lett 783:1–3. https://doi.org/10.1016/j.cplett.2021.139069

UNFCCC (2021) National adaptation plan of peru. https://www4.unfccc.int/sites/NAPC/Pages/national-adaptation-plans.aspx

UNTC (1992) United nations framework convention on climate change. https://treaties.un.org/Pages/ViewDetailsIII.aspx?src=TREATY&mtdsg_no=XXVII-7&chapter=27&Temp=mtdsg3&clang=_en

Veettil BK, Kamp U (2021) Glacial lakes in the Andes under a changing climate: a review. J Earth Sci 32(6):1575–1593. https://doi.org/10.1007/s12583-020-1118-z

Vicente-Serrano SM, Beguería S, López-Moreno JI (2010) A multiscalar drought index sensitive to global warming: the standardized precipitation evapotranspiration index. J Clim 23(7):1696–1718. https://doi.org/10.1175/2009JCLI2909.1

Viveen W, Zevallos-Valdivia L, Sanjurjo-Sanchez J (2019) The influence of centennial-scale variations in the South American summer monsoon and base-level fall on Holocene fluvial systems in the Peruvian Andes. Glob Planet Change 176:1–22. https://doi.org/10.1016/j.gloplacha.2019.03.001

Vorobevskii I, Kronenberg R, Bernhofer C (2022) Linking different drought types in a small catchment from a statistical perspective—case study of the Wernersbach catchment, Germany. J Hydrol X 15:100122. https://doi.org/10.1016/j.hydroa.2022.100122

Vuille M (2013) El cambio climatico y los recursos hidricos en los andes tropicales. Banco Interamericano de Desarrollo, p 21. http://publications.iadb.org/bitstream/handle/11319/5826/SR2012_VUILLE_FINAL_ESP.pdf?sequence=1

Wang Y, Zhou B, Qin D, Wu J, Gao R, Song L (2017) Changes in mean and extreme temperature and precipitation over the arid region of northwestern China: observation and projection. Adv Atmos Sci 34(3):289–305. https://doi.org/10.1007/s00376-016-6160-5

Wang T, Tu X, Singh VP, Chen X, Lin K, Lai R, Zhou Z (2022) Socioeconomic drought analysis by standardized water supply and demand index under changing environment. J Clean Prod 347:131248. https://doi.org/10.1016/j.jclepro.2022.131248

WMO (2016) Handbook of drought indicators and indices. https://library.wmo.int/records/item/55169-handbook-of-drought-indicators-and-indices

Wongchuig SC, Mello CR, Chou SC (2018) Projections of the impacts of climate change on the water deficit and on the precipitation erosive indexes in Mantaro River Basin, Peru. J Mt Sci 15(2):264–279. https://doi.org/10.1007/s11629-017-4418-8

Wood JL, Harrison S, Wilson R, Emmer A, Yarleque C, Glasser NF, Torres JC, Caballero A, Araujo J, Bennett GL, Diaz-Moreno A, Garay D, Jara H, Poma C, Reynolds JM, Riveros CA, Romero E, Shannon S, Tinoco T, Villafane H (2021) Contemporary glacial lakes in the Peruvian Andes. Glob Planet Change 204:103574. https://doi.org/10.1016/j.gloplacha.2021.103574

Zárate Malpica AH, Miranda Zambrano GA (2017) Impacto del cambio climático en la seguridad alimentaria en zonas campesinas vulnerables de los Andes del Perú. Rev Mex Cienc Agríc 7(1):71–82. https://doi.org/10.29312/remexca.v7i1.371

Zhao R, Wang H, Chen J, Fu G, Zhan C, Yang H (2021) Quantitative analysis of nonlinear climate change impact on drought based on the standardized precipitation and evapotranspiration index. Ecol Ind 121:107107. https://doi.org/10.1016/j.ecolind.2020.107107

Zhuang XW, Li YP, Nie S, Fan YR, Huang GH (2018) Analyzing climate change impacts on water resources under uncertainty using an integrated simulation-optimization approach. J Hydrol 556:523–538. https://doi.org/10.1016/j.jhydrol.2017.11.016

Leveraging Technology to Accelerate Implementation and Scale

Leveraging Leading Edge Digital Models, Developed Under Urban Computing Principles, to Support Flood Risk Assessment in Cities

Katherine Foster, Pourya Salehi, Jannika Aalto, David Corbett, and Pietro Visetti

Abstract The prevalence of flooding, other natural disasters, and extreme weather events are anticipated to increase due to human-induced climate change. Flooding is a worldwide phenomenon with severe human, ecological, and economic impacts. Developing processes to assess and manage the risk of these hazards is a critical global challenge. Addressing this challenge effectively could mitigate significant negative human and economic impacts, and technology has a fundamental role in addressing this challenge. However, the potential of leading-edge technologies needs to be explored in the role that they could play in supporting rigorous urban flood risk assessment processes. Although much progress has been made in applying technology in cities (albeit to differing extents), several unanswered challenges and opportunities, such as technical readiness in the public sector and data availability in the Global South, have yet to be addressed or captured. This chapter explores various established and emerging urban computing and digitalization techniques and how they are currently being combined and applied on the ground in cities worldwide. Based on this review, challenges and opportunities are identified to suggest how cutting-edge digital technologies could be better integrated with flood risk assessment strategies in the future.

K. Foster (✉) · J. Aalto · P. Visetti
Green Digital Finance Alliance (GDFA), Geneva, Switzerland
e-mail: kf@gdfalliance.org

J. Aalto
e-mail: ja@gdfalliance.org

P. Visetti
e-mail: pv@gdfalliance.org

P. Salehi · D. Corbett
ICLEI—Local Governments for Sustainability, Bonn, Germany
e-mail: pourya.salehi@iclei.org

D. Corbett
e-mail: david.corbett@iclei.org

K. Kim and C. Atkin (eds.), *Cross-Cultural Perspectives on Climate Change Adaptation*, Cities and Nature, https://doi.org/10.1007/978-3-031-50365-8_21

Keywords Flood risk assessment · Digitalization · Smart city · Digital twins · Metaverse

1 Introduction

The rate and severity of flooding events have been increasing worldwide in recent years (in conjunction with other extreme weather events) due to the ongoing human-induced process of climate change and associated effects such as sea level rise and storm surges (Angra and Sapountzaki 2022; Coumou and Rahmstorf 2012). Warmer atmospheric conditions lead to more extreme precipitation because warmer air can contain more moisture (Coumou and Rahmstorf 2012). Over the last ten years, weather-related events have resulted in upwards of 24,000 deaths and an average of 190 million people affected annually (CRED & UNDRR 2021). It is forecast that floods and water-related disasters could cost the global economy $5.6 trillion by 2050. Although flooding occurs across urban and rural areas due to the concentration of population, critical infrastructure, and buildings, events can have a particularly severe socio-economic and environmental impact in urban areas. Due to their increasing prevalence and geographic spread—including regions that have not previously experienced flooding—the risk of flooding has become a global issue of increasing significance.

Past metrics to inform flood-related planning, such as the 100- and 500-year flood probabilities, are insufficient when applied in resilience planning—especially as uncertainties increase (Bilskie et al. n.d.; Highfield et al. 2013; Linkov et al. 2014). Traditional approaches to flood risk management have focused on engineering-led flood control and protection measures (such as embankments and dams) to reduce the probability of flooding.

Furthermore, investment in flood control measures has primarily focused on protecting high-income areas and economic hubs. This investment imbalance has made lower-income countries and areas increasingly disproportionately exposed and vulnerable to flooding. Although the engineering-led measures are effective at retaining water, they will be insufficient in response to more frequent and severe events associated with climate change. Traditional measures have also been found to negatively affect ecosystems and reduce mitigation efforts due to the security they offer and the low perception of the risk of failure (Cea and Costabile 2022). There is an urgent need for innovation in this space with approaches that leverage technology to better assess the dynamic risks of flooding events and reduce vulnerabilities through a holistic perspective.

While risk management includes the activities taken to analyze, prepare for, and manage risks through mitigation and adaptation approaches, risk assessment is usually considered a prerequisite of risk management. It entails a process whereby the nature of risk is comprehended, expressed, and evaluated using available knowledge. As the first step in a risk management strategy, the assessment component's rigor and quality can determine the entire strategy's efficacy. Risk assessment is the supporting cornerstone of both the private and public sectors in future-orientated planning and risk management strategies. Although technology has the potential to significantly impact flood risk management and assessment processes, this chapter focuses on flood risk assessment. There are multiple existing methods and frameworks for climate and flood risk assessment. In this chapter, we utilize the 9-step Climate Disaster Risk Assessment, and Green Recovery Planning Framework put forward by Dr. Kim in Chapter Four to survey existing urban computing tools and technologies and suggest the role they could play across each stage of the 9-step framework.

Urban computing is a field of study focusing on computational techniques and technologies to understand and address challenges and opportunities in urban environments. Urban computing processes often involve leveraging large datasets and machine learning algorithms to analyze and model urban systems to generate insights that can inform decision-making and support the development of more effective policies and interventions. When applied to flood risk assessment, urban computing can be used to digitalize the process of assessing risk in cities to better model and understand risks and the potential impact of different strategies to inform better flood mitigation, adaptation, and resilience policies.

This chapter explores how digital models developed under urban computing principles could be leveraged to deliver sophisticated processes of flood risk assessment in cities, allowing for a more holistic approach to managing the associated risks of flooding and challenges of climate change. To achieve this, the chapter begins by providing an overview of existing digital tools and techniques used in flood risk assessment. It outlines their potential benefits and applicability within the introduced 9-step framework. Next, tangible case studies demonstrate how subnational governments leverage digital twins, metaverse, and other tools and techniques to support flood risk assessment. The chapter concludes by exploring the challenges and opportunities of using digital twins and metaverse for flood risk assessment and suggests how these could be addressed and the opportunities captured.

2 Urban Floods and Flood Risk Assessment

Flooding is a devastating natural disaster that directly and indirectly causes severe human, ecological, and economic damage. At the broadest level, the effects of flooding occur due to water temporarily covering or saturating a large area of land that would not usually be covered. Most floods are categorized into three main subtypes: fluvial (or riverine), pluvial, and coastal. Further information on the flood subtypes is provided in the Table 1. Climate change will impact all subtypes as rising sea levels, storm surges, and increased frequency of extreme weather events touch on the main natural drivers of all flooding events.

Short-term direct impacts of flooding events include the overwhelming of critical infrastructure (such as drainage systems, electricity infrastructure, and transport infrastructure including bridges, among others), massive property damage, essential

Table 1 Subtypes of flooding events

Flooding sub-type	Characteristics	Drivers
Fluvial	Occurs when the water level in a body of water such as a river, lake, or stream rises and overflows onto the neighboring land	Excessive rainfall, snowmelt, infrastructure failure, and runoff
Pluvial (Riverine)	Occur in urban and rural environments where the drainage system (either manmade or natural) is overwhelmed by the volume of water. Pluvial floods can be flash floods characterized by a fast torrent of water driven by a high amount of rainfall in a short time on elevated terrain. Due to the speed of this event, flash floods are particularly destructive	Extreme rainfall, local drainage failure
Coastal	Characterized by seawater's overwhelming coastal land areas	Extreme weather events, including high winds, tsunamis, and high tide storm surges

supply chains, food security, and threat to human safety (Cea and Costabile 2022). Longer-term indirect effects include the loss of productivity and business opportunities and longer-term human health and community impacts. In terms of both short-term direct impacts and longer-term indirect impacts, urban areas are particularly vulnerable due to ongoing population growth, the densely built environment, industrial expansion (changing land use), and significant economic activity (Taromideh et al. 2022). In addition to these heightened risks associated with the urban context, multiform, there is the compounding effect of multiform risk which is forecast to escalate with climate change. Compound hazard events occur when a single hazard can trigger multiple events, both as subtypes of the primary event or as completely new ones. For example, the combination of a long-term drought which negatively impacts drainage capacity, heightens the risk of flooding when rainfall does occur. This destabilization of the land may lead to a higher potential for landslides in the future. Kruczkiewicz et al. (2022) argue that due to the complexity of these scenarios, they are challenging to model, and this is an under-capacitated and explored topic within flood risk.

The most vulnerable urban populations disproportionately experience the most negative effects of flooding. Informal settlements, in which housing typically does not comply with codes and regulations, settlement is often not legally recognized, and there is insufficient physical infrastructure and social services are often the worst affected (Tom et al. 2022). Housing availability and rising living costs are two driving factors driving settlement in flood-prone areas. Vulnerable populations often have limited options to settle outside of flood zones (Tellman et al. 2021).

Climate change and the associated increase in extreme weather events lead to an increased risk of natural hazards and impact chains in which one hazard may trigger another immediately or in the future. This phenomenon adds layers of complexity to the assessment process. Climate risk assessments which assess how climate risk impacts people, assets, value chains, infrastructure, and ecosystems, have therefore become necessary for future planning across all scales (GIZ Global Programme on Risk Assessment and Management for Adaptation to Climate Change (Loss and Damage) 2022). Like flood risk assessments, climate risk assessments lay the foundation for effective risk management. Multiple methods exist for compiling Climate Risk Assessments, including the cyclical methodology introduced by GIZ (Fig. 1). Under this framework, the assessment component includes identifying risks and assessing the magnitude of impacts and potential measures. Stakeholder engagement plays a significant role in effectively applying frameworks such as this to contextualize the process and identify and fill potential gaps in data and awareness.

In the case of flood risk assessment and management, in recent years, there has been a transition away from the dominant approach of resistance-based strategies, which focus on civil engineering infrastructure to contain or drain water, such as dams and embankments. Sörensen et al. (2016) argue that these traditional methods

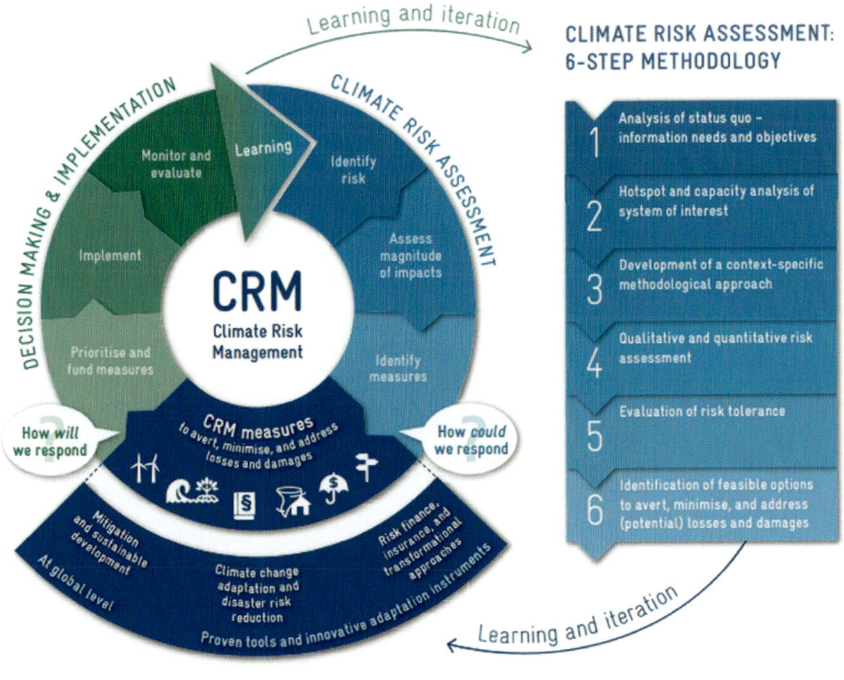

Fig. 1 Climate risk management framework by GIZ (GIZ Global Programme on Risk Assessment and Management for Adaptation to Climate Change (Loss and Damage), 2022)

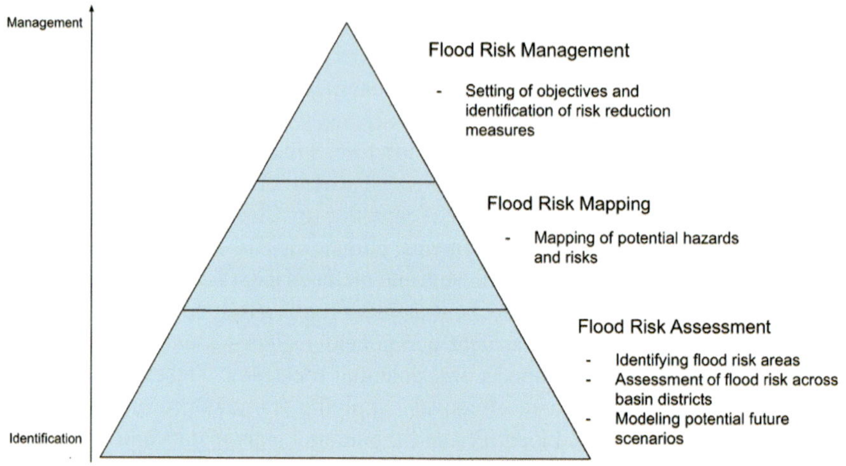

Figure 12.2: Flood Risk Assessment as the foundation for effective Flood Risk Management (Author's Own)

Fig. 2 Flood risk assessment as the foundation for effective flood risk management (author's own)

negatively impact freshwater ecosystems and cannot provide comprehensive protection against floods—which are predicted to intensify in severity and frequency due to climate change. In place of this perspective, a more systemic and collaborative approach is required in which contemporary challenges are acknowledged and technology is leveraged to build resilience to the effects of climate change (Cea and Costabile 2022).

As a result of increased incidents of severe flooding, there has been a push for more future-oriented, inclusive, and standardized approaches to flood risk assessment and management (Fig. 2).

In Europe, this shift from the traditional toward a more risk-based approach has been solidified through the European Floods Directive, 2007/60/EU (Cea and Costabile 2022). The European Floods Directive signifies a paradigm shift in how floods are studied. Within the directive, which was primarily driven by severe floods for the assessment and management of flood risk, the following three-step process, updated every six years, has been established:

1. Preliminary flood risk assessment.
2. Flood mapping.
3. Flood risk management plan.

Flood risk assessment is the first step of the process and involves identifying the areas of potential significant flood risk for each river basin. For this chapter, the 9-step Climate Disaster Risk Assessment and Green Recovery Planning framework proposed by Dr. Kim in Chapter Four is used as a base to provide an overview of the current application of urban computing techniques to flood risk assessment. The 9-step process is provided in detail in Table 2.

Table 2 Dr. Kim's 9-step climate disaster risk assessment and green recovery planning framework

9-step climate disaster risk assessment and green recovery planning framework	
Step 1	Delineate the boundary of landscape district and place unit
Step 2	Delineate boundary of catchment area
Step 3	Identify factors (indicators) to be included in the flood risk assessment exercise
Step 4	Prepare an "inventory map" for each factor showing how it varies over the study area
Step 5	Create composite maps by overlaying two or more inventory maps
Step 6	Analyze the composite maps to make influences relevant to flooding risk assessment
Step 7	Use spatial multi-criteria evaluation—including hazard mapping and vulnerability assessment—to assess overall risk including the probability of damage and loss
Step 8	Identify appropriate hazard reduction measures and remediation strategies and create a comprehensive implementation framework
Step 9	Identify comprehensive monitoring and evaluation framework including MRV and ICT applications

In the section that follows, an overview of digital models, tools, and techniques that fall under urban computing are assessed according to their functionality and application to Flood Risk Assessment—specifically in relation to which of the nine steps they could be applied to.

3 Digital Models, Tools, and Techniques for Flood Risk Assessment

Urban computing offers a powerful means of digitalizing the process of flood risk assessment in cities by leveraging large datasets, machine learning algorithms, and other computational tools, techniques, and technologies to model and analyze flood risk in urban environments. This enables more accurate and comprehensive assessments of flood risk and the increased effectiveness of different flood risk management and reduction strategies. Some of the key ways in which these models can be used include.

1. Real-Time Monitoring and Analysis: Leading-edge digital models can monitor and analyze flood risk in real-time, allowing for a more accurate and up-to-date understanding of flood risk in a given city. This can be particularly useful in situations where there are complex interactions between different physical systems, as it allows for a more complete and accurate understanding of how these systems might behave under different flood conditions.
2. Predictive Modeling: Leading-edge digital models can be used to predict the potential impacts of flooding on a given city based on data on topography, soil type, land use, and weather patterns. This can be useful for identifying areas of

the city with a particularly high risk of flooding and for developing strategies to mitigate this risk.

3. Evaluation of Flood Risk Reduction Strategies: Leading-edge digital models can be used to evaluate the effectiveness of different strategies, such as deploying flood barriers or implementing green infrastructure. This can be useful for identifying the most effective strategies for reducing flood risk in a given city.
4. Communication and Visualization: Leading-edge digital models can be used to communicate and visualize flood risk in a way that is easily understandable to a wide audience. This can be particularly useful for engaging with stakeholders and policymakers and developing effective flood risk reduction strategies (Table 3).

Digital twins and metaverse are considered leading-edge digital models with the potential to support flood risk assessment by providing real-time, interactive, and immersive representations of physical systems. Digital twins are virtual models of physical systems that allow for real-time monitoring and analysis. At the same time, the metaverse is a virtual reality environment that allows users to interact with digital models in a more immersive way. In combination with tools such as BIM, it is possible to simulate and predict the potential impacts of flooding on a given area. This can be done by incorporating data on topography, soil type, and weather patterns into the digital twin or metaverse model and then using the model to simulate how flood waters might flow and affect different areas and how interventions could address this without compromising natural ecosystems. Therefore, when combined, digital tools allow for real-time monitoring and analysis of the system and immersive, interactive experiences. As new data becomes available, the digital model can be updated to reflect the most current understanding of the system enabling a dynamic, accurate, and immersive way of understanding flood risk.

The graph below shows which leading-edge urban computing technologies have the potential to contribute to most of the steps in the 9-step Assessment Framework. Digital twin and metaverse are among the top performers in the graph and are enabled through the data collected using Remote Sensing and Geospatial Information Systems (GIS). As seen in the graph below, no single urban computing tool can independently address every step of the 9-step framework.

In most cases, a combination of urban computing tools is used within a city in a complementary way combining data collection, processing, analysis, and visualization. The following section assesses case studies in which different combinations of tools are used to gain deeper insight into the application of digital twins, metaverse specifically, and other urban computing methods in cities and identify the key solution providers in this space (Fig. 3).

Table 3 Comparative assessment of urban computing tools and techniques

Urban computing models/tools/ techniques	Description/how they work	Potential for flood risk assessment	9-step classification
Digital Twin	A digital twin is a virtual representation of a physical system or process. This term has only recently been adopted in urban planning to enable smart cities (Ketzler et al. 2020)	It can be used to simulate and analyze the performance of a system under different conditions, including flood events. Cities, for instance, can utilize digital twins to simulate numerous flooding processes to enhance flood risk management (Ye et al. 2022)	Step 3, 7, 8, 9
Building Information Modeling (BIM)	BIM incorporates data from architecture, engineering, construction, and facilities management to digitally represent a facility's physical and functional features (Liu et al. 2017)	The potential of BIM for flood risk management can be illustrated with the example of a Flood Risk Information System (FRIS). FRIS is a tool developed by the European Environment Agency that provides information on flood risk, flood hazard, and flood vulnerability for different regions in Europe	Step 3, 8
Inundation Modeling	Inundation modeling is a digital tool that uses computer simulations to predict how floodwater will spread and how high it will reach in a particular area (Teng et al. 2017)	There are different inundation modeling approaches with unique strengths and limitations. They vary in how they address uncertainties in the field. The specific model should be selected based on the specific issue that needs to be solved (Teng et al. 2017)	Step 7
Hydrodynamic Modeling	Hydrodynamic modeling is a tool used to simulate the movement and behavior of water in a given area	Hydrodynamic modeling is an important tool in identifying flooding risk hotspots and has been widely adopted for flood simulations in urban areas. It can be used to predict flood risk and assess the effectiveness of different flood prevention measures (Singh et al. 2020; Zhang et al. 2021)	Step 7, 8
Geospatial Information System (GIS)	GIS is a tool that allows users to create and analyze spatial data, including data related to flood risk	GIS can be used to visualize flood risk in different areas. In particular, flood inundation mapping and risk management are two areas that benefit greatly from the application of GIS (Wang and Xie 2018)	Step 1, 2, 4, 5, 7

(continued)

Table 3 (continued)

Urban computing models/tools/ techniques	Description/how they work	Potential for flood risk assessment	9-step classification
Remote Sensing	Remote sensing involves using satellites or aircraft to collect data about the Earth's surface	Remote sensing provides data for mapping water resources and making up-to-date measurements. In addition, it can measure hydrological fluxes, including river stages. Thus, it can be used to identify areas at risk of flooding and to monitor flood events as they occur (Wang and Xie 2018)	Step 1, 2, 4, 5, 7
Artificial Intelligence and Machine Learning	Machine learning is a branch of Artificial Intelligence that seeks to address the shortcomings of traditional statistical and physical models (Maspo et al. 2020)	Machine learning has the potential to improve flood risk assessment significantly while acknowledging the issues of bias and ethics. Machine learning algorithms can be used to analyze large datasets related to flood risks, such as weather data, topographic data, and historical flood data. This can help identify patterns and predict future flood events more accurately (Wagenaar et al. 2020)	Step 8
Data Analytics	Data analytics techniques, such as statistical analysis and data mining, can be used to extract meaningful insights from large datasets related to flood risk	Data analytics tools and techniques can potentially inform flood-related decision-making and risk management strategies (Donratanapat et al. 2020)	Step 4, 7, 8
Cloud Computing	Cloud computing has profound implications for the provision of Internet services such as servers, storage, databases (Botta et al. 2016)	Cloud computing platforms can be used to store and process large amounts of data related to flood risk, making it more accessible and easier to analyze (Han and Mozumder 2022)	Step 8

(continued)

Table 3 (continued)

Urban computing models/tools/ techniques	Description/how they work	Potential for flood risk assessment	9-step classification
Internet of Things (IoT)	The Internet of Things (IoT) is a telecommunication network of interconnected devices that contain embedded systems such as sensors that can perform numerous services and tasks (Ali et al. 2015)	IoT devices, such as sensors and smart devices, can be used to collect real-time data about flood events and conditions, which can be used to inform flood risk assessment and management (Mangukiya and Sharma 2022)	Step 3, 8
Metaverse	Metaverse is a virtual world that exists in the digital realm and comprises interconnected virtual spaces	While the metaverse is not a specific tool or model developed specifically for flood risk assessment, it has the potential to support flood risk assessment in several ways. For example, the metaverse could be used to create virtual simulations of flood events, which could be used to test and analyze different flood prevention and mitigation strategies. The metaverse could also be used to create virtual representations of real-world cities and infrastructure, which could be used to visualize and analyze flood risk in different areas. Additionally, the metaverse could be used to collect and analyze data related to flood risks, such as weather data, topographic data, and historical flood data. This data could be used to inform flood risk assessment and management efforts (Allam et al. 2022)	Step 7, 8, 9

4 Case Studies

Cities worldwide are already integrating urban computing tools, techniques, and technologies into Climate and Flood Risk Assessment. Table 4 contains a contextual review of case studies that leverage digital twin and metaverse tools as well as other complementary forms of urban computing to support urban flood risk assessment.

Whereas Table 3 identifies the potential of urban computing tools and techniques in relation to flood risk assessment, Table 4 reports how these technologies are

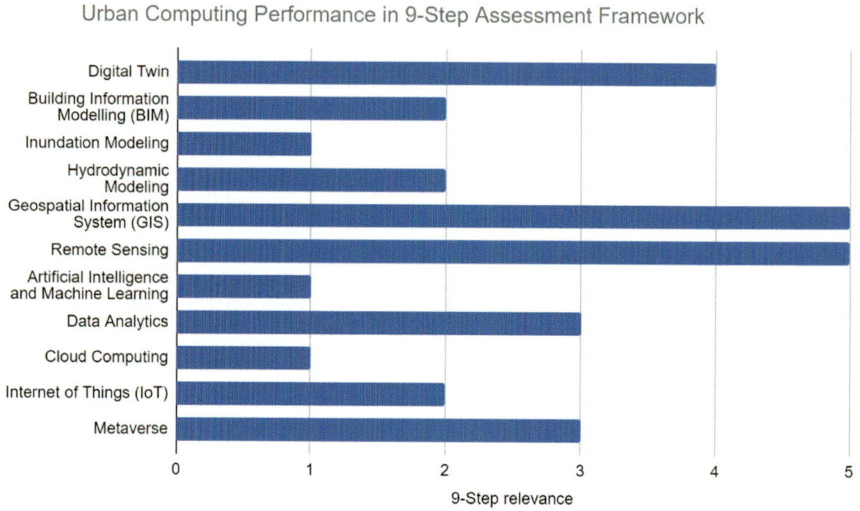

Fig. 3 The potential of urban computing tools to be applied to the 9-step assessment framework

currently being used on the ground. The cases demonstrate that combinations of actors are collaborating to apply urban computing models to various use cases, from infrastructure prediction to traffic and waste management and extreme weather and disaster risk management. The cases show a strong overlap between the potential of the technologies and their current application in cities. The application of urban computing models has progressed from initial applications largely reliant on the leadership of multinational IT companies and narrow in scope toward a more developed point on the continuum. At this point, there is greater collaboration across government, academia, civil society, and business, and the range of application of technology is broader, with more cases incorporating elements of flood and disaster risk assessment in their functionality. However, there remains uncaptured potential in the use of technology. As new technologies emerge, this gap will broaden until the models of collaboration and application are further developed to catch up with what is technically feasible. Therefore, it is imperative that in parallel to technological and scientific breakthroughs, adequate attention is provided to evolving partnerships of collaboration, governance, and civic engagement to ensure that the potential of new technologies can be captured.

The cases provide insights into common parameters in successfully developing and applying the identified urban computing solutions. The drivers or purpose of the use cases covered the 9-step assessment framework from assessing and delineating risk and planning for mitigation, recovery, and resilience, preparing for real-time emergency response and movements, as well as for citizen education and information. Most reported cases were developed before metaverse platforms were readily available on the market. While some commercial digital twin platforms were available on the market or being developed in a parallel temporal track to these cases, they

Table 4 Case studies using urban computing tools and techniques in cities

Case and location	Description	Implementer body/funder	Developer/provider	Types of Urban Computing (technology and data)	Purpose
Newcastle Digital Twin, United Kingdom (2019)	The world's first digital twin created of a city to allow testing of city infrastructure in the face of emergencies	City Water Utility (Northumbrian Water)	Newcastle University, City Water Utility (Northumbrian Water)	Digital Twin, IoT data (from existing sensors)	Infrastructure prediction, extreme weather response and planning, human movement tracking, city planning
Rotterdam Digital TwinCity, Netherlands (2019)	Rotterdam's digital twin is built on a three-tiered framework that comprises data sources, a generic platform, and applications			Digital Twin, IoT data	Urban planning, traffic and waste management efficiency, emergency response
Port of Rotterdam Digital Twin, Netherlands (2019)	The Port of Rotterdam created a digital twin of the port and is committed to becoming the first digital port by 2030, and the smartest port in the world	Rotterdam Port Authority	Private Companies (IBM, Axians, Cisco)	AI, Digital Twin, IoT data	Efficiency, cost reduction, and planning
Singapore Digital Twin, Singapore (2022)	Singapore created the world's largest digital twin of the entire city nation. A metaverse version is expected as the next step in the project	Singapore Land Authority	Private Company (GPS Lands Singapore), Company (Vizzio Technologies)	Digital Twin, AI, remote sensing data	Resource and risk management, planning, RE planning, coastal protection

(continued)

Table 4 (continued)

Case and location	Description	Implementer body/funder	Developer/provider	Types of Urban Computing (technology and data)	Purpose
MOSE Project, Venice, Italy (2003–2025)	The project aims to protect Venice from flooding and prevent further sea level rise. The project involves the installation of 78 mobile gates that separate the Venetian Lagoon from the Adriatic Sea	City of Venice—Venice Water Authority, Consorzio Venezia Nuova	Private Company (IN.TE.SE. Costruzioni d'acciaio),	IoT data linked physical infrastructure	Climate change adaptation and resilience, coastal protection
Lisbon Digital Twin, Spain	Lisbon has created a digital twin for urban flood simulation and for paving the way for the development of a drainage master plan to increase flood resilience [27]	City of Lisbon	Private Company (Bentley Systems)	Digital Twin	Urban flood simulation, drainage master plan, planning for mitigation, resilience and recovery
Takamatsu, Japan	Takamatsu, a port city in Japan, has implemented an IoT-based disaster management system that visualizes emergency information in real-time	City of Takamatsu	FIWARE Foundation	IoT data, meteorological data, dashboard for multi-layer data visualization	Flood mitigation and emergency response in real-time for human/city infrastructure, disaster management

(continued)

Table 4 (continued)

Case and location	Description	Implementer body/funder	Developer/provider	Types of Urban Computing (technology and data)	Purpose
Helsinki Digital Twin, Finland	Helsinki's digital twin integrates IT services and open data to provide real-time insight into the city's operations	Helsinki City Agencies	Helsinki 3D+, Helsinki City Environment Division, Forum Virium Helsinki, Private Company (Bentley)	Digital Twin, Open data, IT services	Citizen decision-making for low carbon activities (mitigation), flood assessment for city
Toa Baja Disasters Project, Puerto Rico (2020),	The project created a spatial flood risk and susceptibility model for Toa Baja, Puerto Rico. The project aims to provide the city with information for disaster mitigation strategies and a GIS tool to assess flood risk	NASA DEVELOP City	MIT Urban Risk Lab Municipality of Toa Baja, ResilientSEE	Open-source remote sensing data (NOAA, NLCD, ESA and Sentinel-1C-SAR) data	Public awareness of flood risk, flood risk management, disaster mitigation
Jakarta, Indonesia	PetaBencana.id is in Jakarta, Indonesia, is the first online tool to create real-time maps of urban flooding using crowd-sourced social media reports and emergency data from government agencies	Indonesia National Government, MIT	Researchers (MIT Urban Risk Lab), Jakarta branch of the National Disaster Management Agency (Badan Nasional Penanggulangan Bencana—BNPB)	IoT data (existing hydraulic sensors), low-tech citizen inputs (social media content, mainly Twitter), government emergency data (water levels and flood locations)	Flood risk management, emergency response, civic information co-management

(continued)

Table 4 (continued)

Case and location	Description	Implementer body/ funder	Developer/provider	Types of Urban Computing (technology and data)	Purpose
Pennsylvania,, United States	Digital twin is being implemented for a chemical plant construction project in Pennsylvania, United States. The company is using drone technology to generate 2D and 3D images of the construction site on a weekly basis. The collected data is then used to create high-resolution rainfall predictions and automated 2D flood simulations that can be accessed through cloud-based web services and a Web-GIS platform. The project's ultimate goal is to reduce the impacts of flooding on "machinery, critical infrastructure, and people." (OECD 2018)	Corporate (Shell Chemical)	Procured private software Company (Bentley)	Private drone technology, Web-GIS Platform (not real-time), 2D Maps—GIS	Real-time monitoring of construction, inventory control, emergency response management
Calgary Digital Twin	A digital twin for the city of Calgary was created to simulate the effect of flood hazards and reduce the damage from future flooding events	McMaster University and City of Calgary	McMaster University	IoT data, remote sensing data, demographic information, and interoperability translators	Infrastructure risk modeling and prediction, disaster preparedness and risk mitigation planning, climate change resilience

were largely not a direct fit for purpose in urban applications. Therefore, in many cases where platforms were provided in case cities, the process was highly collaborative to adapt the project to the local context. Without a collaborative and inclusive process with local stakeholders, there is a risk of technical solutions being applied with few changes across diverse urban contexts. This technocentric approach could risk exacerbating existing issues and vulnerabilities and addressing the symptoms of issues as opposed to their root causes by not engaging adequately with complex local processes.

Many of the cases examined featured collaboration between universities, researchers, and private companies, with only one direct procurement of a service provider in a corporate use case. National, regional, or city-based government agencies, including specific public utilities or departments, were central in many cases to the instigation, funding, and providing digital and data inputs for the initiatives. Many cases leveraged existing digital infrastructure and data inputs and were developed on top of existing 2D maps, some retaining simplicity of user interface in basic digital format (for example, static maps layered with real-time data inputs from existing sensors such as those used for ground, traffic, and air quality, open platforms, weather service platforms, and even citizen data). Most cases utilized a combination of urban technologies to layer up new or data inputs such as open data, satellite, drone, and even citizen data on top of existing digital infrastructure and to develop a suitable interface such as an app or building on existing services and platforms such as planning tools. The approach of leveraging and layering existing data, inputs, and infrastructure also enabled the successful adaptation of technologies to the local context and within existing parameters. For example, sensor infrastructure and data such as for traffic and air pollution already met privacy governance parameters, and in the case of leveraging citizen data input, the governance is covered through the ICT application policies.

The focus on human-centered design and contextualized development—bridging purpose and existing data and digital infrastructure—is a fundamental parameter across the use cases. Universities and research collaborations were essential to developing many of these layered and contextualized solutions, and even where private companies were included, the solutions were highly collaborative and contextualized. Even where private companies and platforms were procured or engaged, government agencies and public utilities' roles were essential in initiating, hosting, and developing the contextualized approaches.

5 Challenges and Opportunities

In examining the challenges and opportunities, it is important to recognize the temporal element in the existing cases and the development of the digital twin and metaverse platforms. This section outlines several challenges and opportunities observed from research and case studies regarding integrating technology in flood risk assessment processes.

5.1 *Public Sector Capacity and Collaborative Development of Solutions*

Technology has the potential to play a vital role in enabling governments and relevant stakeholders to transition from static and reactive toward a dynamic and proactive approach to governance and the addressing of challenges and risks. This transition is imperative in the context of climate change, where the issues being faced require actions to be undertaken in a coordinated, sustained, fast and flexible manner as never before. This is especially true in response to the increasing prevalence of extreme weather events and associated natural disasters. Over the past two decades, governments have advanced at differing adoption rates to integrate information and communication technologies (ICTs) into governance and service delivery (Manoharan and Ingrams 2018). Public sector innovation and in-house utilization of technology require capacity-building. During the COVID-19 international pandemic, there was a surge in e-government worldwide driven by the need to provide uninterrupted services without in-person interaction (Mensah et al. 2022). Moreover, the selected case studies demonstrated the importance of collaboration between research institutions, businesses, local governments, and civic society to leverage collective resources (knowledge, finance, access) to address local issues such as risk from flooding events.

Integrating technologies such as digital twins and metaverse at the subnational level requires capacity-building. As illustrated in the use cases, a key avenue to achieve this is establishing and strengthening collaborative partnerships between research institutions, businesses, and local governments toward leveraging collective resources (knowledge, finance, access) toward addressing local issues such as risk from flooding events. Subnational governments are uniquely positioned to develop these connections with the support of independent organizations as intermediaries. A second critical component in developing capacity is decentralization through multilevel governance approaches. Multilevel governance refers to vertical (different levels of government) and horizontal integration (different local stakeholders (nongovernment)) to enable efficient policymaking, service delivery, and cohesive leadership (UNHABITAT 2022).

The use cases demonstrated the emergence of partnerships and collaborations featuring the engagement of multiple stakeholders from across business, academia, and government in urban computing projects. However, a legacy of a strong focus and reliance on multinational service providers for urban computing solutions remains. In the dominant approach to establishing digital twin projects, subnational governments would most likely use a specialized vendor to generate the model. Digital twins and other urban computing techniques are highly specialized, and building or applying them with in-house expertise, even if desired, would require significant mid to long-term commitments to develop the required software engineering teams and capabilities internally. There is a large cost associated with this. As a result, it is not currently a viable option for many cities worldwide from a budgetary perspective—especially in the Global South. A second aspect of capacity required for local governments, even when tendering a project to an external vendor, would be to develop a better technical understanding of how the technology is applied and maintained so that the tendering criteria can be defined, and submissions can be assessed effectively to ensure the government entity gets value for money and a solution that responds to their needs. It is clear from the cases that new networks of collaboration are forming around the application of new technologies in cities, with governments taking a more active role in developing and maintaining digital platforms. The emergence of consortia-based approaches to the application of urban computing technologies represents an opportunity for less technocentric and more human-centered uses of technology in cities. However, to unlock the full potential of these partnerships, it is imperative that the public sector builds the capacity to engage with the application of technology in their jurisdiction and that a strong link between research and practice is established so that expert insights can shape the use of technology to be ethical and respond to local needs and parameters.

5.2 Technical Readiness, Local Contextual Integration, Interoperability, Harmonization, and Standardization at the Global Scale

The state of digital transformation in government is critical in successfully integrating urban computing models, such as digital twins, into existing governance processes. Enabling disruptive new technologies through public sector innovation has significant potential to accelerate progress toward international targets such as the Sustainable Development Goals. Specifically, converting the potential of technology into tangible, measurable, and consistent outcomes is a key challenge (Dener et al. 2021). The GovTech Maturity Index by the World Bank uses the following four areas of focus to measure maturity:

1. Supporting core government systems.
2. Enhancing service delivery.
3. Mainstreaming citizen engagement.
4. Fostering GovTech enablers to support public sector modernization (Dener et al. 2021).

Building digital skills and innovation in the public sector are critical for transitioning to a more data-driven approach to decision-making and action. This transition requires significant investment and time to achieve and a clear plan to build this capacity which is enabled by strong governance mechanisms and political will. Although some breakout regions exist in Latin America, the highest technical readiness levels are unevenly distributed in the Global North. Large-scale barriers to technology integration are more evident in Sub-Saharan Africa and South Asia, where significant resources are required to address the digital divide, infrastructure, and governance issues (Dener et al. 2021).

The application of technology in cities is a dynamic and rapidly changing landscape. Solution providers are actively competing for large-scale urban projects that leverage technology toward issues from public safety to traffic management to risk assessment. Commercial platforms and vendors are not able to offer one-size-fits-all solutions. While powerful and versatile, some dynamic elements cannot be or are not currently captured by all technologies. Furthermore, there is a severe lack of data originating in and focused on the Global South, which may impact the reliability and applicability of certain commercial solutions.

With new technologies and applications emerging regularly, a stack of systems develops both in the IoT and the city's physical infrastructure. For the city's systems to be scalable and sustainable, the technologies used within the stack need to be cohesive and chosen according to their efficacy and applicability to a given context. The viability of different urban computing methods differs across contexts depending on budget constraints and data availability, among other factors. Technologies need to be able to integrate with existing infrastructure and data. Therefore, IoT and AI will be essential to interpreting and combining existing infrastructure with new layers and inputs.

Fragmentation in the technologies and data used can undermine the overall system's effectiveness and prevent future integration of new technologies. Assumptions about readiness, capacity, and need can augment the fragmentation and further delay integration. Achieving harmonization of different technologies requires an overview of the local infrastructure and context and the setting and implementation of standards to ensure that future technologies can be integrated in an ethical and human-centered way. Ideally, this should be completed by the subnational government and aligned with the standards set by the national and international scales.

5.3 Data Availability in the Global South

The COVID-19 pandemic accelerated digital transitions in the private and public sectors. This has led to a significant increase in the availability of data stacks and big data that can be leveraged for urban computing, potentially altering how many processes are completed. However, according to the Digital Evolution Index produced by the Fletcher School at Tufts University and Mastercard, there are only a handful of main Global North regions in which data is generated, and value is created. The result is a severe lack of data originating in and focused on the Global South. As platforms are being funded and developed largely in the global north, this can potentially increase the digital divide and impact burden on the Global South. As urban computing develops, the opportunity exists to bridge the gaps and move from scarcely employed and limited modeling from publicly available weather prediction data to interactive and real-time data for the Global South, which is bearing the brunt of the climate crisis (Pörtner et al. 2022). For example, in a 2021 study, Tellman et al. (2021) identified previously unidentified increases in flood exposure in Southern Asia, Southern Latin America, and the Middle East using satellite data.

The case studies demonstrate the importance of contextualized tailored approaches in leveraging and layering up on existing digital and data infrastructure to ensure fit for purpose and relevance to local contexts. For the Global North and Global South, this requires leveraging both high and low-tech solutions and designing people-centered approaches and user-friendly interfaces to enable citizen inclusion for use and input. Social identities such as gender, class, race, and age are associated with different levels of disaster vulnerability, adaptive capability, and even digital connectivity and technological access. In the case of the Global South, emergency response, in particular, has a gender dimension, where linguistic, cultural, and social factors lead to higher vulnerability and a lack of access to technology, communication, and critical information (Shah et al. 2022).

Data availability in many Global South contexts means that many significant benefits of technologies for addressing challenges such as flooding cannot yet be realized. However, many innovations in the Global South focus on addressing human needs and leveraging available resources within existing constraints. Data could be a significant resource in creating actionable knowledge that empowers local decision-makers, stakeholders, and citizens. Satellite and remote sensing data could play a significant role in addressing the lack of spatial data for Global South contexts. However, traceable ground truth observations are needed to leverage this data to validate satellite observations and ensure reliable data (Loew et al. 2017). Addressing this issue requires collaboration with the Global North to share data-capturing activities and technologies with the Global South to help generate quality and actionable data globally. Training and capacity-building on the ground in Global South contexts is essential to ensure that there is expertise on the ground to generate accurate data and to apply it to local issues independently. In addition, the development of interfaces and data sourcing, such as citizen observation, is an important element that should not be overlooked as platforms develop. An interesting observation from the case studies is that data privacy and governance may not be a significant obstacle where solutions layer up or leverage data inputs and digital infrastructure developed in alignment with privacy regulation (such as the European cases aligned with GDPR). This may shift as civic and private inputs, and interfaces are developed and integrated.

The generation of context-specific data in the Global South is one of the key opportunities for the future of urban computing toward disaster and risk management applications and broader citizen-centered smart city initiatives. However, as data is collected, the value will be created, and new markets will emerge for solution providers to enter the space. A consortia-based approach must be used to shape urban computing projects that emerge in these geographies to ensure they engage with local complexities and serve local needs. This can only be achieved when the project processes are designed to value and incorporate the inputs from local experts and citizens to shape the technical outcomes and use them as a tool for sustained and inclusive engagement on local development issues.

5.4 *Collaborative, Context-Specific, and Inclusive Solution Development*

The application of technologies to urban issues can lead to a technocratic and engineering-led approach that fails to engage with the political, cultural, social, gender, economic, and environmental intricacies of the context. The dangers of such an approach are that the root causes of issues can be neglected or the specific social make-up of the city (for example, the importance of the local informal economy for livelihood security), and this can lead to the exacerbation of existing issues (Agyeman and McLaren 2017). It is, therefore, important that the socio-economic context of a city is included in the Flood Risk Assessment and the application of technology such as digital twins. Effective citizen participation processes have a vital role, such as using ICT to leverage citizen science.

The role of citizen participation in governance has changed significantly from unidirectional consultation toward greater citizen control and inputs at the initial phase of decision-making (Arnstein 1969). Despite general progress in integrating citizen participation in decision-making, designing processes that lead to more effective policies and long-term outcomes remains challenging. Technology provides an opportunity to address this challenge and include citizens and stakeholders in decision-making, design, and data inputs, as demonstrated in the case studies by combining sensor technologies and ICTs such as mobile applications. An example of this would be the establishment of "citizen observatories" (Wehn et al. 2015). In the Flood Risk Assessment phase of risk management, citizens could play a critical role in generating and validating data collected through remote sensing (Fritz et al. 2017). Especially in densely populated and complex settlements such as informal settlements where spatial data is particularly lacking. Crowdsourcing processes to ensure the quality of data could enable more accurate information used to raise public awareness, engage citizens in decision-making processes and promote behavioral change. A lack of data undermines the potential for meaningful citizen engagement as the evidence required to explore potential scenarios effectively with citizens is unavailable. Data collected and validated in collaboration with citizens could be leveraged to enable better and more inclusive citizen engagement processes in decision-making through data visualization to simulate scenarios and explain potential trade-offs.

5.5 *Scientific Studies of These Technologies and Their Application (Future Research Required)*

In general, there is a lack of independent scientific research regarding the application of digital twins, metaverse, and other urban computing methods to Urban Flood Risk Assessment. Most information available regarding the case studies is produced by the researchers, collaborators, and vendors who develop or supply the models to subnational governments. Future research must be undertaken that independently studies the application process of these models, their efficacy, impact, and maintenance. The production of in-depth and rigorous case studies would enable setting standards and criteria for success in applying these technologies to address issues such as flood risk.

6 Conclusion

This chapter has surveyed the potential of leading-edge technologies, such as digital twins and metaverse, in enhancing urban flood risk assessment processes. Data inputs, layering, and systems integration are the foundations to flood risk assessment and the development of effective management strategies. Applying urban computing technologies in this phase could significantly improve the overall process. The 9-step Climate Risk Assessment framework introduced by Dr. Kim in Chapter Four was used as the basis for a contextual review of the potential role of urban computing tools and techniques in Flood Risk Assessment. Through the case studies and research, it was observed that urban computing tools have the potential to contribute to all steps of the framework. Following this, digital twins and metaverse were in focus through a contextual review of case studies utilizing and developing these technologies. It was found that most projects were focused on and developed in the Global North and that multinational IT companies participated in many of the cases. However, the cases also demonstrated the emergence of a consortia-based approach to solution development featuring collaborations with government agencies or public utilities, and researchers. Moreover, we observed that many solutions leveraged existing technology and data infrastructure, including 2D maps, traffic, ground, air pollution, and traffic sensors, with some layering with open data platforms, drone, satellite, and citizen observation data. The case study reviews informed the identification of challenges and opportunities regarding the integration of leading-edge technologies in urban flood risk assessment, and potential ways to address them were suggested.

There remains a significant gap in data availability and digital readiness in government across Global North and South contexts. Addressing this gap is critical for digital solutions to enable cities in the Global South to generate and leverage data to inform urgently-required and locally relevant action. Digital maturity in subnational governments will continue to be a decisive factor in the integration of technology. Addressing this requires significant political will and alignment across the scales of government through multilevel governance and stakeholder and citizen inclusion to establish standards for the meaningful integration of technology and build the internal capacity to maintain an active role in using technology toward data-driven decision-making.

References

Agyeman J, McLaren D (2017) Sharing cities. Environment 59(3):22–27 (Informa UK Limited)

Ali ZH, Ali HA, Badawy MM (2015) Internet of Things (IoT): definitions, challenges and recent research directions. Int J Comput Appl Technol 128(1):37–47 (Foundation of Computer Science)

Allam Z, Sharifi A, Bibri SE et al (2022) The metaverse as a virtual form of smart cities: opportunities and challenges for environmental, economic, and social sustainability in urban futures. Smart Cities 5(3):771–801 (Multidisciplinary Digital Publishing Institute)

Angra D, Sapountzaki K (2022) Climate change affecting forest fire and flood risk—facts, predictions, and perceptions in Central and South Greece. Sustain Sci Pract Policy 14(20):13395 (Multidisciplinary Digital Publishing Institute)

Arnstein SR (1969) A ladder of citizen participation. J Am Inst Planners 35(4):216–224 (Routledge)

Bilskie M, Del Angel D, Yoskowitz D et al (n.d.) Future flood risk exacerbated by the dynamic impacts of sea level rise. https://doi.org/10.21203/rs.3.rs-63173/v1

Botta A, de Donato W, Persico V et al (2016) Integration of cloud computing and internet of things: a survey. Future Gener Comput Syst FGCS 56:684–700

Cea L, Costabile P (2022) Flood risk in urban areas: modelling, management and adaptation to climate change. A review. Hydrology 9(3):50 (Multidisciplinary Digital Publishing Institute)

Coumou D, Rahmstorf S (2012) A decade of weather extremes. Nat Climate Change 2(7):491–496 (Nature Publishing Group)

CRED & UNDRR (2021) 2020 The non-COVID year in disasters

Dener C, Nii-Aponsah H, Ghunney LE et al (2021) GovTech maturity index: the state of public sector digital transformation. World Bank Publications

Donratanapat N, Samadi S, Vidal JM et al (2020) A national scale big data analytics pipeline to assess the potential impacts of flooding on critical infrastructures and communities. Environ Model Softw 133:104828

Fritz S, Fonte CC, See L (2017) The role of citizen science in Earth observation. Remote Sens 9(4):357 (Multidisciplinary Digital Publishing Institute)

GIZ Global Programme on Risk Assessment and Management for Adaptation to Climate Change (Loss and Damage) (2022) Scoping study: a comparison of climate risk assessment methods to support informed decision-making. GIZ

Han Y, Mozumder P (2022) Risk-based flood adaptation assessment for large-scale buildings in coastal cities using cloud computing. Sustain Cities Soc 76:103415

Highfield WE, Norman SA, Brody SD (2013) Examining the 100-year floodplain as a metric of risk, loss, and household adjustment. Risk Anal Official Publ Soc Risk Anal 33(2):186–191 (Wiley)

Ketzler B, Naserentin V, Latino F et al (2020) Digital twins for cities: a state of the art review. Built Environ 46(4):547–573

Kruczkiewicz A, Cian F, Monasterolo I et al (2022) Multiform flood risk in a rapidly changing world: what we do not do, what we should and why it matters. Environ Res Lett ERL [Web site] 17(8):081001 (IOP Publishing)

Linkov I, Bridges T, Creutzig F et al (2014) Changing the resilience paradigm. Nat Climate Change 4(6):407–409 (Nature Publishing Group)

Liu X, Wang X, Wright G et al (2017) A state-of-the-art review on the integration of Building Information Modeling (BIM) and Geographic Information System (GIS). ISPRS Int J Geo-Information 6(2):53 (Multidisciplinary Digital Publishing Institute)

Loew A, Bell W, Brocca L et al (2017) Validation practices for satellite-based Earth observation data across communities. Rev Geophys. https://doi.org/10.1002/2017rg000562

Mangukiya NK, Sharma A (2022) Flood risk mapping for the lower Narmada basin in India: a machine learning and IoT-based framework. Nat Hazards 113(2):1285–1304

Manoharan AP, Ingrams A (2018) Conceptualizing E-government from local government perspectives. In: State and local government review. https://doi.org/10.1177/0160323x18763964

Maspo N-A, Harun ANB, Goto M et al (2020) Evaluation of machine learning approach in flood prediction scenarios and its input parameters: a systematic review. IOP Conf Ser Earth Environ Sci 479(1):012038 (IOP Publishing)

Mensah IK, Adams S, Adjei JK et al (2022) Drivers of E-government adoption amidst COVID-19 pandemic: the Information Adoption Model (IAM) approach. Inf Dev 38(4):494–509 (SAGE Publications Ltd)

OECD (2018) Embracing innovation in government: global trends 2018. OECD Paris

Pörtner H-O, Roberts DC, Tignor M, Poloczanska ES, Mintenbeck K, Alegría A, Craig M, Langsdorf S, Löschke S, Möller V, Okem A, Rama B (2022) Climate change 2022: impacts, adaptation and vulnerability. Working Group II Contribution to the IPCC Sixth Assessment Report. IPCC

Shah AA, Ullah A, Khan NA et al (2022) Gender perspective of flood early warning systems: people-centered approach. WATER 14(14):2261 (Multidisciplinary Digital Publishing Institute)

Singh RK, Kumar Villuri VG, Pasupuleti S et al (2020) Hydrodynamic modeling for identifying flood vulnerability zones in lower Damodar river of eastern India. Ain Shams Eng J 11(4):1035–1046

Sörensen J, Persson A, Sternudd C et al (2016) Re-thinking urban flood management—time for a regime shift. Water. https://doi.org/10.3390/w8080332

Taromideh F, Fazloula R, Choubin B et al (2022) Urban flood-risk assessment: integration of decision-making and machine learning. Sustain Sci Pract Policy 14(8):4483 (Multidisciplinary Digital Publishing Institute)

Tellman B, Sullivan JA, Kuhn C et al (2021) Satellite imaging reveals increased proportion of population exposed to floods. Nature 596(7870):80–86

Teng J, Jakeman AJ, Vaze J et al (2017) Flood inundation modelling: a review of methods, recent advances and uncertainty analysis. Environ Model Softw 90:201–216

Tom RO, George KO, Joanes AO et al (2022) Review of flood modelling and models in developing cities and informal settlements: a case of Nairobi city. J Hydrol Reg Stud 43:101188

UNHABITAT (2022) Multilevel governance for SDG localization

Wagenaar D, Curran A, Balbi M et al (2020) Invited perspectives: how machine learning will change flood risk and impact assessment. Nat Hazards Earth Syst Sci 20(4):1149–1161 (Copernicus GmbH)

Wang X and Xie H (2018) A review on applications of remote sensing and Geographic Information Systems (GIS) in water resources and flood risk management. WATER 10(5):608 (Multidisciplinary Digital Publishing Institute)

Wehn U, Rusca M, Evers J et al (2015) Participation in flood risk management and the potential of citizen observatories: a governance analysis. Environ Sci Policy 48:225–236

Ye X, Du J, Han Y et al (2022) Developing human-centered urban digital twins for community infrastructure resilience: a research agenda. J Plann Literature:08854122221137861 (SAGE Publications Inc)

Zhang H, Wu W, Hu C et al (2021) A distributed hydrodynamic model for urban storm flood risk assessment. J Hydrol 600:126513

Distributed Ledger Technology (DLT) to Drive Flood Reduction Solution Markets and Financing

Si Chen

Abstract As climate change brings greater risks of flooding and natural disasters, insurance is a critical mechanism for both proper urban planning and development to reduce risks and for funding recovery after disasters. The high cost and complexity of traditional insurance, however, has severely limited its adoption, especially in developing economies. In this chapter, we look at how distributed ledger technology (DLT) could be used to structure insurance mechanisms that help protect against disasters. We propose a combination of automated risk transfers using technology and local groups that work with policyholders to mitigate and adapt to climate change.

Keywords Blockchain · Distributed ledger technology · DLT · Web3 · Insurance · Climate adaptation · Flooding · Flood insurance · Mangrove insurance · Crop insurance · Insurance adoption · Reinsurance · Digital measurement reporting verification · Digital MRV · Urban planning

1 Introduction

When floods or disasters strike, developed economies have a big advantage for recovery: Insurance.

In developing economies, aid could be the only source for funding the recovery, but it is often unreliable. If the victims of the disaster could not get donors' attention, very little aid could be coming. Even when they do, sometimes the promised aid shows up late, and sometimes not at all.

In developed economies, however, insurance is available for recovery after disasters, thanks to both premiums paid by policyholders and capital from investors. Payouts are based on contracts, not donations or goodwill. After disasters happen, insurers send out adjusters to determine the amount of damage and offer payments to policyholders. Thus, even though it could be a slow and costly process, property insurance is the key to recovery and rebuilding. Furthermore, because developed

S. Chen (✉)
Open Source Strategies, Inc., Pasadena, CA 91106, USA
e-mail: sichen@opensourcestrategies.com

K. Kim and C. Atkin (eds.), *Cross-Cultural Perspectives on Climate Change Adaptation*, Cities and Nature, https://doi.org/10.1007/978-3-031-50365-8_22

economies rely on insurance for protecting against risks such as storms and flooding, insurance serves as a powerful market signal for responsible urban planning and development.

In this chapter, we'll look at the role of insurance for climate adaptation and risk mitigation; why its penetration is so low in developing economies, and how distributed ledger technologies ("DLT"), combined with other new technologies, could help scale the risk and insurance markets for climate adaptation.

2 Role for Insurance in Climate and Flood Risk Financing

Insurance is a transfer of risk. Those who wish to be protected against risks pay insurers to protect them. As such, insurance performs several important functions for helping both individuals and society as a whole to manage risks such as natural disasters and floods. Insured policyholders pay regular premiums in exchange for large payments when disasters happen. This turns catastrophes that could wipe out savings, businesses, and whole communities into known and manageable costs. When disasters happen, the insurance payout is known in advance, which helps the policyholders plan for rebuilding and recovery. Thus, insurance is first and foremost a savings mechanism for disaster recovery.

On a larger scale, insurance pools or diversifies the risks of individual policyholders. When there are large numbers of policyholders insured across a range of risks, the insurer evens out its exposure to each insured policyholder and risk. This turns large, occasional payments for risky events into regular and predictable outflows, reducing the amount of risk capital the insurer must hold against its risks. That, in turn, reduces the cost to insure the risks.

Finally, to reduce payouts and increase profits, insurers are naturally incentivized to help policyholders reduce their risks. They may do so by providing discounts for practices which reduce the risk of loss. For example, insurance companies may provide discounts on fire insurance policies if businesses or building owners install smoke alarms and sprinkler systems. Insurance companies also provide research and data which become part of standards that lower risks, such as the Insurance Institute for Business and Home Safety's regular review of building codes in US states that face high hurricane risks (Wright 2021). Alternatively, if insurance companies deem the risk to be too great, they could charge a high premium or decline to issue policies altogether. This, in turn, creates a powerful market signal against high risk projects moving forward.

A functioning insurance industry provides the foundation for building up the investment capital and entrepreneurial risk taking that are necessary for people and communities to succeed. Entrepreneurs are more comfortable investing their time and capital when they could purchase insurance, instead of hoarding their resources in low return activities "just in case something happens" (Swiss Re Institute 2017). More importantly, when there is insurance available for businesses and assets, lenders and investors are more willing to finance them. The entire US housing finance industry,

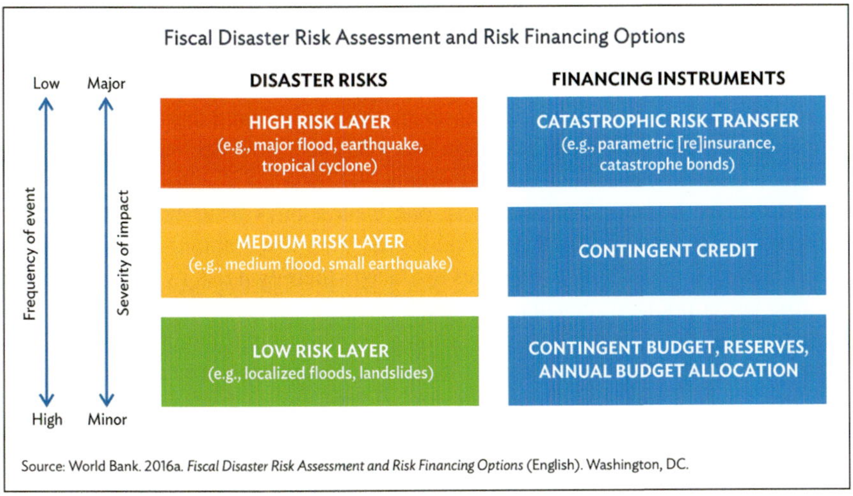

Fig. 1 Insurance as a risk financing option (Surminski et al. 2019)

for example, relies on insurance to protect lenders against damage to the houses that serve as collateral for home loans. A corollary is that insurance then becomes the enforcer of urban planning. Cities may publish zoning regulations and building codes against building in high risk areas, but the lack of available insurance and mortgage financing would actually prevent houses from being built there (Fig. 1).

Insurance should be in place to protect against large but rare risks that would exceed the budgets or even borrowing capacities of the affected individuals, businesses, and communities. There are four types of property insurance mechanisms which could be combined to do so:

- **Indemnity policies** are traditional insurance policies where the insurer protects a particular asset from damage or loss, such as flood damage to a building. The policy could be based on an agreed cash value or replacement cost at the time of loss. In either case, the insurer would pay if there is actual damage to the asset. When the policyholder files a claim to notify the insurer of damage, a claims adjuster would be sent out to evaluate the asset and determine the amount that the insurer would pay. While this could be slow and costly, it has the advantage of being linked to the actual asset. There is no "basis risk" between the insured event and the loss, such as for example buildings in low lying areas incurring flood damage even though there was little rainfall at a weather station up the hill.
- **Parametric policies** are contracts which make payouts based on measured data, such as wind speed or rainfall at a particular agreed upon sensor. There is no need to send a claims adjuster or assess the damage to any particular property, so the payments could be made quickly—sometimes just days after a disaster. Such policies also allow very flexible use of payouts, since they do not have to be tied to the cleanup or rebuilding of a particular asset. The downside is that

such policies could have significant basis risk between the metric being measured for payout (i.e., rainfall at a weather station) and actual damage suffered by a particular property (i.e., flooding at a building). This, in turn, makes parametric policies less useful to the ultimate insured. One study found that every kilometer of distance between a weather station and a farmer's field decreased demand for crop insurance by 6.4% (Swiss Re Institute 2017).

- **Reinsurance** is an agreement where one insurer buys insurance from another insurer to protect itself from potential losses. This is a common arrangement in the industry for insurance companies to manage their own risk by spreading them across larger pools of reinsurance capital.
- **Insurance-linked securities**, also known as catastrophe bonds, are securities which promise a higher payout if no loss event occurs but whose payout is significantly lower if one does. These securities allow insurers to tap into the larger capital markets to raise capital for insurance risks.

As climate change increases the risk of floods and other disasters, insurance is increasingly becoming an important part of adaptation strategies. In recent years, demand for insurance in hurricane-prone areas such as Puerto Rico and the Caribbean has outpaced the capacity of traditional insurers to meet them. As a result, insurers have turned to a combination of parametric insurance policies and insurance-linked securities based on parametric weather data to raise additional capital from outside the industry (Evans 2020).

One example of parametric policies being used for climate adaptation is the insurance of natural assets which help protect properties against storms and flooding. For example, mangrove forests have massive aerial root networks that help reduce soil erosion and protect against surging storm water. A model developed by The Nature Conservancy showed that when mangrove forests are sited near high value infrastructure, such as airports, bridges, and ports, or high value real estate such as homes and condos, their values could be well into millions of dollars per hectare. Even when capped at $50,000 per hectare, the value of mangroves could be eight to fifteen times their costs of restoration in the Caribbean regions (Beck et al. 2020). Studies based on past storms also show their mitigation value:

> In Florida, one study found, mangroves alone prevented $1.5 billion in direct flood damages and protected over half a million people during Hurricane Irma in 2017, reducing damages by nearly 25% in counties with mangroves. Another study found mangroves actively prevent more than $65 billion in property damage and protect over 15 million people every year worldwide. (Dunsavage 2020)

The Nature Conservancy and the government of the Mexican state of Quintana Roo pioneered the use of insurance to protect coral reefs and beaches along 160 km of the coast. The government collected fees from the area hotels, businesses, and residents, deposited them into a trust fund, which then purchased a parametric policy:

> The insurance is triggered if wind speeds in a designated area are recorded in excess of 100 knots. The maximum payout over the 12-month period, or the Annual Aggregate Limit, was $3.8 million (USD). The ultimate payout of the insurance product is based on the maximum recorded wind speed during a storm event: (i) a payout equal to 40% of the maximum payout

> (~$1.5 million) if wind speeds are between 100 and 130 knots; (ii) a payout equal to 80% of the maximum payout (~$3.0 million) if wind speeds are between 130 and 160 knots; and (iii) a payout equal to 100% of the maximum payout (~$3.8 million) if wind speeds are in excess of 160 knots. (Beck et al. 2020)

The parametric policy cost nearly $230,000 (5 million pesos) and paid out $800,000 (17 million pesos) when Hurricane Delta struck Cancun and the Yucatan in October 2020 (Dunsavage 2020). The payout happened even though the actual storm did not cause as much damage as initially feared (Sieff 2020).

Properly used, insurance could offer many of benefits for storm and flood risk mitigation, including:

- Protecting assets, including homes, businesses, and critical infrastructure
- Encouraging and financing risk mitigation before disaster, by
  - Concentrating responsible development in areas of lower risk
  - Building up infrastructure such as levees, mangroves, etc.
  - Improving building codes and construction standards to reduce losses
- Financing recovery after disasters with quick payouts.

Yet despite all these benefits, insurance is still an underutilized mechanism. Globally, $4 trillion has been lost to extreme natural disasters over the past 40 years, but only $1.1 trillion were recovered through insurance (Brown 2018). In 2021, only $20 billion of $90 billion in flooding losses worldwide were insured. In addition, much of the public infrastructure such as roads, railways, bridges, and storm dykes and levees, are uninsured (Munich Re, n.d.). Further, the share of insurance in developing economies is much lower than that of developed economies. For example, in 2016, only 11% of losses from natural catastrophes were covered in developing economies, versus 42% in developed economies (Swiss Re Institute 2017). In Asia as of 2019, more developed economies such as Japan and Australia had ten times as much insurance as Indonesia, Philippines, and India. Thus, these economies lack the savings, borrowing capacity, and insurance capacity for disaster recovery (Fig. 2).

Insurance faces many barriers to adoption in both developed and developing economies, but some of these are especially strong in developing economies where they are less established:

- **Legal and Regulatory Framework**: It's not always clear what types of insurance contracts are valid and how insurance companies are regulated. Governments may interfere with insurance markets for political gain, such as requiring them to write contracts which are not economical or pay outside their contracted liabilities.
- **Social Taboos**: In some places, protecting against disasters before they happen is considered a sign of weakness or going against nature. In other cases, prohibitions against charging interest means that insurance must be organized as mutual aid, such as Takaful insurance in Islamic countries.
- **High Cost of Transactions**: Today, every step of the insurance industry is costly and labor intensive. The sales process is long, the underwriting and pricing of risks are highly complex, and the claims payment process with paper invoices

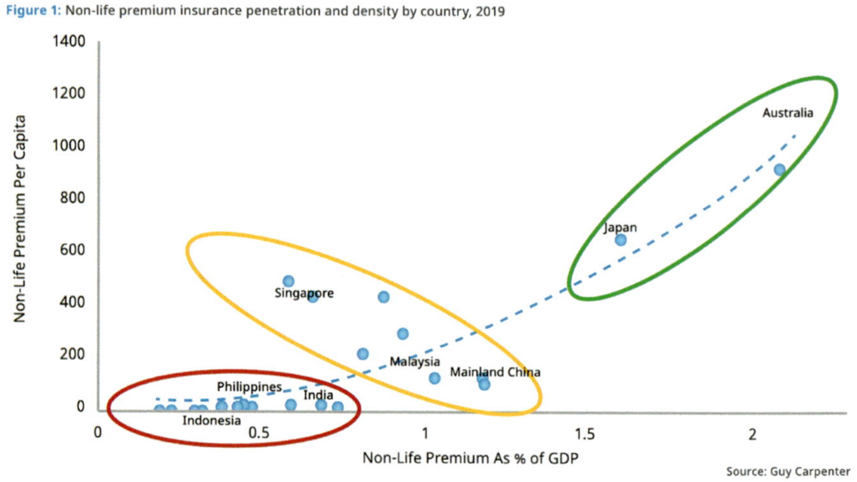

Fig. 2 Non-life insurance premium across Asia (Guy Carpenter 2022)

and checks is slow. All these costs make insurance uneconomical in developing economies, where the value of assets and amounts paid out are small.

- **Lack of Trust in Payoffs**: People are naturally skeptical when strangers ask for money in exchange for promises. Unless there is something to show that the insurer would remain solvent and that the process for getting paid is fair, buying insurance becomes a leap of faith few are willing to take.
- **Lack of Insurable Transactions**: The insurance industry claims that a key barrier is that people don't understand insurance, and the answer is more education. But in reality, human beings are optimists and generally prefer to buy lottery tickets instead of insurance. Therefore, insurance must be linked to other transactions, such as loans for homes or businesses, for us to buy them. In developing economies, these transactions are often not available, so there is nothing to insure.
- **Lack of Data to Enable Underwriting of Risks**: Finally, in many regions, there is no historical data of floods and storms, no accurate maps of developed cities and infrastructure, and no risk models for pricing risks. Unless such data and models were available, insurance is simply not possible.

New digital technologies such as satellite imaging, Internet of Things (IOT) devices, and mobile payments are being used to address these challenges for insurance in developing economies. For example, the Kenyan Livestock Insurance Program (KLIP) provides an indexed insurance to pastoralists' animals which is paid based on satellite images of the land. These images measure how dry the land is based on its color from green to yellow. If a threshold of dryness is detected from the images, the pastoralists receive a payout, regardless of the condition of the herds. In 2017, around 14,000 pastoralists were using this insurance program (Swiss Re Institute 2017).

Similarly, the Kilimo Salama Crop Insurance Program in East Africa uses weather station readings to insure farmers against drought. When a farmer purchases seed, they are given a contact number in the bags. The farmer could text the number and provide the insurer with their GPS location and date. Then, if weather stations later measure a shortfall in rain, the insurer would make a payment to the farmer's mobile phone account. At the end of 2016, more than 1 million farmers were using this crop insurance program, which was renamed ACRE Africa (Swiss Re Institute 2017).

More recently, the African Risk Capacity (ARC) is developing a pan-Africa flood risk model using satellite images to offer parametric flood insurance to member states. This model, called the ARC Flood Extent Depiction Model (AFED), will use microwave sensors from satellites to map floods. Microwave radiation naturally emits from the Earth, and a decrease in this radiation is a sign of flooding. Combined with ground elevation data, the model would produce georeferenced images for whether areas throughout Africa were flooded during the day or not. In 2017, the AFED was piloted in Ghana, the Gambia, and Cote d'Ivoire (Africa Risk Capacity 2022).

Distributed ledger technologies, or "DLT," could build upon such initiatives and help scale insurance by lowering transaction costs while creating a transparent and trusted insurance mechanism. To see how, let's first take a look at DLT itself.

3 The Basics of Distributed Ledger Technologies

A distributed ledger is a way for multiple parties to store identical copies of the same data without relying on a central authority. Instead of relying on a bank, government, or tech company, data on a distributed ledger could be stored across multiple computers, such as the computers of all the participants in a transaction or market. While there are several distributed ledger algorithms for decentralized data storage, such as directed acyclic graphs, the blockchain is the best known. In a blockchain, the data is broken down into a series of blocks of fixed length. A set of algorithms ensure that the data is stored correctly on each computer by using a hash function, which computes a summary value, to make sure the data is the same each time. The blocks are then linked or chained to each other (hence, "blockchain").

The best known example of a blockchain is Bitcoin. It is a distributed ledger which is used to record ownership of a digital asset, which in this case is a "cryptocurrency" that could be traded between many people. The magic of Bitcoin is that there is no central authority, such as a bank or an exchange, to maintain records of who owns how many Bitcoins. Instead, anybody could start their own Bitcoin node and hold a record of all the transactions of Bitcoin. When people buy and sell Bitcoins with each other, a group of miners compete for the right to enter the new transactions into the ledger. They do so by solving complex mathematical puzzles, which consume a lot of expensive computing power. The first miner to solve the puzzle gets to enter the transaction on the ledger and is in return rewarded with a Bitcoin as payment. This mechanism, known as a "proof of work" consensus algorithm, thus sets a high cost for who could enter new transactions into the blockchain. Bitcoin has grown since its introduction in 2009 and today stores nearly $400 billion in market value of assets for people around the world.

Since the introduction of Bitcoin, many important improvements have been made to web3 technologies. First, there are smart contracts, which are programs that could run on a distributed ledger. This opens up a whole new world of not just storing digital tokens of value but also transacting on the ledger using pre-programmed code. Today, virtually all major blockchains support smart contracts, which are used for everything from trading to lending to decentralized voting.

Second, while the proof of work consensus algorithm of Bitcoin has successfully supported a decentralized network, it is slow, expensive, and consumes tremendous amounts of energy. As a result, newer distributed ledgers are shifting to an alternative consensus algorithm such as "proof of stake." Instead of requiring miners to solve complex but useless mathematical problems to enter new transactions on the network, parties stake their assets to validate the new transactions. If the new transactions are later repudiated, the stakers would lose those assets. The advantage is faster, lower cost transactions with significantly lower energy and climate footprint. For example, while Bitcoin consumes a staggering 1400 kWh of electricity for each transaction, many proof of stake algorithms consume less energy than Google searches or sending emails (Digiconomist, n.d.). Given all these advantages, it's no surprise that Ethereum, the largest public blockchain after Bitcoin, switched to proof of stake in 2022.

Finally, while many advocates envision a future where decentralization dethrones the centralized authorities such as governments, corporations, and Big Tech, those same centralized authorities have also seen the benefits of distributed ledgers. They are turning to permissioned distributed ledgers, which are accessible only by known parties, for collaborating with their business partners. For them, enterprise distributed ledgers and web3 technologies offer a faster way to transact than the alternatives while preserving proprietary data.

Distributed ledger storage such as the blockchain are the key components of a whole set of web3 technologies that support a decentralized Internet, including cryptographic wallets and self-sovereign identities. For the rest of this chapter, "blockchain," "distributed ledger," "DLT," and "web3" will be used synonymously to refer to the decentralized storage of data and all the other technologies that support the decentralized Internet.

While distributed ledger technologies continue to evolve, some key benefits are already becoming clear:

- **Lower Transaction Costs**. One recent study showed that the cost to issue green bonds on the blockchain is 90% less than traditional alternatives (HSBC and Sustainable Digital Finance Alliance 2019).
- **Simplified Transaction Process**. Transactions through smart contracts on blockchains are digitized and standardized, reducing many of the redundancies and complexities of legacy processes.
- **Transparency and Visibility**. Once smart contracts and transactions are on blockchains, they are available for all parties to inspect if the blockchain is public

or authorized parties if the blockchain is permissioned. This does not mean that blockchains are an open book, however. Instead, mechanisms such as public/private key encryption and zero knowledge proofs allow blockchains to hold an audit trail which can be used to verify transactions without revealing sensitive data.
- **Settlement in Real Time**. Transactions are settled in real time, as soon as they are written on the blockchain, as opposed to the days required for traditional settlement.
- **Access to the Global Market 24/7**. Blockchains are global and 24/7, allowing parties from all over the world to transact with each other.

Thanks to these benefits, distributed ledger technologies are already finding success in many fields. For example, artists and musicians are building communities of their fans by offering them access to their creative process and selling digital non-fungible tokens (NFT's) linked to their work. Major corporations in industries ranging from food, pharmaceuticals, electric batteries, minerals, and plastics recycling are using blockchains to trace their products through their supply chains. Mortgage lenders are streamlining the application process with blockchains that connect originators, appraisers, credit bureaus, and underwriters to funding to reduce the cost and improve the speed of making loans. In the climate sectors, there are projects such as digital Measure, Reporting, and Verification (MRV) of carbon emissions, tokenization of carbon credits and offsets for trading and investing, and meta registries to prevent double counting of claims.

4 DLT for Flood and Adaptation Insurance

The insurance industry has also started to look at DLT as a way to improve the speed and lower the costs of their business processes. For example, distributed ledgers could help detect fraudulent claims filed across multiple insurers. It could also reduce the time to settle claims between insurers by transacting in real time on the ledger, instead of using paper checks. This could settle auto insurance claims three to five times faster while saving the industry up to $200 billion a year, or 25–50% of its operating costs. Allianz, for example, is using the blockchain to settle auto insurance claims across 23 countries. The health insurer Anthem is planning to move the healthcare data of 40 million customers to a blockchain based system to help process the huge amount of requests for customer healthcare data while allowing customers to control access to that data. Finally, a consortium of insurers has explored using the blockchain for reinsurance through an industry-funded startup b3i, with the goal of reducing the need for entering and reconciling transactions and increasing the speed of transactions and settlements (CB Insights 2022).

More fundamentally, though, DLT offers a way to create an entirely digital and decentralized alternative to traditional insurance. Insurance contracts are risk transfers with some common elements: The policyholder and the insurer enter into a contract, which specifies what event triggers the payout, how the triggering event is measured, the amount of payout, and a mechanism for ensuring the insurer could pay. For example, a typical homeowner's insurance policy may specify that if there is a fire, the insurer would pay the homeowner the cost to repair or rebuild the house. The amount is determined by a claims adjuster. Finally, government regulators audit insurers and require them to hold minimum levels of capital against their policies. Rating agencies such as A.M. Best also review the insurers and publish ratings of their claims paying ability.

Distributed ledgers offer mechanisms for setting up the same elements digitally. As an example, let's first look at the steps required to mirror some real world events using DLT and then transact based on them:

Step 1. Verifiable Credentials. The most basic step for transactions to take place is to be able to identify the parties in the transaction and to know whether a party authorized it or not. Distributed ledgers typically use public and private key cryptography for these purposes. Any party which may record data or transact, including machines, people, or organizations, has a public key with a matching private key. When a transaction is presented, the party signs it with their private key, and their corresponding public key is then published to identify the signing party. The private key must be kept secret, as anyone with access to it could sign on behalf of a party and have full access to their assets. To keep the private key secret, hardware wallets or cloud-based vaults could be used. Finally, to secure transactions and prevent accidental loss, social or multi-signature wallets could be set up so that multiple parties could sign transactions together.

Step 2. Measurement of Real World Events. The first step for any digital process is to use sensors to measure real world events. For example, solar panels can measure the amount of electricity generated. Weather sensors can measure the amount of rainfall or wind speed. Internet of Things (IOT) devices now promise a wealth of real-time data.

Step 3. Recording of Data. Once we have measurements, they must be kept for reporting and analysis. Storing data on distributed ledgers makes them immutable once recorded, so that multiple parties could review them without concern that the data could become lost or tampered with. Depending on the amount, frequency, and sensitivity of the data, they could be stored on permissioned or specialized ledgers that meet those particular requirements.

Step 4. Verification of the Data. Once the data is available for review, they need to be verified before they could trigger events on distributed ledger, such as exchanges of funds or certifications. This could be done either with a traditional auditor or with an "oracle," which is a decentralized service that allows multiple parties to validate the data. A traditional auditor may be preferred if the data is complex and highly technical, and finding qualified reviewers is difficult. A decentralized oracle, on the other hand, offers a more open model for reviewing and validating data. A party could stake tokens of value and validate the data in return for a fee. If the results are questioned, another party could challenge them to trigger supervisory review. If the challenge is successful, the reviewer would lose the tokens of value they had originally staked.

Steps 2 through 4 are the DLT implementation of a process often referred to as Digital Measurement, Reporting, and Verification (MRV). Traditionally, MRV has involved a combination of remote sensors, in-person field measurements, life cycle analysis models, and databases supported by mostly spreadsheets and paper documents. Digital MRV takes traditional MRV to the next level by leveraging distributed ledgers to store volumes of data from IOT devices and sensors. The combination creates an immutable audit trail of data to support conclusions, improving the trust in the results of MRV, while simultaneously significantly reducing the costs and improving the time to deliver results (Baumann 2020).

Step 5. Tokenization of Data. Once the data is verified, the next step is to create tokens on the ledger to represent the measured results. Tokens could be fungible or non-fungible. Fungible tokens denote quantities which are tradeable, such as the value of currencies. They could also be used to record quantities such as the amount of renewable electricity generated or the amount of Greenhouse Gas (GHG) emissions removed. Non-fungible tokens, or NFTs, are linked to unique assets, so that two NFTs are not directly exchangeable for each other. For example, an NFT could represent ownership of a unique work of art, or it could be a certificate of compliance for a particular facility.

Step 6. Smart Contracts. Once we have tokens, we could use smart contracts, or programs which execute on the ledger, to operate on them. For example, a smart contract could automatically issue a payment based on a fungible token of the amount of energy generated by a solar panel. Alternatively, the payment could be made based on an NFT demonstrating compliance with certain goals such as reducing its energy use.

Step 7. Staking. Tokens of value could be "staked" on a smart contract, so that they are held by contract and could be used as needed, for example, to pay one party under some predefined events. In exchange, the contract pays a return on the staked tokens. Staked tokens could be used to guarantee the performance of the parties, much like collateral for futures contracts. For example, a smart contract for the purchase of renewable energy may require cryptocurrencies such as stablecoins, which have constant value in traditional currencies such as US dollars, to be staked to assure the seller that the buyer is able to complete the purchase. Staking offers the advantage of transparency: All parties could see how much value has been staked to assure the performance of the contract.

Step 8. Protocols. Tokens and smart contracts are just the building blocks that could support a variety of protocols built on top of them. For example, if a smart contract advances payment for tokens of renewable energy generated, then a protocol could be set up to fund the purchase of renewable energy equipment in exchange for those payments. Another protocol could be set up to use tokens of energy use to transfer payments from facilities which use a lot of energy to those that use less. The variety and complexity of possible protocols are virtually unlimited, thanks to the flexibility of smart contracts and the ingenuity of developers.

Structuring insurance for flood risk and their mitigation on distributed ledgers would follow these steps to mirror real world events, such as flooding, and then manipulate them through staking, tokens, and transactions. These mechanisms could create both a combination of parametric and indemnity policies. For example, there could be a Layer 1 insurance policy, which could be similar to a parametric insurance policy based on commonly agreed metrics such as rainfall or wind speed, and a Layer 2 insurance policy, which could be similar to an indemnity insurance policy based on the value and losses of actual assets. Layer 1 could tap into the global reinsurance or insurance-linked securities markets to raise capacity and make quick payments in the event of disasters. Layer 2 could be administered locally by groups familiar with the assets, who could price the policies based on their knowledge of their values and their mitigation measures. They could be for profit local insurance companies, such as in the USA, or mutual or community groups similar to African savings circles or Korean *kye*. The choice of for profit or mutual insurance would be based on the cost to sell and administer the policies, as well as the cultural acceptance of the formats.

In either case, the administrators of the Layer 2 policies would purchase coverage from the Layer 1 policies to lay off their overall risk. They would also assume the basis risk between standard metrics such as rainfall or wind speed and the potential for loss at specific assets. Thus, they would be incentivized to encourage mitigation measures, including both nature-based solutions such as mangroves and artificial ones such as dykes, levees, raising or floating assets. Where it is unprofitable to underwrite the risks, they would be incentivized to raise prices, decline providing coverage, or encourage moving out of disaster prone areas. Finally, the Layer 2 administrators would need to manage both moral hazard, where the availability of insurance may encourage riskier behavior, and outright fraudulent claims. In the USA, insurers rely on contractual exclusions, which may deny payouts if policyholders do not meet required behavior for reducing risks, and criminal penalties against insurance fraud. In other markets, social reputation may help produce the proper behavior (Table 1).

Layer 1 insurance protocols could be structured with DLT following the steps discussed above:

Step 1. Verifiable Credentials. The insurer and the insured parties would all be issued cryptographic accounts.

Step 2. Measurement of Events. This would happen with agreed upon sensors, such as official weather stations, or specific weather stations set up and maintained by trusted third parties.

Step 3. Recording of Data. A distributed ledger could be set up to record the data from the weather stations.

Step 4. Verification of Data. An oracle or decentralized verification board would be set up to review the data to make sure it is accurate. This should include people who are technically familiar with the measurements and people who could verify that the devices used are functioning correctly in the field.

Step 5. Tokenization. The data could be tokenized as an NFT which denotes the amount of rainfall or wind speed at the weather station.

Table 1 Layer 1 and Layer 2 insurance using blockchain

	Layer 1	Layer 2
Purpose	Parametric policy with fast payout tied to measured events, such as wind speed or rainfall	Indemnity policy with payout tied to specific assets and losses. Linked to Layer 1
Verifiable credentials	Cryptographic accounts issued to all parties	Cryptographic accounts issued to all parties
Measurement of events	Sensors maintained by official or third-party weather stations	Documentation of assets before and after loss claims. Appraisals and repair estimates for the assets
Data recording	Ledger for direct recording of data from sensors	Ledger for recording assets, appraisals, and loss claims
Data verification	Decentralized oracle	Auditor or decentralized oracle
Tokenization	NFT records data	Fungible token for amount of loss
Smart contract	Makes payouts based on data recorded on NFT	Makes payouts based on tokenized loss
Staking	Payer stakes funds	Payer stakes funds
Protocols	Connect premiums, payouts, insurance-linked securities	Connect Layer 2 policies to Layer 1 policies

Step 6. Smart Contract. The smart contract is the heart of the insurance policy. It would execute payout in exchange for the NFT. Once the contract is paid, the NFT is marked as used and could not be used again.
Step 7. Staking. The payer on the contract would stake cryptocurrencies or stablecoins to assure other parties that it is able to make the payments when needed.
Step 8. Protocols. Protocols could be set up to fund the insurance contract. For example, policy holders would pay their premium to enable the insurance smart contract. A protocol could be set up to link a separate smart contract for reinsurance or to set up an insurance-linked bond, where an investor would stake their funds for either a high return if there is no insured event or a loss if there is.

For Layer 2 policies, the steps would be:

Step 1. Verifiable Credentials. Again, all parties would be issued cryptographic accounts.
Step 2. Measurement of Events. The relevant events for an indemnity policy would be the assets and the loss of values. Thus, measurements would include photographs of assets at the time the insurance policy began and after the disaster, as well as appraisals of their values and reports of claims adjusters.
Step 3. Recording of Data. A distributed ledger could be set up to record information about the assets and their appraisals. Since this may include sensitive information, it should be placed in permissioned ledgers and only shared between the insurer, the insured, and any appraiser and claims adjusters.

Step 4. Verification of Data. Given their complexity, analyzing assets and verifying losses would require parties with highly specialized knowledge. Overreliance on one party, however, creates the potential for fraud. On the other hand, a decentralized oracle could introduce a trusted process that uses multiple parties to evaluate losses and resolve disagreements. DLT offers the possibility to combine the two, where multiple trusted parties could review and sign off on the assets appraisals and loss claims.
Step 5. Tokenization. Once resolved by the previous step, losses could be tokenized. Since it is a monetary amount, it could be a fungible token.
Step 6. Smart Contract. The smart contract executes the insurance agreement. With indemnity policies, the payout could be a percentage of the amount of loss with a deductible.
Step 7. Staking. The seller of the insurance would stake cryptocurrencies or stablecoins of value to assure buyers that it is able to perform when needed.
Step 8. Protocols. An additional protocol would link funds from Layer 1 parametric policies to Layer 2 indemnity policies. Policy premiums paid by Layer 2 insureds could automatically be linked to purchases of Layer 1 parametric policies. When a disaster happens, payouts from Layer 1 policies would automatically be transferred to the treasury of the Layer 2 policy and made available. This would increase the transparency and trust of the Layer 2 policies while avoiding missed or lost payments or any other event that could lead to denial of coverage after a disaster.

A "dual token" model could be used to separate the insurance capital pool from operating expenses of the insurance contracts. Under such a model, all funding and disbursement of risk pools, including the premiums paid by the insureds and the payouts from the insurer, are made with a token for the risk pool. Each interaction with the contract, such as entering into an insurance agreement, would require the use of a separate token. The smart contracts of the insurance policies could be written so that it could not use any funds from the treasury of the risk pool tokens to pay for operating expenses (Mussenbrock and Karpischek 2017).

Building out insurance contracts using DLT offer several key advantages versus the traditional mechanisms of the industry, including:

- As a fully digital, highly automated platform, it has much lower costs than the existing processes. In particular, a Layer 1 policy operates mostly with software. A Layer 2 policy requires local administration and know-how, but even then it should be much less expensive to set up and operate than with traditional systems.
- This low cost makes it possible to insure assets of small value, such as those found in developing economies.
- The payout could be made by cryptocurrencies such as stablecoins, thus allowing anyone with a digital wallet to receive payouts, even when they have no access to bank accounts.
- Immediate payouts are possible under Layer 1 policies. In cases where Layer 2 indemnity policies could not be set up due to the costs or lack of know-how, Layer 1 policies provide at least some protection from losses.

- Greater transparency and trust from having insurance smart contracts and insurance risk pools that are both openly available for review. This helps any party, be it regulators or policyholders, to determine whether the contract is written to make the needed payouts and whether it has the funds to do so.

5 Implementation and Challenges

The insurance industry has sponsored several initiatives based on some of the ideas discussed here. For example, California Extreme Heat Concept, the Mexican Smallholder Farmer Climate Risk Initiative, The Cool Capital Stack, and B-READY are all piloting the use of parametric policies to protect against extreme heat, droughts, floods, and typhoon exposures (Carpenter 2022). The blockchain startup Etherisc has piloted weather index microinsurance to protect Kenyan farmers against excessive rainfall or drought (Eyase 2021).

Scaling such initiatives to the insurance industry, though, would need to overcome many challenges, especially if DLT is involved.

The first and foremost are the regulatory and social requirements. Insurance is a highly regulated industry. So, while the smart contracts as described above could be used by anyone, if they are marketed as a form of insurance, they may be subject to insurance regulations. One solution is to have a regulated insurance company provide an umbrella or wrapper on top of the smart contracts. The insured parties would enter into legally binding insurance contracts with the insurance company, which would then use the smart contracts to administer the policies (Mussenbrock and Karpischek 2017).

In some cases, though, that might not be enough. For example, homes or businesses may have been built without zoning or permits, and the owners may not have clean legal title. As a result, it may not be clear what an insurance policy would cover, if a policy on such a property would be legally binding, or if it would even be legal to write such a policy in the first place. In other cases, government policies may push insurers to write policies even when it's uneconomical. For example, as climate change has increased storm and flooding risks in coastal areas, the US states of Florida and Louisiana have introduced government-run insurance programs to make insurance available to homeowners at lower rates than private sector insurers would provide. Altogether, these issues make it difficult to link DLT smart contracts and actual insurance policies.

This then leads us to the question of how insurance would be sold. Even in developed economies, selling insurance is time-consuming and labor intensive. In developing economies where many people may not even be aware of insurance or trust their performance, it could be even more difficult, especially when amounts are small. Expectations of government bailouts in the event of disasters have also hampered insurance sales. Studies in the USA, for example, showed that for every $1 of government disaster relief, insurance purchase was reduced by $6 (International Insurance Society 2018). The solution could be to link insurance purchases to other transactions, such as the purchase of seeds by farmers, supplies by businesses, or obtaining any type of loan. Further, mobile technologies and payment networks such as M-PESA in Kenya could reduce the costs of insurance and allow small premium policies to be paid as the insureds have the money (Swiss Re Institute 2017).

Third, the integration of traditional and digital insurance runs up against an industry locked into legacy systems and paper-based processes. New and innovative technologies such as digital MRV and blockchains have had a difficult time taking hold. For example, while the startup b3i was successful technically in executing reinsurance contracts through the blockchain, it ultimately ceased operations. Commenting on this turn of events, Christian Mumenthaler, Group Chief Executive Officer (CEO) of Swiss Re noted:

> You would need all insurance companies to basically create smart contracts at the beginning, at the origin. And then, based on that you could of course then construct a digital reinsurance contract that can be traded afterwards. And then you have the full efficiency end-to-end, and as claims come in, you would automatically pass them through to the reinsurer, for example. So, that's very visionary, but it would imply that all insurers have to switch all IT systems and create smart contracts.
>
> But, insurance companies, of course, have a huge amount of different data systems, and you just don't get to the efficiency you need if you just start with that. It's only part of the value chain. (Gallin 2022)

Finally, creating insurance markets in developing economies would require both data and risk models to accurately price the risks involved. Such models should support both current risk profiles and allow analysis of the effects of various mitigation strategies. The lack of risk models, accurate meteorological data, and registers of assets which needed to be insured were key challenges when a group of 20 Caribbean nations tried to purchase reinsurance against hurricane risks after Hurricane Ivan in 2004. Eventually, a grant from the Japanese government was needed to develop a risk model. In the future, open source development of risk models would create a public good that would enable both public and private insurance programs to create products for developing economies, without proprietary ownership that could either close off access or make them prohibitively expensive (Jarzabkowski et al. 2020). One example of an open source insurance risk model is the Oasis Loss Modeling Framework (available at oasislmf.org), which supports the development and running of catastrophe risk models. Part of the Global Risk Modeling Alliance, its development is driven by insurers and reinsurers including Allianz, Swiss Re, and Zurich. It currently supports over 90 models from 18 suppliers, and it offers an open data standard but does not include any data.

6 Conclusions

With climate change already happening at a rapid pace, adaptation and mitigation will become increasingly important. Insurance will be one of the key mechanisms for adapting to natural disasters such as floods and hurricanes. A functioning insurance market could encourage proper adaptation and mitigation and provide capital for rebuilding after disasters. When integrated into financial products such as mortgages and business loans, insurance helps to reinforce urban planning against floods and natural disasters that may be difficult to enforce otherwise.

Given the lack of insurance in developing countries and slow pace of change in the insurance industry, the blockchain could offer an accelerated alternative to forming risk markets. It could bring in international capital through contracts tied to measured weather events, while helping local groups, whether mutual aid groups or for profit insurance companies, protect assets against loss at much lower operating costs than traditional insurers.

After the fall of cryptocurrencies and the spectacular collapse of FTX in 2022, as well as the failure of b3i, some may be ready to write off the value of blockchain in the insurance industry altogether. This is premature. A more holistic view would recognize that insurance is just one of several mechanisms, along with futures, swaps, and securitization, for transferring risk. DLT is naturally suited for transferring risks between parties, and it could do it at far lower costs than traditional finance or insurance. It should therefore only be a matter of time before others attempt to use blockchain to transfer and mitigate risks.

More importantly, the decentralized and open source philosophy of blockchain could help solve the central conundrum of the insurance industry: information asymmetry. Many insurers want to believe that proprietary risk models and data give them a competitive advantage. In reality, though, proprietary risk models and data limit profitability and growth while imposing high costs on the insurance industry as a whole. Information asymmetry creates distrust in insurers and their products. As a result, it is very difficult and expensive to sell insurance. At the same time, lack of transparency leads to regulation, again at high costs to the insurers. Finally, the reliance on proprietary models has saddled the industry with expensive and high cost legacy business processes.

The alternative may be a decentralized risk market based on open data sets and loss models. Imagine a world where anyone could trade rainfall as easily as stocks and futures. Restaurants and concerts could "go long" rainfall and get paid if there is more rain, while golf courses could "go short" and get paid if there is less rain. Farmers could buy straddles, or combinations of options which pay if there is too little or too much rain. Traders from around the world could then participate in the market, making it deep and liquid. Both for profit insurers and nonprofit mutual aid groups could then tap into these weather markets to offer insurance policies, as well as help their policyholders reduce their risks in the event of rain or drought. All the transactions could happen on DLTs, and the entire industry could innovate at the speed of financial markets rather than insurance companies.

The technologies for this to happen already exist. What's missing in the field of risk transfer is the data and modeling to support a marketplace. Mobile phones and IOT devices could provide the data, and open source loss modeling could provide the modeling. If there is actuarial know-how in the developing economies, then the combination could lead to climate risk mitigation using DLT. The lack of insurance markets in developing economies due to high costs may make it the perfect breeding ground for such disruptive change.

References

Africa Risk Capacity (2022) River flood. African Risk Capacity. Accessed 19 Sept 2022. https://www.arc.int/river-flood

Baumann T (2020) What is digital MRV? Digital measurement, reporting and verification to enhance climate actions and sustainability. LinkedIn, 8 Dec 2020. Accessed 15 Sept 2022. https://www.linkedin.com/pulse/what-digital-mrv-measurement-reporting-verification-tom-baumann-%E5%8C%85%E8%AD%BD%E6%96%87

Beck MW, Heck N, Narayan S, Menendez P, Torres-Ortega S, Losada IJ, Way M, Rogers M, McFarlane-Connelly L (2020) Reducing Caribbean risk: opportunities for cost-effective mangrove restoration and insurance. axa-contento-118412.eu. Accessed 13 Sept 2022. https://www-axa-com.cdn.axa-contento-118412.eu/www-axa-com%2Ff83724d2-fcd0-4a41-bde9-e0330a501d07_tnc_mangrove+insurance_final+hi.pdf

Brown JR (2018) The natural catastrophe protection gap: measurement, root causes and ways of addressing underinsurance for extreme events—abstract. International Insurance Society. Accessed 13 Sept 2022. https://www.internationalinsurance.org/sites/default/files/2018-04/The%20Natural%20Catastrophe%20Protection%20Gap.pdf

Carpenter G (2022) Fulfilling a legacy of societal risk management. Marsh McLennan. Accessed 31 Jan 2023. https://www.marshmclennan.com/content/dam/mmc-web/insights/publications/2022/november/Race%20to%20Reslience%20Report_COP27%20FINAL.pdf

CB Insights (2022) How blockchain is disrupting insurance, 19 July 2022. https://www.cbinsights.com/research/blockchain-insurance-disruption/

Digiconomist (n.d.) Bitcoin energy consumption index. Digiconomist. Accessed 14 Sept 2022. https://digiconomist.net/bitcoin-energy-consumption

Dunsavage J (2020) Triple-I blog | Hurricane Delta triggered coral reef parametric insurance. Insurance Information Institute, 2 Nov 2020. Accessed 13 Sept 2022. https://www.iii.org/insurance industryblog/hurricane-delta-triggered-coral-reef-parametric-insurance/

Evans S (2020) Aon seeing "demand surge" for parametric risk transfer products. Artemis.bm, 14 Dec 2020. Accessed 13 Sept 2022. https://www.artemis.bm/news/aon-seeing-demand-surge-for-parametric-risk-transfer-products/

Evans S (2021) Swiss Re CorSo settles parametric hurricane Ida claim. Artemis.bm, 10 Sept 2021. Accessed 13 Sept 2022. https://www.artemis.bm/news/swiss-re-corso-settles-parametric-hurricane-ida-claim/

Eyase J (2021) Reimagining agriculture insurance using blockchain technology—ACRE Africa. ACRE Africa, 11 Mar 2021. Accessed 31 Jan 2023. https://acreafrica.com/reimagining-agriculture-insurance-using-blockchain-technology/

Gallin L (2022) B3i conceptually interesting, but required end-to-end view: Swiss Re execs. Reinsurance News, 29 July 2022. Accessed 19 Sept 2022. https://www.reinsurancene.ws/b3i-conceptually-interesting-but-required-end-to-end-view-swiss-re-execs/

Guy Carpenter (2022) Natural catastrophe protection gap in Asia calls for collaborative innovation. Guy Carpenter. Accessed 13 Sept 2022. https://www.guycarp.com/content/dam/guycarp-rebrand/pdf/Insights/2022/2022.7-Protection-Gap-Publish-final.pdf

HSBC and Sustainable Digital Finance Alliance (2019) Blockchain. Gateway for sustainability linked bonds. HSBC Centre of Sustainable Finance, 24 Sept 2019. Accessed 14 Sept 2022. https://www.sustainablefinance.hsbc.com/mobilising-finance/blockchain-gateway-for-sustainability-linked-bonds

International Insurance Society (2018) The natural catastrophe protection gap: measurement, root causes and ways of addressing underinsurance for extreme events—abstract. International Insurance Society. Accessed 16 Sept 2022. https://www.internationalinsurance.org/sites/default/files/2018-04/The%20Natural%20Catastrophe%20Protection%20Gap.pdf

Jarzabkowski P, Chalkias K, Clarke D, Iyahen E, Stadtmueller D, Zwick A (2020) Insurance for climate adaptation: opportunities and limitations. Insurance Development Forum. Accessed 19 Sept 2022. https://www.insdevforum.org/wp-content/uploads/2020/08/Insurance-for-Climate-Adaptation-Opportunities-and-Limitations.pdf

Munich Re (n.d.) Flood risk: underestimated natural hazards. Munich Re. Accessed 13 Sept 2022. https://www.munichre.com/en/risks/natural-disasters-losses-are-trending-upwards/floods-and-flash-floods-underestimated-natural-hazards.html

Mussenbrock C, Karpischek S (2017) White paper. Webflow, 1 Nov 2017. Accessed 15 Sept 2022. https://uploads-ssl.webflow.com/6243075ff83d08a79dc7b307/624edb8a4b12178635a629c6_etherisc_whitepaper_1.01_en.pdf

Sieff K (2020) Hurricane Delta hits Cancún, but resort city escapes major damage. The Washington Post, 7 Oct 2020. Accessed 13 Sept 2022. https://www.washingtonpost.com/world/the_americas/hurricane-delta-mexico-cancun-cozumel/2020/10/07/184494aa-081f-11eb-859b-f9c27abe638d_story.html

Surminski S, Panda A, Lambert PJ (2019) Disaster insurance in developing Asia: an analysis of market-based schemes (EWP 590). Asian Development Bank, Sept 2019. Accessed 13 Sept 2022. https://www.adb.org/sites/default/files/publication/528026/ewp-590-disaster-insurance-developing-asia.pdf

Swiss Re Institute (2017) Sigma/insurance: adding value to development in emerging markets. Swiss Re. Accessed 13 Sept 2022. https://www.swissre.com/dam/jcr:0c654530-302f-4442-b01e-585d11ea3beb/sigma4_2017_en.pdf

Wright R (2021) As hurricane season begins, new homes are being built at a record rate—but not strongly enough. The Hill, 9 June 2021. Accessed 13 Sept 2022. https://thehill.com/opinion/energy-environment/557630-as-hurricane-season-begins-new-homes-are-being-built-at-a-record/?rl=1

Case Study Synthesis and Transformational Approaches to Adaptation and Green Recovery

Case Study Findings and Synthesis of Learnings

Catherine Atkin and Kwi-Gon Kim

Abstract Book co-editors summarize learnings from the thirteen case studies focused on flooding hazard risk assessment and green recovery planning in regions across Africa, South and Southeast Asia, Middle East, and South America. Case study authors engaged in flood risk assessments and local context variable analyses including an exploration of the socio-cultural, economic, institutional, and technological dimensions that define adaptive capacity and green recovery planning and implementation. The case study studies provide strong evidence of the devastating effects on people, animals, and places of climate change, extreme weather, and flooding hazards. They also reveal the current challenges that Global South communities face in creating robust risk assessments including the limited availability of robust data sets, analytical tools, and technology solutions at the local level.

Keywords Global South flood risk assessment · Climate risk management · Climate risk indicator analysis · Flooding hazard reduction measures · Structural and non-structural reduction measures

1 Introduction

The thirteen case studies focused on Global South communities in Africa, South America, South and Southeast Asia and the Middle East tell a powerful story about the increasing adversity that people and animals are experiencing as climate change exacerbates rural, urban, riverine, and flash flooding hazards and disasters globally. The authors, professionals and practitioners from Global South countries pursuing their master's degree in Climate Change Policy from KOICA/Hallym University have leveraged available data and insights and applied new tools developed through

C. Atkin (✉)
Stanford CodeX Climate Data Policy Initiative, Stanford, USA
e-mail: catherine.atkin@codex.stanford.edu

K.-G. Kim
Seoul National University, Seoul, South Korea
e-mail: kwigon@snu.ac.kr

K. Kim and C. Atkin (eds.), *Cross-Cultural Perspectives on Climate Change Adaptation*, Cities and Nature, https://doi.org/10.1007/978-3-031-50365-8_23

their coursework and the case studies featured in this book. These case studies contribute to a deeper understanding of the status of hazard risk assessment and comprehensive climate risk management and green recovery and provide critical insights across the two key learning dimensions introduced in Chap. "Overview of Case Studies: CommonElements and Framework for Analysi (Catherine Atkin and Kwi-Gon Kim)"

1. What are the key findings from the case studies regarding flood risk assessment—including hazard, exposure, and vulnerability analysis—and green recovery planning?
2. How are the currently available climate risk management planning tools and frameworks used and what does it tell us about opportunities and limitations of their current application and deployment?

2 Case Study Flood Risk Assessments

While robust disaster and hazard risk assessment is a critical foundation for comprehensive and successful disaster reduction management and planning, the case studies reveal the current challenges that Global South stakeholders face in developing comprehensive assessments. A review of the case studies demonstrates that lack of available risk assessment data and insights for use in affected communities.

Key findings include:

- **Project Site Selection, Boundary Delineation and Profile**: Authors chose their project site and boundaries based on a range of factors including history of flood disaster, strength of available data, and personal and professional connection. While there is a variation in the type of flooding hazard, urban and riverine floods were the most common focus of case study authors. The geographic boundaries of the case studies vary with a focus on neighborhoods, cities, and regions that may be rural or a mixture of rural and urban areas. Publicly available maps as well as a variety of GIS tools were used to delineate site boundaries.
- **Description of Historical and Prospective Flood Hazard Data**: Historical flood hazard data varied significantly depending on country and location. While case studies in the five countries designated as Least Developed Countries by the United Nations—Ethiopia, Bangladesh, Cambodia, Afghanistan, and Uganda—generally lacked more granular hazard data than other country locations, this was not always the case. In countries like Bangladesh and Cambodia where international organizations have ongoing development efforts, robust historical and projected data is more readily available.
- **Community Vulnerability Assessment**: Although most case study authors were able to generalize about the sensitivity of communities to the adverse effects of flooding on people and systems, only a few case study authors had access to the necessary data to undertake a robust place-based vulnerability framework analysis to examine the social, natural resource, and structural vulnerabilities to

flooding impacts. Similarly, there was limited systematic assessment of the adaptive capacity of communities including the ability of systems and institutions to mitigate flooding impacts. See Sect. 4 for discussion of local context variables by a subset of case study authors.

3 Green Recovery Planning

Case study authors used their risk assessment results to identify hazard reduction measures using a widely accepted approach that groups hazard reduction measures as structural and non-structural.

Non-structural measures use knowledge, practice, or agreement to reduce disaster risks and impacts through policies and laws, public awareness raising, training, and education.

Structural measures are physical constructs to reduce or avoid possible impacts of hazards, or the application of engineering techniques or technology to achieve hazard resistance and resilience in structures or systems (UNDRR).

For the purposes of this analysis structural measures as further divided between "soft" solutions that are nature-based and "hard" solutions that involve infrastructure development and land engineering.

Table 1 includes data on the frequency with which a comprehensive set of structural and non-structural measures were identified in the case studies featured in Chaps. "Flood Risk Analysis: Causes and Landscape-Based Mitigation Strategies in Dire-Dawa City, Ethiopia"–"Proposed Study Design for Spatio-Temporal Drought Assessment Using Standardized Precipitation Evapotranspiration Index (SPEI) and Adaptation in the Mantaro Valley of Peru".

Key findings include:

- Authors identified 55 non-structural and 69 structural hazard reduction measures.
- Non-structural hazard reduction measures: Some of the most frequently identified hazard reduction measures include:
  - Flood monitoring and early warming frameworks and systems
  - Community flood awareness and preparation
  - Comprehensive land use planning and zoning
  - National disaster and climate risk management policies
- Soft structural hazard reduction measures: Some of the most frequently identified hazard reduction measures include:
  - Wetlands restoration
  - Upper watershed restoration
  - Natural drainage path restoration
  - Riparian and mangrove vegetation restoration
  - Tree forestry planting and development

Table 1 Frequency data for structural and non-structural hazard reduction measures identified in case studies

Non-structural Measures	
Flood monitoring and early warning systems	10
Community flood awareness and preparedness	7
Comprehensive land use planning and zoning	6
National disaster and climate risk management policy	5
Strengthen government coordination and capacity	2
Flood response and rescue scenario planning	5
Improved risk assessment	3
Comprehensive resilience and adaptation master planning	3
Solid waste management planning	3
Public acquisition of floodplain lands	2
Economic development planning	2
Flood insurance programs	2
Capacity building for community and stakeholders	2
Regional planning and regulation	1
Crop change and alternative land use	1
Leverage new technologies	1
Total frequency of non-structural measures	55
Soft structural measures	
Other	14
Wetlands restoration	3
Upper watershed restoration	2
Natural drainage path restoration	2
Removal of barriers	2
Land infiltration capacity enhancement	2
Riparian and mangrove vegetation restoration	3
Tree forestry planting and development	2
Rainwater harvesting	2
Agroforestry systems	1
Parks development	1
Soil conservation	1
Total frequency of soft structural measures	**35**
Hard structural measures	
Damn and reservoir construction and maintenance	5
Levees and river embankment enhancement	4
Drainage system engineering and maintainance	4
Roads and transportation infrastructure enhancement	5

(continued)

Table 1 (continued)

Water systems development	2
Traditional and modified irrigation systems	2
Diversions, floodways, and canal development	4
Early warning systems and mobile networks	1
Resettle and rebuild shelter and businesses	1
Retention and detention ponds	1
Terracing and bunds	2
Percolation tanks	1
Oceanfront protection measures	1
Construct livestock dips	1
Total frequency of hard structural measures	34

- Hard Structural hazard reduction measures: Some of the most frequently identified hard structural measures include:
  - Damn and reservoir construction and maintenance
  - Levee and riverbank enhancement
  - Drainage system engineering and maintenance
  - Road and transportation infrastructure enhancement
  - Water systems development
- The distribution of hazard reduction measures did not vary significantly across countries and regions. Non-structural measures, including the creation of nested policies at the national and local levels focused on disaster risk management including land use planning, community education, and flood monitoring and early warning systems, were front and center regardless of location. As would be expected, local topography informed the kinds of hazard reduction measure that were identified. For example, for those case studies areas along rivers, a key non-structural measure identified was the need for planning reform regarding dwelling placement on the edges of rivers and structural measures identified for consideration included the fortification and expansion of river capacity and improved vegetation restoration.

4 Local Context Variables and Adaptive Capacity Conditions

While case study authors described a range of local context variables that were considered in developing their risk assessments, many case study authors had limited access to stakeholders with knowledge and insights about local systems they would need to identify and assess the local context variables. In addition, few evidenced a clear understanding of how to evaluate an area's adaptive capacity including the economic, social, institutional, and technological conditions that may constrain the development and deployment of flood hazard reduction measures. This is not surprising given that this critical dimension of the conversation is often undervalued in academic discourse and many of the climate risk management frameworks and methodologies lack clear guidance on how to evaluate adaptive capacity.

4.1 Social Conditions

Local Community Knowledge: Several case study authors acknowledged that although community members had deep knowledge and historical experience with flooding that would bring value to resilience building, they were not involved in adaptation planning efforts. For example, authors spoke about how networks of residents communicate with each other when flooding is imminent and have valuable information about when and how flooding may occur in their communities. In addition, residents and villagers have traditional knowledge and experience shared over generations about agricultural production and protection techniques, and local water and crop harvesting technologies that could inform the design and implementation of flooding reduction measures.

Community Engagement: Some case studies highlighted the importance and power of community capacity building and engagement and described ways in which the strengthening local community institutions can drive the success of adaptation projects. In particular, authors described importance of localized planning processes that strengthen buy in and ensure residents are given the opportunity to design, evaluate, and support the implementation of solutions over time.

4.2 Governmental and Policy Conditions

National and Regional Levels: The case studies revealed that most countries do not have robust national adaptation and climate risk management policies and plans that guide national and regional action. While many countries were described as having weak policy environments, authors from some countries—including from the Philippines, Bangladesh, and Uganda—included information in their case studies about significant national disaster relief and resilience building policies.

City Level: Most case study authors reported that study area cities lacked the range of policies—including climate risk management plans and comprehensive land use and economic development—necessary to effectively address flood hazards and build resilience. They also found that even when policies are in place, enforcement of existing housing and commercial standards, for example, may be weak. Recognizing that there may be resistence to policies restricting access to riverbank development authors highlighted the importance of coupling location restrictions with new approaches to construction of flood resistant structures so that local residents can continue to pursue their livelihoods.

4.3 *Economic and Adaptation Finance Conditions*

Local Economy: High levels of poverty characterize the study area communities and not surprisingly, those residents that are most negatively impacted by flooding are typically the least resourced within the community. Case study authors reported that in both urban and rural areas there were very limited funds for basic operations and in village areas the challenges are even greater. These governments raise local revenue but also depend on significant amounts of operational funding from national governments, multilateral development agencies, and international NGOs.

Adaptation and Disaster Funding: The severe lack of funding to address flood and other climate related impacts was mentioned in many of the case studies. Contributors identified numerous promising hazard reduction measures and planning efforts that cannot be implemented because local governments in case study areas have very limited funding for adaptation projects. Those countries that experience flooding disasters may receive relief funds from international agencies and national governments but even those resources are not sufficient to meet community needs.

4.4 *Technological Infrastructure and Access to Solutions*

A number of case studies featured high-value technologies including geospatial mapping, cloud computing, blockchain, and IoT sensors that if available, would significantly enhance the completeness and ease of flood risk assessment and allow for the creation of early warning systems. Basic tech infrastructure gaps including grid access, internet availability, mobile networks, and cell phone access were identified as challenges to unlocking the power of technology to increase impact and scale.

5 Current Disaster Risk Management Methodologies and Their Application

The case studies offer the reader a rich look into the theory and practice of disaster risk management as a field. Are current tools and frameworks able to support robust risk assessment and management efforts? Do they allow for the integration of the full range of adaptation, mitigation, and development imperatives cities and communities face? What are the opportunities for further development and needed reforms?

Disaster and Climate Risk Assessment and Management: As national and international institutions recognize that the negative impacts of climate change are already having devastating impacts across the globe, there has been an uptick in the development of new and more comprehensive risk assessment and disaster management methodologies. The UN Habitat's Vulnerability and Risk Assessment and Action Planning framework (UN Habitat 2020) and the Climate Disaster Risk Assessment framework (GIZ 2022) developed by Deutsche Gesellschaft für Internationale Zusammenarbeit (GIZ) GmbH are a few examples of the risk assessment models used by case study authors. Many of the case studies used the Nine Step Conceptual Framework for Climate-Disaster Related Risk Assessment and Green Recovery Planning developed by Dr. Kwi-Gon Kim and presented in Chap. "Nine-Step Conceptual Framework for Climate Disaster Risk Assessment and Green Recovery Planning" as the framework for their analyses.

One of the weaknesses of many of the leading frameworks is their singular focus on risk assessment. Risk assessment should be coupled with forward looking management including identification and deployment of hazard reduction measures that can mitigate future hazards. The expansion of the frame to include climate and disaster risk management is an important development and Dr. Kim's framework is an important addition to the field because it connects risk assessment with green recovery planning and implementation.

As methodologies are expanded to incorporate go-forward planning and implementation of hazard reduction measures, there must be a concerted effort to create the kinds of tools, guides, and capacity building that will be necessary to ensure needed progress. For example, while many case study authors were able to identify a wide range of potential structural and non-structural hazard reduction measures, they did not assess the efficacy and feasibility of various measures.

Co-benefits and Integration of Adaptation and Mitigation: While there is increasing discussion of the need for integrated solutions that allow communities to advance connected strategies, the tools and frameworks used by case study authors did not support an integrated, co-benefit approach to adaptation and mitigation planning. There have been very few efforts to look comprehensively at adaptation planning and city-level mitigation efforts and develop holistic and connected planning process and methodologies that incorporate mitigation opportunities and impacts into the assessment and evaluation of potential hazard reduction measures. The case study authors benefited from Dr. Kim's inclusion of this topic into their coursework and a number identified co-benefit hazard reduction opportunities including the implementation of nature-based solutions including wetlands, mangrove vegetation restoration, tree forestry planting, and soil conservation that offer adaptive protection and also sequester carbon as well as land use planning and building code reforms that both reduce flooding hazards and incentivize the use of low-carbon materials.

References

GIZ (2022) Global programme on risk assessment and management for adaptation to climate change (loss and damage). In: Scoping study: a comparison of climate risk assessment methods to support informed decision-making. GIZ

UN Habitat (2020) Climate change vulnerability and risk: a guide for community assessments, action planning and implementation. https://unhabitat.org/sites/default/files/2020/05/climatechange_vulnerabilityandriskguide.pdf

United Nations Office of Disaster Risk Reduction (website). https://www.undrr.org/terminology/structural-and-non-structural-measures

Co-benefits of Combining Adaptation and Mitigation Frameworks: A Cross-Cutting Approach

Kwi-Gon Kim

Abstract The case studies show that adaptation can accomplish much more than simply mitigating and preventing future disasters. It also presents major opportunities to achieve a larger multipurpose agenda and put countries featured in the case studies on a new green, low-carbon and resilient path to climate change, adaptation and green growth. Adaptation, mitigation and climate-friendly urban development hand in hand can create powerful synergies that meet global Sustainable Development Goals, Paris Agreement Goals, Green New Deal Goals and goals of post-COVID-19 recovery. This chapter deals with an integration platform which is needed to address this large and critical step to achieve sustainability, climate safety and resiliency. A particular emphasis is put on climate change and development: The evolution of an intersecting policy domain, which incorporates adaptation and mitigation solutions.

Keywords Co-benefits · Green growth · Sustainable Development Goals · Mitigation interventions · Synergies · Trade-offs · Circular material flow · Land-based options · Sustainable urban planning · Coordinated policies · Ethics · Paris Climate Action Plan · Climate impacts · Sustainable lifestyles · Climate innovation

1 IPCC Guidelines for the Linkages Between Mitigation, Adaptation and Sustainable Development

Low ambitious policies increase the risk of lock-in buildings in carbon for decades while well-designed and effectively implemented mitigation interventions, in both new buildings and existing ones if retrofitted, have significant potential to contribute to achieving SDGs in all regions while adapting buildings to future climate (IPCC-AR6-WGIII-Summary for Policy-makers. Pdf 5 MB, 2021).

K.-G. Kim (✉)
Seoul National University, Seoul, Korea
e-mail: kwigon@snu.ac.kr

K. Kim and C. Atkin (eds.), *Cross-Cultural Perspectives on Climate Change Adaptation*, Cities and Nature, https://doi.org/10.1007/978-3-031-50365-8_24

According to the report, the linkages between mitigation, adaptation and sustainable development involves:

- Accelerated and equitable climate action in mitigating, and adapting to climate change impacts which is critical to sustainable development.
- Trade-offs of individual options which could be managed through policy design. The Sustainable Development Goals (SDGs) adopted under the UN 2030 Agenda for Sustainable Development can be used as a basis for evaluating climate action in the context of sustainable development
- Human-induced climate change is a consequence of more than a century of net GHG emissions from unsustainable energy use, land use and land use change, lifestyle and patterns of consumption and production. Without urgent, effective and equitable mitigation actions, climate change increasingly threatens the health and livelihoods of people around the globe, ecosystem health and biodiversity.
- There are both synergies and trade-offs between climate action and the pursuit of other SDGs. Synergies and trade-offs depend on the development context including inequalities, with consideration of climate justice. They also depend on means of implementation, intra- and inter-sectoral interactions, cooperation between countries and regions, the sequencing, timing and stringency of mitigation actions, governance and policy design. Maximising synergies and avoiding trade-offs pose particular challenges for developing countries, vulnerable populations and indigenous peoples with limited institutional, technological and financial capacity, and with constrained social, human and economic capital. Trade-offs can be evaluated and minimised by giving emphasis to capacity building, finance, governance, technology transfer, investments and development and social equity considerations with meaningful participation of indigenous peoples and vulnerable populations.
- There are potential synergies between sustainable development and energy efficiency and renewable energy, urban planning with more green spaces, reduced air pollution and demand side mitigation including shifts to balanced, sustainable healthy diets. Electrification combined with low GHG energy, and shifts to public transport can enhance health, employment, and can elicit energy security and deliver equity.
- In industry, electrification and circular material flows contribute to reduced environmental pressures and increased economic activity and employment. However, some industrial options could impose high costs.
- Land-based options such as reforestation and forest conservation, avoided deforestation and restoration and conservation of natural ecosystems and biodiversity, improved sustainable forest management, agroforestry, soil carbon management and options that reduce CH_4 and N_2O emissions in agriculture from livestock and soil, can have multiple synergies with the SDGs.
- The sustainability of bioenergy and other bio-based products is influenced by feedstock, land management practice, climatic region, the context of existing land management and the timing, scale and speed of deployment.

- Ecosystem restoration and reforestation sequester carbon in plants and soil, and can enhance biodiversity and provide additional biomass, but can displace food production and livelihoods, which calls for integrated approaches to land use planning, to meet multiple objectives including food security. However, due to limited application of some of the options today, there are some uncertainties about potential benefits.
- Figure 1 presents a summary of the chapter-level assessment for selected mitigation options. Blank cells indicate that interactions have not been assessed due to limited literature. They do not indicate the absence of interactions between mitigation options and the SDGs. Confidence levels depend on the quality of evidence and level of agreement in the underlying literature assessed by the sectoral chapters. Where both synergies and trade-offs exist, the lower of the confidence levels for these interactions is used. Some mitigation options may have applications in more than one sector or system. The interactions between mitigation options and the SDGs might differ depending on the sector or system, and also on the context and the scale of implementation. Scale of implementation particularly matters when there is competition for scarce resources.

There is a strong link between sustainable development, vulnerability and climate risks. Limited economic, social and institutional resources often result in high vulnerability and low adaptive capacity, especially in developing countries. Several response options deliver both mitigation and adaptation outcomes, especially in human settlements, land management and in relation to ecosystems. However, land and aquatic ecosystems can be adversely affected by some mitigation actions, depending on their implementation. Coordinated cross-sectoral policies and planning can maximise synergies and avoid or reduce trade-offs between mitigation and adaptation (high confidence).

- Sustainable urban planning and infrastructure design including green roofs and facades, networks of parks and open spaces, management of urban forests and wetlands, urban agriculture and water-sensitive design can deliver both mitigation and adaptation benefits in settlements. These options can also reduce flood risks, pressure on urban sewer systems, urban heat island effects and can deliver health benefits from reduced air pollution. There could also be trade-offs. For example, increasing urban density to reduce travel demand could imply high vulnerability to heat waves and flooding.
- Land-related mitigation options with potential co-benefits for adaptation include agroforestry, cover crops, intercropping and perennial plants, thus restoring natural vegetation and rehabilitating degraded land. These can enhance resilience by maintaining land productivity and protecting and diversifying livelihoods.
- Restoration of mangroves and coastal wetlands sequester carbon, while also reducing coastal erosion and protecting against storm surges, thus, reduce the risks from sea level rise and extreme weather.
- Some mitigation options can increase competition for scarce resources including land, water and biomass. Consequently, these can also reduce adaptive capacity,

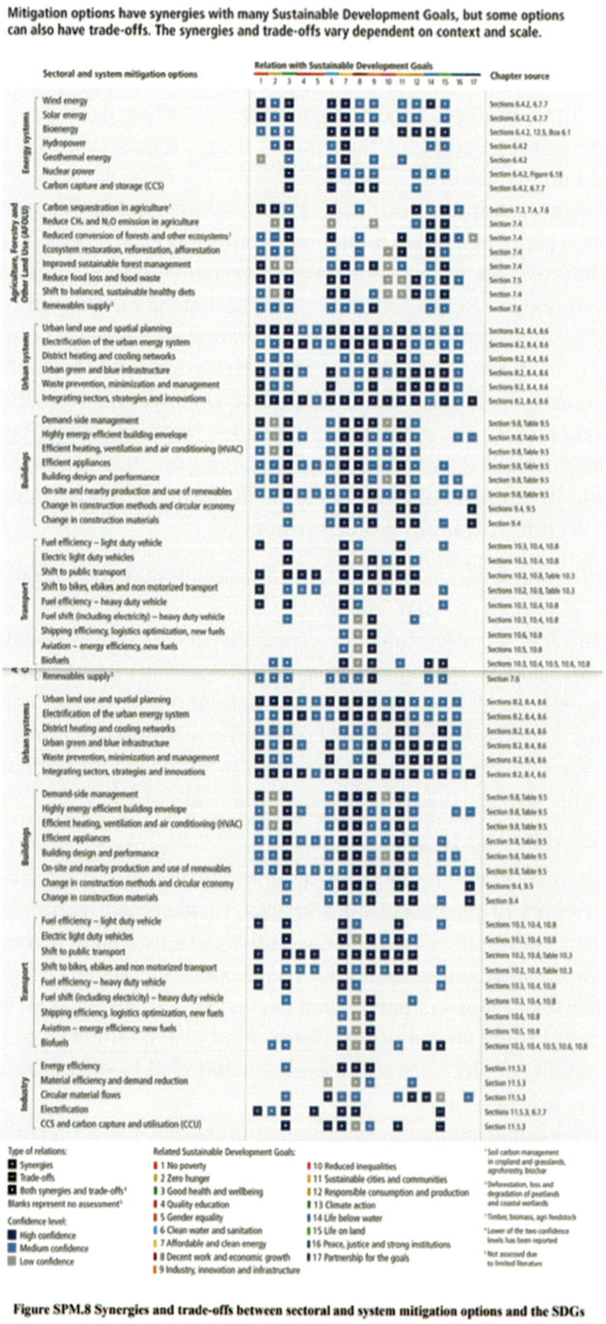

Fig. 1 Synergies and trade-offs between sectoral and system mitigation options and SDGs. *Source* IPCC (2022) Climate Change 2022: Mitigation of Climate Change, IPCC-AR6-WGIII-Fullreport.pdf, p. 54

especially if deployed at larger scale and with high expansion rates thus exacerbating existing risks in particular where land and water resources are very limited. Examples include the large-scale or poorly planned deployment of bioenergy, biochar and afforestation of naturally unforested land.

- Coordinated policies, equitable partnerships and integration of adaptation and mitigation within and across sectors can maximise synergies and minimise trade-offs and thereby enhance the support for climate action. Absence or limited resources in social and institutional systems can lead to poorly coordinated responses, thus reducing the potential for maximising mitigation and adaptation benefits, and increasing risk.
- Enhanced mitigation and broader action to shift development pathways towards sustainability will have distributional consequences within and between countries. Attention to equity and broad and meaningful participation of all relevant actors in decision-making at all scales can build social trust, and deepen and widen support for transformative changes.
- Countries at all stages of economic development seek to improve the well-being of people, and their development priorities reflect different starting points and contexts. Different contexts include social, economic, environmental, cultural or political conditions, resource endowment, capabilities, international environment and history. The enabling conditions for shifting development pathways towards increased sustainability will therefore also differ, giving rise to different needs.
- Ambitious mitigation pathways imply large and sometimes disruptive changes in economic structure, with significant distributional consequences, within and between countries. Equity remains a central element in the UN climate regime, notwithstanding shifts in differentiation between states over time and challenges in assessing fair shares. Distributional consequences within and between countries include shifting of income and employment during the transition from high to low emissions activities. While some jobs may be lost, low emissions development can also open more opportunities to enhance skills and create more jobs that last, with differences across countries and sectors. Integrated policy packages can improve the ability to integrate considerations of equity, gender equality and justice. Inequalities in the distribution of emissions and in the impacts of mitigation policies within countries affect social cohesion and the acceptability of mitigation and other environmental policies. Equity and just transitions can enable deeper ambitions for accelerated mitigation. Applying just transition principles and implementing them through collective and participatory decision-making processes is an effective way of integrating equity principles into policies at all scales, in different ways depending on national circumstances.
- Broadening equitable access to domestic and international finance, technologies that facilitate mitigation, and capacity, while explicitly addressing needs can further integrate equity and justice into national and international policies and act as a catalyst for accelerating mitigation and shifting development pathways.
- The consideration of ethics and equity can help address the uneven distribution of adverse impacts associated with 1.5 °C and higher levels of global warming, in all societies. Consideration of climate justice can help to facilitate

shifting development pathways towards sustainability, including through equitable sharing of benefits and burdens of mitigation, increasing resilience to the impacts of climate change, especially for vulnerable countries and communities, and equitably supporting those in need.

(Source: IPCC-AR6-WGIII-Summary for Policy-makers.pdf).

WRI (17 August 2022) has suggested advancing gender and social equity through nature-based solutions (NBS). DEWpoint identifies three types of co-benefits such as mitigation-development co-benefits, adaptation-sustainable development co-benefits and adaptation-mitigation co-benefits (DEWpoint 2011).

2 An Integrated Approach to Air, Earth, Fire and Water

In Fig. 14.2, three characteristics of flood risk management including avoidance, control and reduction can be seen from the mitigation perspective.

Paris Climate Action Plan (2018) describes how to make sure that socio-economic issues are prioritised as part of Paris's climate efforts (Fig. 2). The city conducted a study on the socio-economic impacts of the energy, and climate and ecological transition in order to specifically identify the impacts of the Climate Plan, whether positive or negative. A large of individual, economic, social and political impacts were analysed for the main actions in the Climate Plan. Their effects on employment, well-being, equity and balance across the city were also examined.

The study showed that through its actions, the Climate Plan strengthens social links between citizens and helps reduce insecurity related to energy, food, housing and transport. The goal of these measures is to reduce social divides and encourage Parisians to adopt more sustainable lifestyles.

3 Implications of Case Studies for Benefits for SDGs

Figure 3 depicts digitally streamlined steps of combined climate innovation planning of adaptation (9 steps) and mitigation (13 steps). In the middle of the diagram, climate-friendly and resilient path to climate innovation hand in hand describes forward-looking, advanced climate innovation planning towards a new "green" growth and development agenda for climate innovation.

Figure 3 is a model example of multilayer climate solutions to meaningfully reduce GHG emissions, strengthen community to climate change and promote green development.

Appendix B may help to digitalize the streamlined steps shown in Fig. 14.3.

The case studies described in the previous chapters showed numerous measures (structural and non-structural) forward to adapt and to build resilience. They present

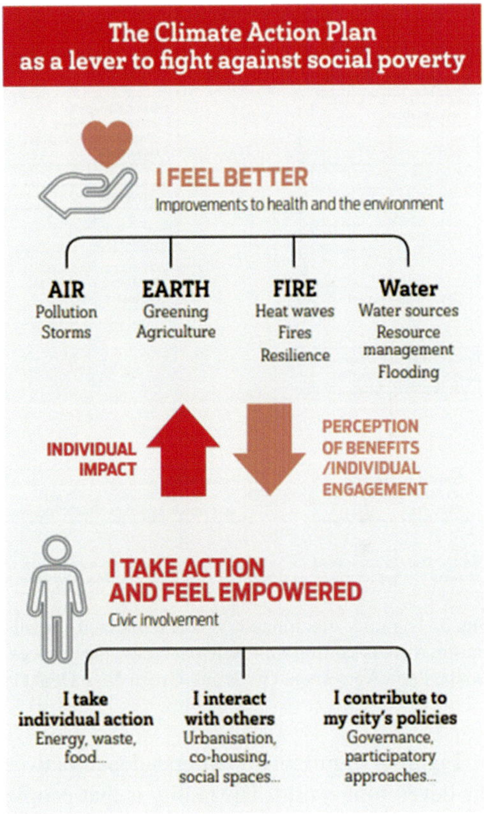

Fig. 2 Climate action plan as a lever to fight against social poverty. *Source* City of Paris (2018) Paris Climate Action Plan: Towards a carbon neutral city and 100% renewable energies, p. 71

a detailed blueprint for action by offering innovative, adaptation and resilience ideas, solutions and policy recommendations in the context of different local variables.

As these measures protect people and communities from floods and other impacts, they also help lift people out of poverty, raise incomes and living standards, fight diseases, create jobs, reduce inequality, mitigate the risk of conflicts and give voice to the vulnerable. These realisable results, in turn, further increase resilience to climate impacts and contribute to achieve SDGs.

Table 1 shows how the adaptation and mitigation measures and benefits can be related to SDGs as an illustration.

More in specific, PBL (2022) discusses the SDGs related to water. Major climate adaptation challenges include water security issues with respect to increases water scarcity, drought and flood risk, and increasing water temperatures affecting water quality and biodiversity. With its link to human health and well-being, clean water and sanitation, food production, sustainable cities and communities and the quality of ecosystems, water is directly and indirectly also linked to many of the SDGs (PBL Netherlands Environmental Assessment Agency 2022). Figure 4 depicts the SDGs related to water.

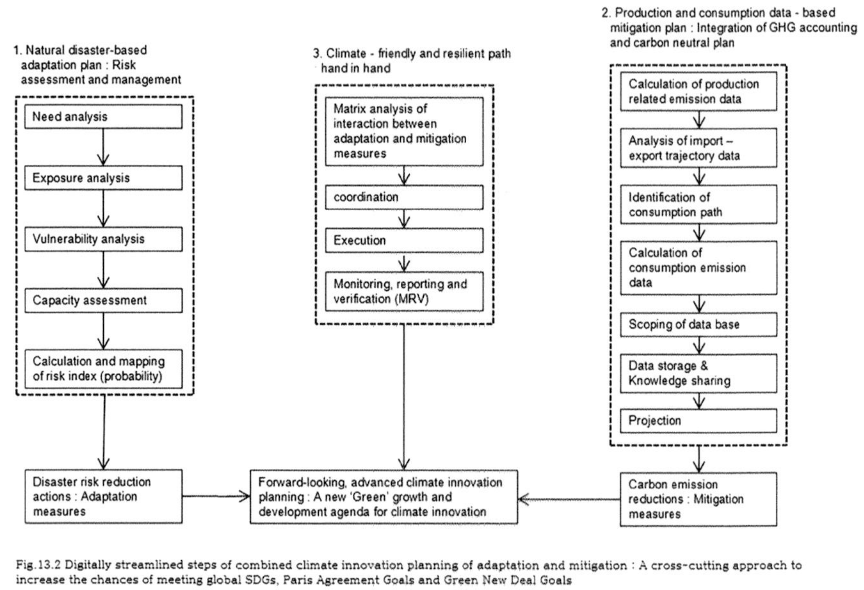

Fig. 3 Digitally streamlined steps of combined climate innovation planning of adaptation and mitigation: A cross-cutting approach to increase the chances of meeting global Sustainable Development Goals, Paris Agreement Goals and Green New Deal Goals

Figure 5 highlights interdependence between climate solutions and real lives in the developing world. The reality is that people live their lives in integrated, holistic ways, and therefore it doesn't make sense to think that we can address climate change in a silo, separate from, for instance, addressing poverty. The figure shows global potential emission reduction or carbon storage of each solution group (Patterson 2022).

Table 1 Adaptation and mitigation measures and benefits for the SDGs: an illustration

Sector	Sub-sector	Measures	Benefits	SDG 1	2	3	4	5	6	7	8	9	10	11	12	13	14	15	16	17
Adaptation	Urban development	• Structural measures • Non-structural measures (Refer to Sect. 3, Chap. 4 of this book for structural and non-structural measures).	• To protect People and communities from floods	✗	✗	✗	✗	✗	✗	✗	✗	✗	✗	✗	✗	✗	✗	✗	✗	✗
			• To lift people out of property, hunger and undernourishment	✗						✗										
			• To give voice to the vulnerable		✗															
			• To raise incomes and living standards																	
			• To fight diseases										✗							
			• To create jobs																	
			• To reduce inequality													✗				
			• To mitigate the risk of conflicts																	✗

(continued)

Table 1 (continued)

Sector	Sub-sector	Measures	Benefits	SDG 1	2	3	4	5	6	7	8	9	10	11	12	13	14	15	16	17
			• To trade carbon credit								✗									
Mitigation	• Urban Systems • Buildings	• Sustainable urban planning and infrastructure design • Electrification • Circular economy • Land-based options • Bioenergy and bio-based products • Ecosystem restoration and reforestation • Land-related mitigation options • Considerations of equity, gender equality and Justice • Broadening of equitable access to finance and technologies • Considerations of ethics	• To reduce flood risks • To reduce urban heat island effects • To deliver health benefits • To sequester carbon • To maintain land productivity • To protect and diversify livelihoods • To build social trust • To deepen and widen support for transformative changes • Equitable sharing of benefits and burden of mitigation				✗													✗

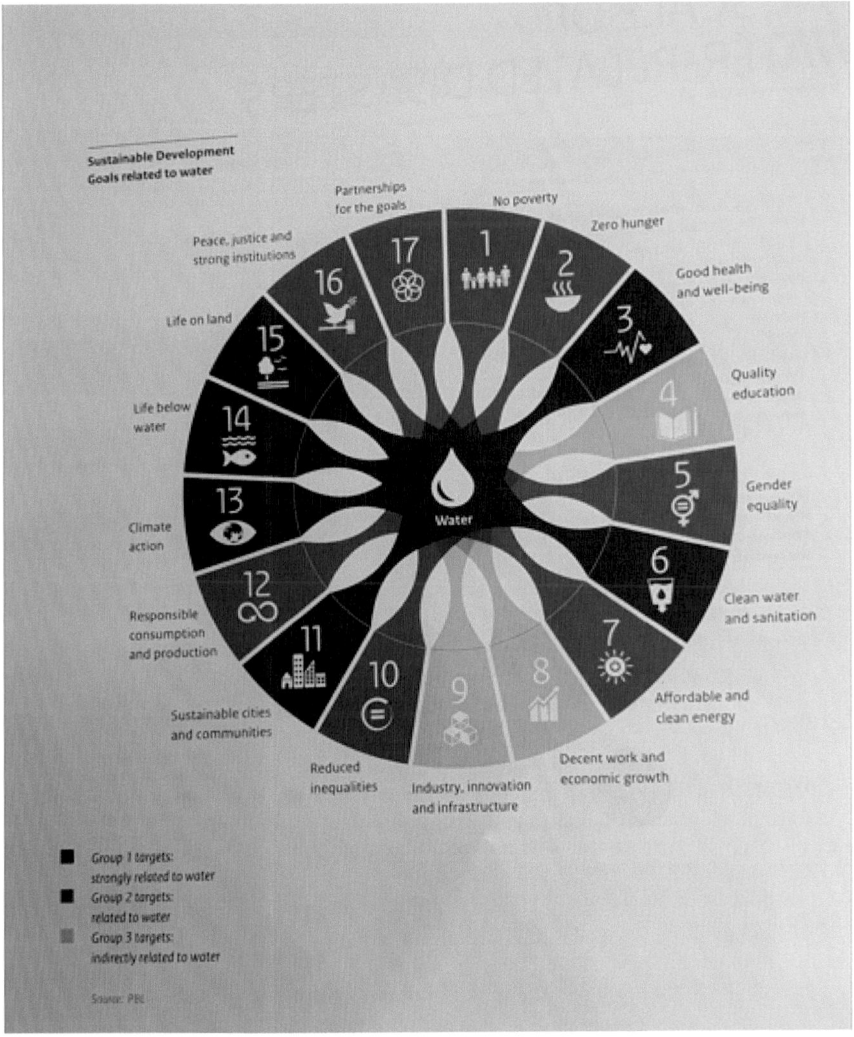

Fig. 4 Sustainable Development Goals related to Water. *Source* PBL Netherlands Environmental Assessment Agency, Hague, p. 13

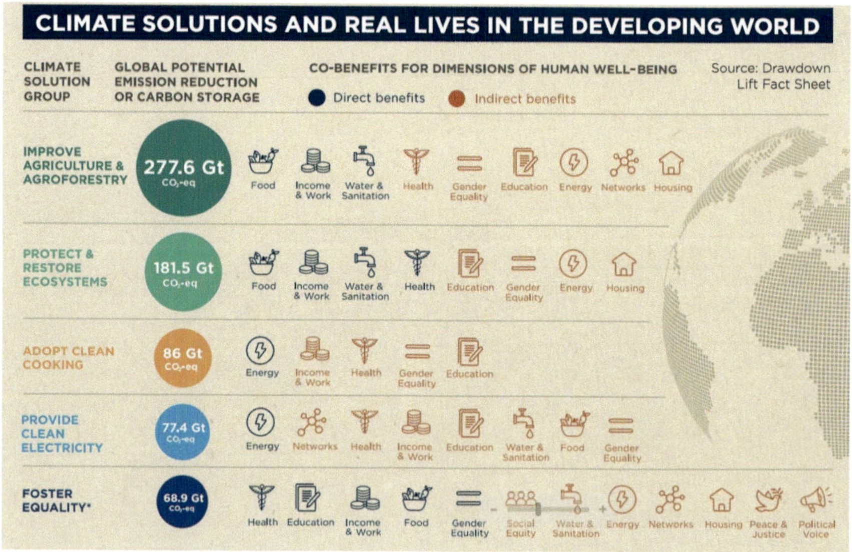

Fig. 5 Interdependence between climate action and human well-being. *Source* Patterson, K P (2022) "We cannot address climate change in a silo." OPEC FUND QUARTERLY 3/2022, The OPEC Fund for International Development, pp. 38–41

References

DEWPoint-A0406-Jan 2011-Co-Benefits-of-adaptation-vl-l.pdf, Co Benefits of adaptation, mitigation and development: ICF background paper prepared for DfID (2011)

Patterson KP (2022) We cannot address climate change in a silo. OPEC QUARTERLY 3/2022, The OPEC Fund for International Development, pp 38–41

PBL Netherlands Environmental Assessment Agency (2022) The geography of future water challenges, Hague

The Critical Role of Community-Driven Adaptation and Resilience Building: Bright Spots and Lessons from the State of California

Catherine Atkin and Saharnaz Mirzazad

Abstract Across the globe policymakers and institutional stakeholders see that approaches where outsiders deploy top-down solutions for communities facing complex local challenges and where climate change, poverty and weak systems overlap, consistently fail to create the desired impact. While public engagement is often part of local processes, recognition of the importance and power of community voice and agency to address power imbalances and systems of exclusion requires deeper and more fundamental shifts in the design of public initiatives. Governments and stakeholders across the world need to rethink the role of communities in problem identification and solution making and accelerate reforms and investments to center community voice and agency. California's Transformative Climate Communities program is a powerful example of government valuing participatory decision making and community-led resilience planning and action. After an introduction to community engagement in the context of key California adaptation and resilience policies, the chapter showcases California's Transformative Climate Communities program, features the initiatives on the ground in cities of Fresno and Stockton, and offers concrete advice for other governments as they design and undertake similar initiatives.

Keywords Community-led Adaptation Planning and California · Climate Resilience and Participatory Governance · California Transformative Climate Communities Program · Adaptation and community engagement · Locally led adaptation and California · Collaborative Governance · Spectrum of Public Participation and California

---

The views expressed in this chapter do not necessarily represent the views of the Governor's Office of Planning and Research or the State of California.

C. Atkin (✉)
Stanford CodeX Climate Data Policy Initiative, Stanford, USA
e-mail: catherine.atkin@codex.stanford.edu

S. Mirzazad
California Governor's Office of Planning and Research, Sacramento, USA

K. Kim and C. Atkin (eds.), *Cross-Cultural Perspectives on Climate Change Adaptation*, Cities and Nature, https://doi.org/10.1007/978-3-031-50365-8_25

1 Overview

California, the United States' most populous state in the USA and the 5th largest economy in the world, is a powerful laboratory for innovation and learning about how to address critical challenges that undermine the sustainability of diverse communities. One of these challenges is the devastating impacts to people, ecosystems and the economy that climate change-related hazards—in the form of drought, wildfires, flooding and sea level rise—are having across California. Given this, the state is continuing its history of global environmental leadership with the development of comprehensive approaches to climate adaptation and resilience building that provide powerful insights and best practices for the USA and countries around the world.

State and local governments work in California are increasingly focused on the imperative that systems must be reformed, and programs designed to ensure that all communities of California experience equitable outcomes. These efforts are guided by the reality that California is a state with wide disparities based on income and race and that negative impacts of climate change are felt most acutely in low-income communities with high concentrations of people of color where there is a long histories of disinvestment, disenfranchisement, and structural racism. This means not only creating policies and funding that ensures that historically marginalized communities receive the targeted attention and resources required to address long standing inequities but ensuring they also have a voice in defining and meeting their own needs and aspirations.

California's Transformative Climate Communities (TCC) program is designed to support community-led transformation through the development and infrastructure projects that achieve major environmental, health, and economic benefits in California's most disadvantaged communities and is a powerful example of government doing business differently. California's 6-year experience in implementing the TCC program provides rich insights into the state's effort to move from community collaboration to more fully enfranchising community voices in community-led resilience planning and action that can inform similar efforts by governments across the USA and the globe.

The chapter begins by introducing a framework to evaluate community engagement as well as reviewing the research on community-driven models. After an introduction to key California adaptation and resilience policies, the chapter showcases the Transformative Climate Communities program, features the initiatives on the ground in cities of Fresno and Stockton, and offers concrete advice for other governments as they design and undertake similar initiatives.

2 The Continuum of Community Engagement

As the value and importance of community engagement gets more attention and focus, there is still confusion about what community engagement means and what are the strategies that must be implemented to achieve the impacts desired.

The International Association for Public Participation's "IAP2 Spectrum of Public Participation" IAP2 Spectrum of Public Participation (2004) is a framework that has been used across multiple contexts including the development of adaptation and resilience initiatives as a way for local governments and stakeholders to be intentional about the kind of community engagement they desire. Understanding community engagement as a continuum defined by the level of decision-making power affected communities have—from traditional inform and consult approaches to those that allow for shared leadership and community-driven decision making—is a useful construct.

Traditionally, most efforts at community engagement in the context of adaptation planning focus on gathering the interests, insights, and feedback of community members to inform the initial plan development. At the same time there is increasing recognition of the limitations of this approach and need to support community-led climate adaptation and resiliency building efforts during both the planning and implementation phases of projects. California's Transformational Climate Communities program was designed to be community led and to address traditional power inequities by incorporating communities in all facets of the process, supporting needed capacity building and giving communities final decision-making authority (Fig. 1).

3 Summary of equitable planning consideration in the 2020 adaptation planning guide

When communities are involved in climate risk management their role is often limited to the "inform" or "consult" approaches described in Fig 1. There is robust evidence that this extractive approach can leave communities feeling disempowered and disconnected from the process (Dias et al. 2018). The research on the efficacy and impact of approaches where community is enfranchised and given more decision making authority is robust with researchers identifying greater success in achieving outcomes, reduced vulnerability, and better community outcomes (Johnston et al. 2022). At the same time, there is a recognition that more research needs to be done to identify the key practices and program design elements of community-led initiatives (Geekiyanage et al. 2020).

Leading researchers in the United Kingdom performed a meta-analysis based on a structure literature review to evaluate the current challenges and barriers to community-driven decision making that can be used to assess the design and evaluation of community driven initiatives including California's Transformative Climate

IAP2 Spectrum of Public Participation

IAP2's Spectrum of Public Participation was designed to assist with the selection of the level of participation that defines the public's role in any public participation process. The Spectrum is used internationally, and it is found in public participation plans around the world.

INCREASING IMPACT ON THE DECISION →

	INFORM	CONSULT	INVOLVE	COLLABORATE	EMPOWER
PUBLIC PARTICIPATION GOAL	To provide the public with balanced and objective information to assist them in understanding the problem, alternatives, opportunities and/or solutions.	To obtain public feedback on analysis, alternatives and/or decisions.	To work directly with the public throughout the process to ensure that public concerns and aspirations are consistently understood and considered.	To partner with the public in each aspect of the decision including the development of alternatives and the identification of the preferred solution.	To place final decision making in the hands of the public.
PROMISE TO THE PUBLIC	We will keep you informed.	We will keep you informed, listen to and acknowledge concerns and aspirations, and provide feedback on how public input influenced the decision.	We will work with you to ensure that your concerns and aspirations are directly reflected in the alternatives developed and provide feedback on how public input influenced the decision.	We will look to you for advice and innovation in formulating solutions and incorporate your advice and recommendations into the decisions to the maximum extent possible.	We will implement what you decide.

Fig. 1 IAP2 Spectrum of Public Participation, International Association for Public Participation

Communities. In their 2020 International Journal of Disaster Risk Reduction article: "Assessing the state of the art in community engagement for participatory decision making in disaster risk-sensitive urban development," the authors identified 48 barriers associated with context and community capacity, infrastructure for community engagement, and current decision-making processes. They also found that "[B]y synthesizing the current research, the study found that these barriers can potentially be overcome through attitude transformation and capacity building of both community and professionals, investment in community engagement, and changes to present stakeholder engagement processes and policies" (Geekiyanage et al. 2020) (Fig. 2).

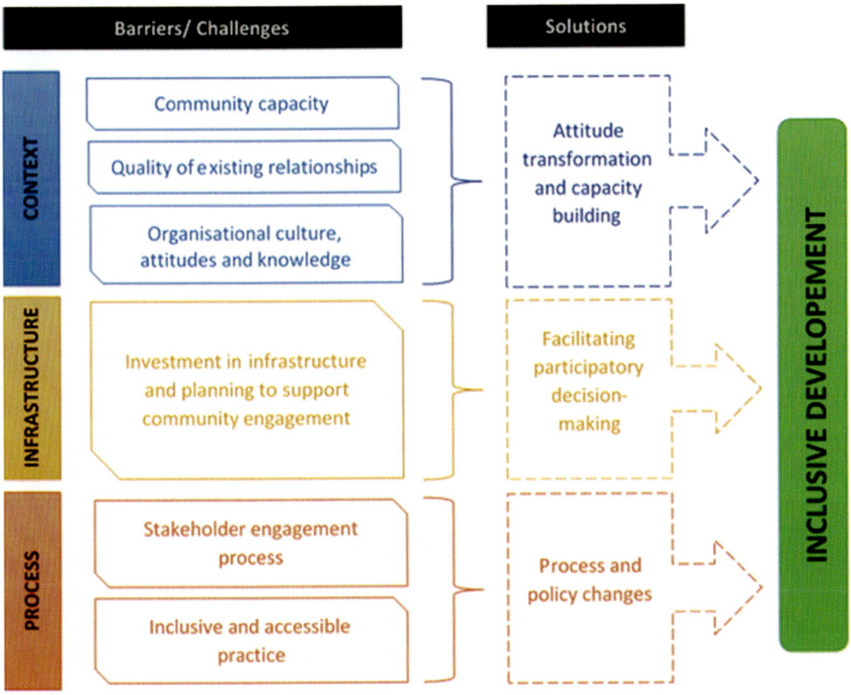

Fig. 2 Conceptual model for achieving inclusive developments through addressing associated barriers and challenges. *Source* https://www.sciencedirect.com/science/article/pii/S2212420920313492?via%3Dihub#bib29

4 How Equity and Community Engagement Is Built into California Adaptation and Resilience Policies

The launch of California's Transformative Climate Communities Program as an experiment in government driven community-led program design was just one of a range of state-level adaptation and resilience policy and practice changes of the past decade. This section outlines some of the other key institutional building blocks—including policies, tools and frameworks—that are in place in California and describes how equity and community are addressed.

4.1 California Climate Adaptation Strategy and Coordination

The California Climate Adaptation Strategy, mandated by Assembly Bill 1482 (Gordon 2015), links together the state's existing and planned climate adaptation efforts and is organized around outcome-based priorities to (1) strengthen protections for climate vulnerable communities, (2) bolster public health and safety to protect against climate risks, (3) accelerate nature-based solutions and strengthen climate resilience of natural systems, (4) make decisions based on the best available climate science, (5) partner and collaborate to leverage resources, and (6) build a climate resilient economy (https://climateresilience.ca.gov/priorities/strengthen-protections/build-capacity.html).

It is significant that the state identifies a focus on climate vulnerable communities as a top priority for the state. The plan also has as a success metric associated with this priority of "increased participation and leadership in climate adaptation decision-making processes and projects by climate vulnerable communities and community organizations" and identifies the Governor's Office of Planning and Research—Strategic Growth Council, Environmental Protection Agency, Natural Resources Agency, and Coastal Commission as agencies responsible for achieving this outcome (https://climateresilience.ca.gov/priorities/strengthen-protections/build-capacity.html). Following the Adaptation Strategy, SB 246 established the Integrated Climate Adaptation and Resiliency Program (ICARP) in 2016 to serve as a cornerstone in California's approach to climate adaptation and resilience. ICARP's mandate is to develop a unified and comprehensive response to climate change impacts, effectively coordinating state and local efforts to accelerate action (https://opr.ca.gov/climate/icarp/).

4.2 State Level Government Agency Resilience Planning and Policy

Executive Order B-30-15 directed State agencies to integrate climate change into all planning and investment, including accounting for current and future climate conditions in infrastructure investment (https://www.library.ca.gov/wp-content/uploads/GovernmentPublications/executive-order-proclamation/39-B-30-15.pdf). The Office of Planning and Research convened a Technical Advisory Group with members from nearly every State agency, local and regional governments, non-governmental and community-based organizations, and the private sector. One element of their work was the production of "Planning and Investing for a Resilient California: A Guidebook for State Agencies" that provides high-level guidance on what future conditions to plan for and how State agencies should approach planning differently in light of a changing climate (https://opr.ca.gov/docs/20180313-Building_a_Resilient_CA.pdf).

The Executive Order also included a first in the nation mandate that because "climate change will disproportionately affect the state's most vulnerable people," all "State agencies' planning, and investments shall…protect the state's most vulnerable populations." The guidebook's four-step process outlines specific actions agencies should take to strengthen community voice and agency and also includes a Resiliency

Guidebook Equity Checklist to "assist State agencies to ensure that plans and investments identify and protect the State's most vulnerable populations" (https://opr.ca.gov/climate/docs/20180312-Equity_Checklist.pdf). Included in this checklist are the following questions focused on community engagement and decision making:

Have you designed a process to collaborate with vulnerable populations?

- Which best practices for meaningful engagement are you implementing? (see Community Engagement Best Practices)
- Are you sharing as much decision-making power as feasible (see the Spectrum of Public Participation)?
- Is there a need for formal agreements to define collaboration, and if so how will you develop them (see example in Community Engagement Best Practices)?

4.3 Local General Plan Requirements

All cities and counties in California are required to develop a General Plan, each local government's blueprint for meeting the community's long-term vision for the future. The Office of Planning and Research has the responsibility to adopt and periodically revise the State General Plan Guidelines that cities and counties must use to develop and update their plans. As a result of legislation (SB 379 and SB 1035) California requires local cities and counties to update the safety elements of their general plans to address climate adaptation and resilience as they are relevant to their jurisdiction. The General Plan Guidelines include a Climate Change Adaptation and Resilience section of the Safety Element and requires that local jurisdictions must include a vulnerability assessment identifying the risks that climate change poses to the local jurisdiction, a set of goals, policies, and objectives based on a vulnerability assessment for the protection of the community and a set of feasible implementation strategies to carry out the goals, policies, and objectives (https://opr.ca.gov/docs/OPR_COMPLETE_7.31.17.pdf).

While the General Plan Guidelines include modest requirements for community engagement, the Community Engagement and Outreach section does outline best practices including the use of the continuum of community engagement framework and highlights examples of locales that have moved to a more community-led—or empowerment—approach (https://opr.ca.gov/docs/OPR_COMPLETE_7.31.17.pdf). The Climate Change Adaptation and Resilience section of the Safety Element requires local governments to use the Adaptation Planning Guide developed by the Governor's Office of Emergency Services in collaboration with the Office of Planning and Research to analyze risk and address goals and measures for climate change adaptation and resilience. The Adaptation Planning Guide centers equity and community engagement with a focus on collaborative and empowering approaches to community involvement. Table 1 shows how they embed equitable planning approaches into the four phases of their planning guide (https://resilientca.org/topics/climate-equity/#equitable-planning-and-community-engagement).

Table 1 Summary of equitable planning considerations in the 2020 California's Adaptation Planning Guide

Phase 1 Explore, Define, Initiative	Phase 2 Assess Vulnerability	Phase 3 Define Adaptation Frameworks & Strategies	Phase 4 Implement, Monitor, Evaluate, and Adjust
• It is essential to bring a diversity of stakeholders when assembling the core planning team. These organizations and individuals have lived experiences, local knowledge, and ideas that the local government may not know or anticipate. Community members most affected by climate issues, when invited to lead on solutions, can drive more effective implementation. • From the start, it is important to empower community participation to create a sense of co-ownership in the planning process. Outreach and engagement should be considered carefully and budgeted appropriately at this stage. • It is helpful if all core planning team members are trained in outreach before engaging stakeholders. Topics such as cultural competence and humility, racism, and systems of injustice may be unfamiliar in the climate risk context, but it is critical to understand those power dynamics when working with vulnerable populations. • It is also useful at this stage to develop internal learning goals and questions to create a space for continuous improvement on equity goals in process and implementation.	• The goal of outreach and engagement in Phase 2 is to collaborate with the community members both within the organization and externally to identify neighborhood strengths, challenges, assets, and climate change effects. • Vulnerability assessments should be driven by community priorities and reflect knowledge and innovative ideas from different population groups and neighborhoods. • Examples of engagement here include interviews and focus groups, storytelling approaches, community-based participatory research, asset mapping, and design strategies. • Outreach and engagement in Phase 2 ensures that community members have opportunities to share their expertise and provide valuable lived experiences of climate vulnerabilities and make sure the assessment more accurately reflects specific local conditions. A robust balance of public contributions towards both quantitative and qualitative data helps ensure a more thorough and accurate vulnerability analysis.	• During the development of adaptation strategies, the project team and community members should think about how strategies can be realistic, aligned with equity principles, and how they can be transformative, addressing social inequities as well as climate change impacts. • There are many creative ways to involve stakeholders in brainstorming and drafting adaptation strategies. This outreach is meant to build on information about climate change effects and previous involvement with the vulnerability assessment to develop solutions. • The project team should be prepared to receive extensive feedback on the adaptation strategies and to interpret how community members understand and value each of the strategies presented. • The entity leading the outreach process should also post the outcomes of voting and workshop summaries online or in other publicly accessible locations for transparency.	• Each step of Phase 4 incorporates outreach and provides a list of sample actions. Long-term implementation cannot be effective without close collaboration with, and direct feedback from, the community. • It is important to rely on frontline, vulnerable, and tribal community representatives to determine implementation options, and leverage their local expertise to carry out the strategies in partnership. • Community members and organizations often see and experience firsthand when actions are not working or are causing harm. Including these groups in assessing and monitoring data and choosing next steps is critical to ongoing effectiveness. This ensures that community members are partners if the results of monitoring and evaluation require a change in approach. • This phase should involve collaboration with the community to update strategies and program implementation based on lessons learned from monitoring, and regular measurement of defined equity-related indicators.

4.4 *Regional and City Adaptation and Resilience Infrastructure and Initiatives*

California is home to robust regional and local sustainability related planning and action efforts. Cities across the state have developed comprehensive adaptation plans and are bringing multiple stakeholders, including community members, to the table. Regional structures like the Alliance of Regional Collaboratives for Climate Adaptation (Fig. 3) brings together collaborative networks from across the state to build regional resilience and acts as a powerful driver of collaboration as well as a learning and practice exchange platform. The State California Coastal Commission undertakes local and regional efforts along California's coast focused on resilience in the face of sea level rise. National networks like the Urban Sustainability Directors Network have a strong presence in California and have taken a leading role in driving a focus on equity and community voice with their Guide to Equitable, Community-Driven Climate Preparedness Planning Guide (https://www.usdn.org/uploads/cms/documents/usdn_guide_to_equitable_community-driven_climate_preparedness-_high_res.pdf) and ICLEI USA has developed a Sustainable Communities Adaptation Suite of tools that includes concrete guidance on how to center community voice (https://icleiusa.org/wp-content/uploads/2022/09/Adaptation_-First-Steps_USCP.pdf).

5 California's Transformative Climate Communities Program: Unlocking the Power of Community-Led Decision Making

As evidenced in the previous section, California has put in place a robust set of policies, frameworks, and tools along with making multi billion dollar in investments to support adaptation and resilience planning. While a review of the policy elements reflects a strong value for equity and centering-community, the balance and tension between what the State wants to advance and what it can require on the local level are real. In this context, California has used its Transformative Climate Communities (TCC) program to embrace and fund an approach that values the voice and agency of communities as fundamental to addressing disparities and advancing social and economic justice while also advancing climate mitigation and adaptation goals.

Even though there has been increased attention and agreement that community-driven approaches to adaptation and resilience planning and action hold powerful promise, there have been relatively few concrete examples of what this looks like on the ground. In launching the Transformative Climate Communities Program in 2017, the Governor, legislature, and community activists have created a groundbreaking initiative that is delivering positive results in California communities and acting as a national model for advancing community-led, place-based, and integrated resilience and sustainability action.

Stockton and Fresno are two of the eight Transformative Climate Communities. A deeper look at what transpired in these two communities provides the reader with rich insights into the power of community-led governance. It is important to note that the TCC program takes a comprehensive approach to community change where adaptation, mitigation, and community development goals are interconnected. While community projects include climate hazard reduction measures including tree planting, parks, and community gardens that reduce the negative impacts of extreme heat and flooding they also respond to a range of community needs that build broader community resilience and adaptive capacity.

Fig. 3 Alliance of Regional Collaboratives for Climate Adaptation—Map of Regional Networks ARCCA

5.1 Overview to California's Transformative Climate Communities Program

The Transformative Climate Communities (TCC) Program was launched in 2017 as a result of California legislation (AB 2722), a bill that was designed and co-sponsored by the Greenlining Institute and California Environmental Justice Advocates, two powerful advocacy organizations focused on environmental justice and community development. Administered by the Strategic Growth Council (SGC) in partnership with the California Department of Conservation, The TCC Program makes investments in communities that face significant environmental burdens and climate impacts and "empowers the communities most impacted by pollution to choose their own goals, strategies, and projects to reduce greenhouse gas emissions and local air pollution and enact transformational change" (http://www.leginfo.ca.gov/pub/15-16/bill/asm/ab_2701-2750/ab_2722_bill_20160914_chaptered.html). The TCC program was originally funded by the California Climate Investments initiative, it is now funded from the State's General Fund and supports the planning and implementation of neighborhood-level plans that both reduce greenhouse gas emissions and provide other co-benefits including health, environmental, and economic benefits while at the same time minimizing the risk of displacement.

The TCC program has garnered significant national and international attention as it integrates three core design elements:

1. A place-based and community-driven approach toward building resilient and flourishing neighborhoods,
2. A comprehensive systematic approach to infrastructure development and implementation that is centered on co-benefits and efficiency, and
3. A focus on bringing together and developing cross-sector partnerships.

Eight communities have been awarded over $230 million in TCC implementation grants in communities across California including South Stockton, East Oakland, Eastside Riverside, Sacramento's River District, Northeast San Fernando Valley (Pacoima, Los Angeles), Watts (Los Angeles), Downtown Ontario, and Fresno's Southwest, Downtown, and Chinatown neighborhoods. Implementation grants fund a range of project elements in each community including affordable housing units close to transit, electric buses, EV charging sites, bike lanes, solar panels, tree planting, parks, community gardens, workforce trainees, workforce job placement, displacement avoidance initiatives and involve close to 7000 community members. (https://greenlining.org/wp-content/uploads/2021/10/Fighting-Climate-Change-and-Redlining-with-Transformative-Climate-Communities-Final-Report.pdf).

Equity, Community Voice, and Leadership

Creating new approaches to power and partnership with residents in local communities is not a means to an end but a fundamental goal with community members involved at all stages and having decision-making authority. The bill that created TCC, Assembly Bill 2722, ensures that equity and community voice are "baked into" the program by requiring that the Strategic Growth Council "award grants for projects that demonstrate community engagement in all phases" (http://www.leginfo.ca.gov/pub/15-16/bill/asm/ab_2701-2750/ab_2722_bill_20160914_chaptered.html). The programmatic guidelines that have been developed since the launch of the program have further articulated what community-led decision making looks like and created concrete guidance on programmatic elements to advance this goal. Systematic and intentional community engagement is identified as one of six transformative elements in the TCC program guidelines (https://sgc.ca.gov/programs/tcc/resources/). Key programmatic drivers of equitable and empowering community include:

Collaborative Stakeholder Structure: TCC program guidelines state that "[A]applications will be evaluated on the degree to which they incorporate community leadership" and require that applicants outline how communities are going to be included in the leadership structure over time stating that" (https://sgc.ca.gov/programs/tcc/resources/) TCC funded projects must develop Collaborative Stakeholder Structures (CSS) that bring together a broad array of residents, community-based organizations, nonprofit organizations, local government and other stakeholders to build a shared vision of transformation for their community.

Displacement Avoidance: The TCC focuses on community centered investment to increase the quality of life for residents. However, as history indicates, increased amenities may reduce affordability which can undermine community interest in participating in the process. Realizing this core challenge, as one of its main pillars, TCC requires proactive policy and investment to make sure that current residents can afford to live in their future neighborhood. This builds trust with the residents that the investment is to benefit them, not future more well-off residents.

Community Engagement Plan and Recommended Programmatic Activities: The TCC guidelines identifies four transformational elements—including community engagement, adaptation and resilience, workforce development, and displacement avoidance—and requires that a plan be developed for each element, and holds communities accountable for implementing these plans. The Community Engagement Plan is a core feature of the work in each community and the guidelines give concrete examples of recommended activities (https://sgc.ca.gov/programs/tcc/resources/).

5.2 TCC Community Spotlight—Transform Fresno

The Basics

Fresno, located in the San Joaquin Valley in the center of California, is the state's fifth-largest city. It is culturally and ethnically diverse with a majority Latino/Hispanic population and is home to a diversity of immigrant communities. The city has long been subject to environmental pollution and economic disparities and residents suffer from poor health outcomes like cardiovascular disease and asthma. While it has long experienced warm temperatures, climate-related extreme heat is on the rise.

Transform Fresno is a collaborative that came together with stakeholders from a set of Fresno neighborhoods that disproportionately face environmental and economic barriers and challenges. After a community-led process, in 2018 Transform Fresno was awarded $66.5 million dollars in implementation funding. These funds are being used for a range of projects that advance GHG reduction goals while also implementing high-value measures to reduce the negative impacts of extreme heat and

Table 2 TCC recommended activities to ensure meaningful community engagement

Recommended activities to ensure meaningful community engagement	
Activities to Inform Community Stakeholders and to Solicit Stakeholder Input	• Public workshops/meetings • Door-to-door canvassing • House meetings • Established website and/or social media • Distributed flyers or other printed materials • Outreach to existing community groups • Surveys • Focus Groups • Community Working Group
Activities to Engage Community Stakeholders in Development of TCC Proposal	Design charrettes • Community-based participatory research • Participatory budgeting • Convene advisory body, community working group or shared decision-making body • Establish website and/or social media • Community benefits agreements • Additional activities to ensure community stakeholders have an opportunity to influence the TCC Proposal development
Activities to Ensure Community Engagement During Implementation of TCC Project	Public workshops/meetings • Door-to-door canvassing • House meetings • Established website and/or social media • Surveys • Focus groups • Sub-contract with community-based organizations to conduct outreach • Allocate staff positions focused on community engagement • Steering committee relative to the participatory budget established • Advisory body, community working group, or shared decision-making body • Additional activities to provide community stakeholders an opportunity to influence the TCC Proposal development • Maintain community engagement throughout the TCC Implementation Plan

supporting community climate resilience. Funded projects include the greening of public spaces, active transportation, affordable housing, and food waste prevention.

Local Drivers of Community Voice and Leadership

Governance and Collaborative Stakeholder Structure: The structure of the initiative requires regular collaborative meetings and includes a 164-member Community Steering Committee. From the effort's inception participants were part of evaluating community needs, developing projects and solutions with options for implementation projects being voted on by the members that were then sent to the state for review. Led by the City of Fresno, the SCC includes 12 project partners, sixteen Outreach and Oversight Committee members, three community partner organizations and residents, business owners, and property owners in the project area.

Community Engagement Plan (CEP) and Strategies: The City of Fresno along with nonprofit partners and a consultant led the development of the CEP with the goal that "residents, workers, business owners, property owners, and other stakeholders in Downtown, Chinatown, and Southwest Fresno to be knowledgeable of Transform Fresno projects, activities, events, and efforts, and to enable these diverse stakeholders to be active participants in all areas of project planning and implementation. The plan outlines specific methods for the City of Fresno, community partners, and project partners to follow to ensure impactful information sharing and communication, participation, and documentation" (https://innovation.luskin.ucla.edu/wp-content/uploads/2022/06/Transform-Fresno-2022-Progress-Report.pdf). Examples of engagement activities include an innovative participatory budgeting process, a commitment to multilingual translation of meetings, over 150 informational workshops, and an annual stakeholder summit (Fig. 4).

Fig. 4 Residents of Southwest Fresno vote on the Transform Fresno plan

5.3 TCC Community Spotlight—Stockton Rising

The Basics

Stockton is a port city along the San Joaquin River in California's Central Valley. The TCC Project Area focuses on South Stockton where approximately 50,000 residents are primarily Latinx, African American, and Asian. Residents in this community experience high levels of poverty and unemployment and face significant environmental burdens and climate impacts such as extreme heat, drought and inland flooding.

Rise Stockton is a coalition of community groups that came together around the TCC Planning and Implementation Grants including faith-based, community organizing, health, and other nonprofits. After a 2018 Planning Grant, in June 2020, Stockton partners won a $10.8 million TCC Implementation Grant for five projects that focus on community-identified needs and priorities and received another $24 million in 2022 for projects to improve streetscapes, provide solar power, ensure energy and water efficiency upgrades, increase food security, and build out the urban tree canopy (https://greenlining.org/wp-content/uploads/2021/11/Seeding-Environmental-Justice-Coalition-Undo-Legacy-Disinvestment-Case-Study-TCC.pdf).

Local Drivers of Community Voice and Leadership

Governance and Collaborative Stakeholder Structure: In Stockton, the Collaborative Stakeholder Structure includes a small steering committee and a larger Community Coalition. There are also working groups that bring people together to focus on community engagement, workforce development, and the oversight of capital projects. Stockton Rising has also created an innovative requirement that the Community Coalition be brought together when there is any proposed change to the TCC scope of work during the grant period (https://greenlining.org/wp-content/uploads/2021/11/Seeding-Environmental-Justice-Coalition-Undo-Legacy-Disinvestment-Case-Study-TCC.pdf).

Community Engagement Plan and Strategies: During the planning phase, core community partners—including Third City Coalition, Public Health Advocates, Fathers, and Families of San Joaquin and PUENTES—were a focal point for the community strategy and undertook a range of activities including town hall and smaller community meetings, Climate Leadership Forum seminars, surveys, and door-to-door canvassing campaigns presentations to local organizations and groups (https://greenlining.org/wp-content/uploads/2021/11/Seeding-Environmental-Justice-Coalition-Undo-Legacy-Disinvestment-Case-Study-TCC.pdf). The collaborative has created a Community Engagement Working Team to coordinate community engagement during the implementation grant phase which is working to develop community liaisons and the creation of youth empowerment programs (https://greenlining.org/wp-content/uploads/2021/11/Seeding-Environmental-Justice-Coalition-Undo-Legacy-Disinvestment-Case-Study-TCC.pdf).

6 California's Transformative Climate Communities Program: Takeaways and Reflections

6.1 TCC Demonstrates the Possibility and Power of Equity-Centered and Community-Led Decision Making

Transform Fresno, Stockton Rising and the other six TCC collaboratives have built new and powerful approaches to meeting the needs of their communities that have delivered real and positive impacts on the ground. The TCC program is giving power to local communities to determine for themselves what they most need and structures the program in ways that requires and supports community voice and leadership. While there is much to be proud of there have also been challenges along the way, much has been learned and there is more work to be done. State-level staff at the Strategic Growth Council, community advocates, and stakeholders have continued to work together as the program has moved from the planning to implementation phase to address the challenges faced in local communities in delivering on the promise of the program and adjust programmatic guidelines accordingly.

Many of those who were involved on the ground in the eight TCC initiatives spoke about the transformative nature of the experience and described with pride what the community had been able to accomplish through the process. Two community organizers that have been deeply involved in the work of Stockton Rising and Transform Fresno speak eloquently about the vision and impact of the work and provide important learnings and insights:

California's TCC program is a bold and trailblazing effort that should be a national model for integrated community-led sustainability efforts. Many people don't know that the creation of the TCC program was the result of coordinated policy advocacy. The Greenlining Institute partnered with other advocacy organizations including the California Environmental Justice Alliance, to influence the design of the original legislation as well as the development of program guidelines to ensure the program actually made the impact we envisioned. I also had the privilege of working in and with the Stockton community as they developed their vision and made it a reality. Based on our experience, we have to scale this work from both the top-down and the bottom up: build progressive policy and program design to support community leadership and invest in deep and ongoing local capacity building at the local level.

—Emi Wang, Associate Director of Capacity Building, The Greenlining Institute

It was a tremendous privilege and honor working alongside community members to pass the bill that created TCC and subsequently work together ensuring a community-driven plan made it to Strategic Growth Council. The work we accomplished was set by decades of advocacy led by community leaders who continued to show up and hold elected officials and government staff accountable. Seeing the community flex their power to ensure tens of millions of dollars was invested in a community I grew up in was a powerful experience and highlight of my career. TCC is an example of how governments must rethink their roles in adaptation, resilience, and community development, moving from a patriarchal approach that dictates what happens in communities to one where they co-develop and support local power building.

—Grecia Elenes, Former Regional Policy Manager, Leadership Counsel for Justice and Accountability

6.2 Government Perspectives: Mindset Shifts, Best Practices, and Policy Drivers for Replication and Scale

The TCC program is increasingly on the radar of policymakers and government agencies outside of California as an example of a successful government-initiated program and investment that is designed to ensure community voice and community-led decision making. While program adaptations will need to be made based on varying governmental, geographic and community contexts, the takeaways and design elements have broad applicability. As other local, regional, and national governments seek to design and implement more equitable and community-led approaches to adaptation and sustainability efforts, the following are some key policy and programmatic design elements, and mindset shifts:

Require Community Involvement Throughout the Process: The TCC program not only asks applicants to involve the community in creating their vision, but also holds grantees accountable for continuing the partnership through provisions in the grant agreement. Communities must have the power to raise their voices and have recourse if the lead-agencies including local governments are not involving communities. The funders—whether public or private—should put in place a balanced power sharing structure. This is even more important in the context of communities that have experienced systematic disenfranchisement and struggled to get local government to be responsive to their needs.

Center the Government Role as Enabler: Government can play a crucial role in enabling and empowering communities by using the tools in their toolbox to amplify the voice of citizens and respond to their needs. This require creative thinking on the part of government teams and an understanding that the status quo is not always the best approach.

Prioritize and Improve Community Relationships: TCC invests in communities that have had challenging relationships with local governments due to past injustice, neglect, and discriminatory practices. The program offers a solution by having the state act as a mediator between the local government and community, holding both to the same standards outlined in the grant agreement. This changes the dynamic and helps all parties find ways to work together. In the global context, a similar approach could be taken by international funders that seek more equitable outcomes in countries with different governance structures by using funding agreements as a tool for change making.

Provide Continued Support: While the lynchpin of TCC's success is the presence of an ecosystem of active and strong local organizations and institutions, at the outset of the program, TCC communities varied in the presence and strength of on the ground organizations. To address this, TCC provides ongoing technical assistance and capacity building throughout the application and implementation process and staff are assigned to each site and have regular check-ins with grantees to ensure desired outcomes are met. However, some communities may lack social infrastructure for engagement which makes it more challenging to create an inclusive process. Government should invest time and resources in ongoing capacity building to make sure that communities have social infrastructure to mobilize if needed. The Governor's Office of Planning and Research efforts to establish the Office of Community Partnerships and Strategic Communications and developing Regional Climate Collaborative Program are examples of providing needed ongoing support and capacity building.

Engage in Ongoing Learning and Evaluation: By bringing in third-party evaluators from the beginning, the program uses data to analyze its impact and get feedback on how it can better support communities. Structure evaluation efforts in ways that also value and gather knowledge and insights from diverse stakeholders. These evaluations will provide important insight on whether the government investment is effective and identify improvements along the way to increase the effectiveness of the program.

Allow for Programmatic Flexibility on the Ground: The legislation that created TCC is broad and high-level, leaving room for the Strategic Growth Council to collaborate with stakeholders and create guidelines that best meet community needs. Special attention should be given to the development of programmatic guidelines that clearly articulate outcomes but don't act as barriers to participation.

Monitor Progress Through Ongoing Engagement and Feedback: The success of the TCC program depends on continued involvement from stakeholders and advocates. Maintaining active partnerships and incorporating feedback from stakeholders should be built into the program design including updates of program guidelines and processes. Thinking about a public process and transparency measures will increase the excitement and willingness by stakeholders to participate in the process.

Remove Barriers: The TCC program intentionally removes barriers that have prevented grassroot organizations or small businesses from participating in the program, such as requiring expenses to be reimbursed, insurance requirements, and withholding portion of the funding until completion of projects. As a result of the program's success, the SGC is leading the first pilot to change the reimbursement-only model.

Be humble and Embrace a Growth Mindset: Government staff and those involved from outside the communities learned powerful lessons from those on the ground in communities and were challenged to reject their institutional power privileges. Agencies overseeing these programs should articulate needed value and mindset shifts and support staff on this journey.

Maximize Upfront Government Coordination: Implementing infrastructure projects impacts daily life of people on the ground. System thinking in deploying infrastructure projects, including the ways in which projects can complement each other and minimizing the negative impact of the construction process through coordination of different government agencies, is crucial in creating a successful overall outcome.

6.3 Conclusion

The rich lessons and insights emerging from the powerful work being advanced in California's TCC communities are timely. Across the globe policymakers and institutional stakeholders see that approaches where outsiders deploy top-down solutions in communities facing complex local challenges and where climate change, poverty and weak systems overlap, consistently fail to create the desired impact. Recognition of the importance and power of community engagement, voice, and agency to address power imbalances and systems of exclusion requires deeper and more fundamental shifts in the design of public initiatives. Governments and stakeholders across the world need to rethink the role of communities in problem identification and solution making and accelerate reforms and investments to center community voice and agency.

References

AB 2722 PART 4. Transformative climate communities program. http://www.leginfo.ca.gov/pub/15-16/bill/asm/ab_2701-2750/ab_2722_bill_20160914_chaptered.html

California Adaptation Planning Guide, Governor's Office of Emergency Services. https://resilientca.org/apg/intro/

California Climate Adaptation Strategy. https://climateresilience.ca.gov/priorities/strengthen-protections/build-capacity.html
California Governor Executive Order B-30-15 (2015) https://www.library.ca.gov/wp-content/uploads/GovernmentPublications/executive-order-proclamation/39-B-30-15.pdf
Dias N et al (2018) Critical success factors of a bottom-up urban design process to deliver sustainable urban designs. Int J Strat Property Manag 22(4):265–277
Fighting Redlining and Climate Change with Transformative Climate Communities, The Greenlining Institute (2021). https://greenlining.org/wp-content/uploads/2021/10/Fighting-Climate-Change-and-Redlining-with-Transformative-Climate-Communities-Final-Report.pdf
Fresno 2022 Transformative Climate Communities Update, UCLA Luskin Center for Innovation. https://innovation.luskin.ucla.edu/wp-content/uploads/2022/06/Transform-Fresno-2022-Progress-Report.pdf
Geekiyanage D, Fernando T, Keraminiyage K (2020) Assessing the state of the art in community engagement for participatory decision-making in disaster risk-sensitive urban development. Int J Disaster Risk Reduction 51:101847. ISSN 2212-4209. https://doi.org/10.1016/j.ijdrr.2020.101847
Guide to Equitable, Community-driven Climate Preparedness Planning Guide (2017) US sustainability directors network. https://www.usdn.org/uploads/cms/documents/usdn_guide_to_equitable_community-driven_climate_preparedness-_high_res.pdf
IAP2 Spectrum of Public Participation (2004) From the International Association for Public Participation. IAP2
ICLEI—Local Governments for Sustainability USA Sustainable Communities Suite Adaptation—First Steps (2022). https://icleiusa.org/wp-content/uploads/2022/09/Adaptation_-First-Steps_USCP.pdf
Johnston KA, Taylor M, Ryan B (2022) Engaging communities to prepare for natural hazards: a conceptual model. Nat Hazards 112:2831–2851. https://doi.org/10.1007/s11069-022-05290-2
Planning and Investing for a Resilient California: A Guidebook for State Agencies, California Office of Planning and Research. https://opr.ca.gov/docs/20180313-Building_a_Resilient_CA.pdf
Resiliency Guide Equity Checklist, Equity and Vulnerable Communities subcommittee of the Technical Advisory Group. https://opr.ca.gov/climate/docs/20180312-Equity_Checklist.pdf
State of California General Plan Guidelines (2017) Governor's Office of Planning and Research. https://opr.ca.gov/docs/OPR_COMPLETE_7.31.17.pdf
State of California General Plan Guidelines (2017) Governor's Office of Planning and Research, p 26. https://opr.ca.gov/docs/OPR_COMPLETE_7.31.17.pdf
Stockton Case Study, The Greenlining Institute (2021). https://greenlining.org/wp-content/uploads/2021/11/Seeding-Environmental-Justice-Coalition-Undo-Legacy-Disinvestment-Case-Study-TCC.pdf
TCC Program Guidelines, Strategic Growth Council. https://sgc.ca.gov/programs/tcc/resources/

Looking Forward: Building Vibrant, Climate Adaptive, and Resilient Cities and Communities

Catherine Atkin and Kwi-Gon Kim

Abstract In the final chapter of the book the co-editors look back at the span of the book and explore key insight and learnings. They review the ground covered in the book, including the thirteen featured case studies from across the Gobal South and identify new tools and frameworks that support climate risk management, the use of new technologies and the inclusion of community voice for greater impact by expert contributors. They also present a set of policy and practice next steps needed to strengthen the theory and practice of comprehensive climate risk management and green recovery planning including the need to develop new tools to support the development and capacity of in-country professionals and the critical importance of assessing and strengthening the economic, social, institutional, and technological conditions that facilitate or limit the development and deployment of adaptive measures.

Keywords Global South climate risk assessment · Africa climate risk management · South East Asia disaster risk management · Community-driven resilience building · Digitilization of climate risk management · IoT sensor technology disaster management · Nature-based hazard reduction measures · Comprehensive urban planning

1 Introduction

With this book we have attempted to explore current status and approaches in the Global South to adaptation management processes including: vulnerability and risk analysis, disaster modeling and forecasting, flood hazard reduction planning, and the assessment of local context variables and adaptive capacity.

C. Atkin (✉)
Stanford CodeX Climate Data Policy Initiative, Stanford, USA
e-mail: catherine.atkin@codex.stanford.edu

K.-G. Kim
Seoul National University, Seoul, South Korea
e-mail: kwigon@snu.ac.kr

K. Kim and C. Atkin (eds.), *Cross-Cultural Perspectives on Climate Change Adaptation*, Cities and Nature, https://doi.org/10.1007/978-3-031-50365-8_26

Extensive ground has been covered in this book.

- **Chapters** "Climate Disaster Risk Management—Risk Assessment and Green Recovery Planning: Core Concepts and Identification of Current Issues" and "International Initiatives and Current Context for Adaptation Planning and Green Recovery" provide an overview of core concepts associated with climate risk management and discusses emerging issues and the global political and policy context.
- **Chapters** "Nine-Step Conceptual Framework for Climate Disaster Risk Assessment and Green Recovery Planning" and "Methods and Technologies for Developing the Indicators of Climate Disaster Risk Assessment and Green Recovery Planning: The Standard Techniques for a Whole System" introduce a new climate risk management framework created by Dr. Kwi-Gon Kim and a review of methods and technologies for climate disaster risk assessment and green recovery planning that informs the case studies that follow.
- **Chapers** "Flood Risk Analysis: Causes and Landscape-Based Mitigation Strategies in Dire-Dawa City, Ethiopia"–"Proposed Study Design for Spatio-Temporal Drought Assessment Using Standardized Precipitation Evapotranspiration Index (SPEI) and Adaptation Over Mantaro Valley, Peru" showcases the thirteen case studies from across the Global South focused on flooding in cities and regions by authors from these countries.
- **Chapters** "Leveraging Leading Edge Digital Models, Developed Under Urban Computing Principles, to Support Flood Risk Assessment in Cities" and "Distributed Ledger Technology (DLT) to Drive Flood Reduction Solution Markets and Financing" explore the promise of technology to accelerate impact and scale and highlights digitalization and next-gen technologies including AI and DLT.
- **Chapters** "Case Study Findings and Synthesis of Learnings"–"The Critical Role of Community-Driven Adaptation and Resilience Building: Bright Spots and Lessons from the State of California" provides a synthesis of learning from the case studies and features new approaches to co-benefit adaptation and mitigation framework integration and community engagement to drive transformational impact.

This chapter summarize case study insights and learnings, identifes new tools and frameworks presented in the book that contribute to climate risk management (CRM) studies, and outlines a set of policy and practice next steps.

2 Key Case Study Insights and Learning

The thirteen case studies author written by professionals from across the Global South offer important insights and learnings:

- The risk assessments in the case studies tell a story of where and how people, animals and natural environments are experiencing the negative impacts of climate change related flooding hazards. They also reveal the current challenges that Global South professionals face in getting access to the data and insights they need to develop this critical planning tool.
- Case study authors were challenged to systematically identify and evaluate local context variables as part of their analysis. The models and frameworks they utilized do not provide the kind of comprehensive and step-wise guidance necessary to assess the range of issues including social, ecological, institutional, and structural dimensions that must inform risk analysis and green recovery planning.
- Weak policy and enabling environments at the local, regional, and national levels were identified by authors as a fundamental challenge to implementing needed flood hazard reduction reforms. This includes the development of comprehensive adaptation plans as well as land use and economic development regulations.
- There was significant variation in the extent to which case study authors were able to move beyond their climate risk assessments to identify and evaluate potential hazard reduction measures. This aligns with insights from the broader CRM literature about the need for additional focus on identifying solutions sets and the building the capacity to analyze and use them.
- Across every local context case study authors recognized that local governments were severely under-resourced and often facing a wide range of difficult local realities including poverty and stressed health and social services systems that limited their ability to execute on plans that might be developed.
- Although most case study authors reported limited current technology capacity and availability—including broadband infrastructure—authors saw technology solutions as offering great promise, especially for use in risk assessment and the implementation of early warning and disaster response systems.

3 New Tools and Frameworks

The book also features the contributions from global experts who brought forth important new frameworks and tools that policymakers, public sector professionals, and on the ground implementers can use. Some of these include:

- **Nine-Step Conceptual Framework for Climate Disaster Risk Assessment and Green Recovery Planning**: Recognizing the shortcomings of dominant models of climate risk management, including a lack of tools that on the ground implementers need, Dr. Kwi-Gon Kim puts forward a new Chapter "Nine-Step Conceptual Framework for Climate Risk Assessment and Green Recovery Planning". This framework lays out the concrete steps required to implement an indicator approach assessing climate risk as well as undertake green recovery planning.
- **Digital and Next-Gen Technology Use Case Matrix and CRM Application**: Authors from ICLEI and the Green Digital Finance Alliance (Chapter "Leveraging Leading Edge Digital Models, Developed Under Urban Computing Principles, to Support Flood Risk Assessment in Cities") share with readers an exciting range of

tech tools—including Digital Twin, Metaverse, IoT, and AI—that can and should be leveraged for adaptation efforts including a matrix that maps the tools to their potential application to elements of comprehensive CRM. Authors also map the technologies to the nine-step framework developed by Dr. Kim.

- **Blockchain Framework for Flood and Adaptation Insurance**: After sharing basic principles and elements of blockchain technology, author Si Chen (Chapter "Distributed Ledger Technology (DLT) to Drive Flood Reduction Solution Markets and Financing") explores disaster insurance as a critical adaptation tool and strategy and explores how Distributed Ledger Technology (DLT) could be used to structure insurance mechanisms that help protect against disasters and proposes a combination of automated risk transfers using technology and focus on local groups that work with policyholders to mitigate and adapt to climate change.
- **Integrated Climate Adaptation and Mitigation Co-benefit Model**: Recognizing that current models and frameworks for critical mitigation and adaptation efforts are siloed, Dr. Kwi-Gon Kim advances a new integrated model in Chapter "Co-benefits of Combining Adaptation and Mitigation Frameworks: A Cross-Cutting Approach" that combines climate risk management and GHG emissions mitigation frameworks that allows planners and implementers to consider how hazard reduction measures like nature-based solutions can both reduce hazards and also advance decarbonization goals.
- **Policies and Programmatic Design for Community-Led Adaptation and Resilience Building**: Adaptation initiatives too often fall short because those on the ground in communities are treated as passive actors instead of actors and leaders of the planning and implementation process. In Chapter "The Critical Role of Community-Driven Adaptation and Resilience Building: Bright Spots and Lessons from the State of California", Catherine Atkin and Saharnaz Mirzazad tell the story of California's Transformative Climate Communities Program as an example of the power of community-led resilience building and share new best practices and policies to unlock the power of communities in driving successful adaptation and resilience building.

4 Key Takeaways and Future Directions

The rich learnings and insights from the case study authors and other experts tell a story of stakeholders and governments in the Global South struggling to effectively address climate change related flooding disasters and hazards. Developing the adaptive capacity of local ecosystems to change this trajectory necessitates an inventory of the tools and solutions that currentlty exist and identification of existing gaps that will require new solutions.

Climate Risk Management Planning Policy

- Starting at the national level and cascading down through regions and cities there should be a requirement for the development of comprehensive and vertically connected **climate and disaster risk management plans** that include risk assessment, hazard reduction measure identification and standards, targets and benchmarks for progress in green recovery and resilience building.
- At the city and community level, **comprehensive land use planning** is a particularly critical feature of successful climate risk management with a focus ensuring that housing and settlements are located in lower-risk geographies and that economic development and natural resource management is included in plans.
- International and national governments should invest in the further development of **localized data sets and climate risk assessment methodologies and processes** that can be accessed by local governments and stakeholders and that are connected to the design and implementation of hazard reduction measures and solutions.
- Climate risk assessment and green recovery planning efforts must incorporate a systematic approach to **identification and evaluation of local context variables**—including the social, cultural, institutional, ecological and economic dimensions, and realities in local communities—that define a community's vulnerability to climate hazards as well as inform the evaluation and choice of hazard mitigation measures.
- **Community participation and leadership** and on the ground capacity building should be a required component of planning and implementation processes as a condition of funding for not only community vulnerability assessments but also green recovery planning and strategy and project implementation and evaluation.
- All climate risk management plans and standards should include a technology evaluation and **explicit reference to how technology and digitalization** solutions will be leveraged for impact and scale in both risk assessment and green recovery plans.

Green Recovery Implementation

- Regional and local governments must **move beyond climate risk assessments to implementation** including the design, evaluation, implementation, and testing of integrated sets of hazard reduction measures and green recovery strategies.
- Green recovery implementation should include a concerted **focus on strengthening the enabling environment for successful adaptation** by developing and advancing policies and aligning and bolstering institutions and systems that increase the adaptive capacity of communities.
- Local governments should **build the capacity of government professionals across agencies** to collaborate and bring together the necessary institutional and community partners to develop, fund, and execute locally contextualized green recovery plans.
- Given the volatile and dynamic context on the ground, **hazard reduction measures and projects should be subject to ongoing monitoring and evaluation** for learning and allow for iteration and change over time.

- There must be a radical expansion at the international, national, and regional levels of **public, private, and blended financing for adaptation projects in the Global South** so that regions, cities, and local communities can execute on their adaptation plans and successfully mitigate their climate hazard risks.
- International, national, and local climate programs should prioritize and incentivize the **integration and execution of combined adaptation and mitigation planning efforts** that maximize the co-benefits of climate hazard reduction and decarbonization **and new funding should drive co-benefit projects and solutions**.

5 Final Message

We hope that planners, regional and local public sector actors, implementers and impacted communities from across the Global North and South will benefit from the case studies as well as the insights and tools shared by other contributors in ways that will support national, regional, and local efforts to advance adaptation, green recovery, and resilience building. Green growth and sustainable development are the only way forward. This requires an integrated approach to climate risk management and participatory frameworks that allow for new ways of doing business. The issues raised in this book should be seen as provocations and the frameworks put forward should be understood to be prototypes that should be further tested and developed.

Other goals of this book are to deepen readers appreciation of the power of research and analysis done by in-country professionals in Global South countries and critical importance of assessing and strengthening the economic, social, institutional, and technological conditions that facilitate or limit the development and deployment of adaptive measures. Without this focus, climate risk management including the implementation of needed strategies and solutions will be limited in their impact.

Lastly, moving from planning to implementation requires not only smart planning and greater implementation capacity, but the financing required to implement strategies and projects. This cannot be the primary responsibility of local governments but must be provided by international and national governments and private sector actors. There is much work to be done. Our collective knowledge, insights and sustained advocacy will be critical to ensure the development and execution of plans and deployment of solutions needed to build resilient and vibrant communities and environments.

GPSR Compliance

The European Union's (EU) General Product Safety Regulation (GPSR) is a set of rules that requires consumer products to be safe and our obligations to ensure this.

If you have any concerns about our products, you can contact us on ProductSafety@springernature.com

In case Publisher is established outside the EU, the EU authorized representative is:

Springer Nature Customer Service Center GmbH
Europaplatz 3
69115 Heidelberg, Germany

Batch number: 10756376

Printed by Printforce, the Netherlands